10/16    3

*After the Flood*

# After the Flood

## Imagining the Global Environment in Early Modern Europe

## LYDIA BARNETT

Johns Hopkins University Press
Baltimore

This book was brought to publication through the generous assistance of the Jack G. Goellner Publishing Fund.

Johns Hopkins University Press
2715 North Charles Street
Baltimore, Maryland 21218-4363
www.press.jhu.edu

Library of Congress Cataloging-in-Publication Data

Names: Barnett, Lydia, 1981– author.
Title: After the flood : imagining the global environment in early modern
    Europe / Lydia Barnett.
Description: Baltimore : Johns Hopkins University Press, 2019. | Includes
 ·  bibliographical references and index.
Identifiers: LCCN 2018041200 | ISBN 9781421429519 (hardcover : alk.
    paper) | ISBN 9781421429526 (electronic) | ISBN 1421429519 (hardcover :
    alk. paper) | ISBN 1421429527 (electronic)
Subjects: LCSH: Ecology—Religious aspects—Christianity—History of
    doctrines. | Sin—Christianity—History of doctrines. | Deluge—His-
    tory of doctrines. | Theological anthropology—Christianity—History
    of doctrines.
Classification: LCC BR115.N3 B37 2019 | DDC 261.8/809—dc23
LC record available at https://lccn.loc.gov/2018041200

A catalog record for this book is available from the British Library.

*Special discounts are available for bulk purchases of this book. For more
information, please contact Special Sales at 410-516-6936 or specialsales
@press.jhu.edu.*

# CONTENTS

*List of Illustrations  vii*
*Acknowledgments  ix*

Introduction. A Natural History of Sin  1

Chapter One. Before the Flood: Gender, Embodied Sin,
and Environmental Agency  20

Chapter Two. After the Flood: Biblical Monogenism,
Global Migrations, and the Origins of Scientific Racism  50

Chapter Three. Protestant Climate Change: From
Edenocene to Fallocene  89

Chapter Four. The Flood and the Apocalypse: Building the
Republic of Letters  129

Chapter Five. Catholic Climate Change: Heritable Sin and
Strategies of Toleration  160

Epilogue. The Flood Subsides  188

*Notes  197      Index  241*

Horn, *Arca Noae* (1666), frontispiece   6

Woodcut depicting the All Saints' Day Flood, 1570   29

Kircher, *Arca Noë* (1675), map of postdiluvian land and oceans   65

Kircher, *Arca Noë* (1675), map of division of the earth by Noah's descendants   71

Delisle, 1752 map of the North Pacific   80

Burnet, *Theory of the Earth* (1684), antediluvian climactic zones   99

Burnet, *Theory of the Earth* (1684), antediluvian global weather pattern   99

Burnet, *Theory of the Earth* (1684), frontispiece   113

Dryden's 1697 translation of Virgil's *Georgics*, frontispiece   122

Scheuchzer, *Piscium querelae et vindiciae* (1708), fossil fish   147

Scheuchzer, *Kupfer Bibel* (1735), "The final destruction of the Earth by fire"   149

I am so grateful to all of the people who have advised, mentored, inspired, encouraged, and supported me during the research and writing of this book. I cannot possibly list them all here, so let me begin with a blanket thank you to each and every one of them. You know who you are, I hope.

While this book is only distantly related to my dissertation, it would not exist without the support I received from mentors, teachers, and colleagues in graduate school at Stanford University: Paula Findlen, Jessica Riskin, David Como, Giovanna Ceserani, Robert Proctor, Londa Schiebinger, Caroline Winterer, Brad Bouley, Josh Howe, Brianna Rego Lind, Josh Lobert, Carol Pal, Sarah Richardson, Peder Roberts, and most especially the Animals: Noah Millstone, Jeff Miner, Suzanne Sutherland, Corey Tazzara, Liz Thornberry, and Nick Valvo. I also wish to thank my colleagues in Europe who generously shared their expertise with me when I was a young scholar embarking on my dissertation research: Marta Cavazza, Pietro Corsi, Ivano dal Prete, Dario Generali, Urs Leu, Francesco Luzzini, Giuliano Pancaldi, Will Poole, Simon Schaffer, Jim Secord, and Martin Rudwick.

As the dissertation grew into a book and then a different book, I was lucky to have been surrounded by an ever-changing cast of brilliant and supportive colleagues. At the University of Michigan, my thanks go to Eric Calderwood, Hussein Fancy, Dena Goodman, Elizabeth Hinton, Jamie Jones, Elise Lipkowitz, Don Lopez, Laura Miles, Dan Myers, Michelle Phelps, Matt Spooner, and the entire community of scholars in the Society of Fellows. I want to thank everyone in the Bates College History Department, especially Dennis Grafflin, and the many friends and colleagues who made my time in Maine so memorable, including Jen Adair, Brooke O'Harra, Arielle Saiber, Adriana Salerno, and Caroline Shaw. My colleagues at Northwestern University provided the ideal environment in which to bring this book to a close, and I thank them all, most especially Ken Alder, Lina Britto, Corey Byrnes, Sarah Maza, Joel Mokyr, Ed Muir, Susan Pearson, Scott Sowerby, Helen Tilley, Kelly Wisecup, Tristram Wolff, and the Global Early Modernists. I also wish to thank scholars who extended their support at key junctures during the evolution of this project: Fredrik Albritton Jonsson, Alix Cooper, David Sepkoski, and John Tresch. Extra thanks are due to Paula Findlen, whose sage advice and unwavering support at every step of the way, from seminar paper to dissertation to book, was crucial in making this project a reality.

I wish to acknowledge the librarians, archivists, and staff at the Archivio di Stato di Reggio Emilia; the Bancroft Library; the Biblioteca Comunale dell'-Archiginnasio; the Biblioteca dell'Accademia dei Concordi di Rovigo (especially Michela Marangoni); the Biblioteca Estense; the Biblioteca Universitaria di Bologna; the Bibliothèque Publique et Universitaire de Neuchâtel (especially Maryse Schmidt-Surdez); the Bodleian Library; the British Library; Cambridge University Library; Clare College and Trinity College Libraries, Cambridge; the Houghton Library; the Linda Hall Library (especially Bruce Bradley); the Massachusetts Historical Society; the Natural History Museum in London; New College Library, Oxford; the Newberry Library; Northwestern University Special Collections; the Royal Society; the Sedgwick Museum of Earth Sciences; Stanford Special Collections (especially John Mustain); and the Zentralbibliothek Zürich. I thank Carlo Sarti for generously giving me a personal tour of the paleontological collections of the University of Bologna.

For their financial support of this project, I thank the National Science Foundation, the Charlotte W. Newcombe Foundation, the Gladys Krieble Delmas Foundation, the Michigan Society of Fellows, the Max Planck Institute for the History of Science, the Linda Hall Library, and the Weter Fund and the Lane Fund at Stanford University. I am also grateful for the Boyle-Shea Fund Grant that I received from Bates College and an Undergraduate Research Assistant Program Grant from Northwestern University, which allowed me to hire a series of excellent undergraduate research assistants: Jackson Fleming, Christina Kiriakos, Scott Long, Teresa Seel, and Michaela Nakayama Shapiro. I could not have covered so much territory and assembled so many sources in so many languages without their help—I thank them all.

I also extend my thanks to the numerous audiences who listened to parts of this book at various stages of its development and to the readers who took the time to give me valuable feedback on drafts: Brad Bouley, Mitch Fraas, Glenda Goodman, Josh Howe, Ed Muir, Tara Nummedal, Jerry Passanante, Paul Ramírez, Kelly Wisecup, participants in the Michigan Eighteenth-Century Studies Group (especially Elise Lipkowitz), everyone in MPIWG Department II (especially Carla Bittel, Lino Camprubí, Lorraine Daston, Matthew Eddy, and Phillip Lehmann), and members of the History of Science, Technology and Environment Division at the Royal Institute of Technology in Stockholm (especially Arne Kaiser, Peder Roberts, and Sverker Sörlin).

I am deeply grateful to Deborah Coen and Dániel Margócsy for their careful, generous, and incisive reading of the manuscript; they improved the book immeasurably. Thanks are also due to my editor Matt McAdam for being an early champion of the book project, the editorial staff at Johns Hopkins University Press, and Beth Gianfagna for her expert copyediting. I also wish to thank the readers and editors of two of my articles that were formative in the evolution of this project, parts of which are incorporated in chapters 3 and 5: "Strategies of Toleration: Talking across Confessions in the Alpine Republic of Letters," *Eighteenth-Century Studies* 48, no. 2 (2015): 141–57 (copyright Johns Hopkins University Press, 2015) and "The Theology of Climate Change: Sin as Agency in the Enlightenment's Anthropocene," *Environmental History* 20, no. 2 (2015): 217–37.

I have been fortunate to have lived most of my life near the conjunction of land and sea, whether the Atlantic, Pacific, or the Great Lakes. I would like to thank Lake Michigan in particular for inspiration during the final stages of writing and revising.

Finally, heartfelt thanks to my kin: my parents, John, Patricia, Clare, Glenda, Casey, Peter, Eleanor, Sam, Edith, Nell, Erin, Emma, and especially Nick, my first, best, and last reader and my constant source of love and encouragement. This book is dedicated to the memory of my brother Spencer.

*After the Flood*

# A Natural History of Sin

In his widely anticipated encyclical on environmental justice, *Laudato Si'* (*Praise Be to You*, 2015), Pope Francis sought to incorporate the scientific consensus on climate change into a spiritual accounting of humanity's past, present, and future relationship to the global environment. Francis lamented the loss of the "originally harmonious relationship between human beings and nature" as described in the early chapters of the book of Genesis and condemned the increasing tendency of people in the postbiblical era to see themselves as the "masters, consumers, [and] ruthless exploiters" of the natural world. As a result of modern peoples' careless avarice, "the earth, our home, is beginning to look more and more like an immense pile of filth."[1] Describing man-made environmental change as a type of sin, Francis emphasized that the only way forward was to "look for solutions not only in technology but in a change of humanity." Because the "environmental problems" that humanity collectively faces have "ethical and spiritual roots," the search for a techno-fix will necessarily fall short of what is needed; only our renewed ethical and spiritual commitment to the planet and to each other can reverse the dangerous trends that now threaten life on earth.[2]

*Laudato Si'* came in for a great deal of criticism in the spring and summer of 2015, just before and after its release, especially from conservative Catholic thinkers and politicians in the United States. Many of these critics invoked the unbridgeable gap between science and religion in order to preemptively discredit Francis's writings on the environment, insisting that the papacy lay on one side of that bright dividing line and climate science on the other. The American Catholic politician Rick Santorum, then a presidential candidate and former Republican senator from Pennsylvania, cautioned in June 2015 that the Catholic Church was "better off leaving science to the scientists and focusing on what we're really good at, which is theology and morality."[3] Kathleen Hartnett White, a conservative thinker on environmental policy and Texas state official, invoked the specter of Galileo's trial before the Roman Inquisition in order to argue that the pope should leave climate change alone. "The last time the church became so embroiled in science," she wrote in the *Federalist* in June 2015, "was when Pope Urban VIII arrested Galileo in 1632."[4] Santorum and Hartnett White's public records of climate denialism furnish context for their stated concern with maintaining a strict separation between science and religion, suggesting that their statements did not spring from a principled deference to professional scientists on environmental issues but rather from a strategic attempt to weaken the potential force of Francis's strongly worded call to climate action.[5] However, the notion that the pope was transgressing professional and disciplinary boundaries by speaking publicly on the topic of climate change had force well beyond the relatively small circle of American Catholic climate deniers. Mainstream news outlets adopted this framing in their coverage of *Laudato Si's* unveiling. A June 2015 article in the *New York Times* stated that Francis, "by wading into the environment debate," was "seeking to redefine a secular topic, one usually framed by scientific data, using theology and faith."[6] Environmental issues were, definitionally, secular and scientific ones; Francis was doing something novel and transgressive in reframing environmental issues as a matter of moral failings and spiritual renewal.

As this book is concerned to demonstrate, the detrimental impact of the human species on the global environment has been a topic of both scientific inquiry and moral concern in the Catholic world for centuries, as indeed in the Protestant world. Unbeknownst to most if not all of the people involved in the 2015 controversy over *Laudato Si'*, Francis was drawing on a centuries-old tradition of exploring environmental issues within a Christian theological framework in which the mutual vulnerability of global humanity and the global environment was mediated through their mutual subjection to God.[7] From the late sixteenth century to the early eighteenth century, a research tradition flourished in western Europe that viewed sin as a world-historical force capable of ruining the climate and planet and thereby shaping the future course of human history. At the height of the Catholic Reformation (known also as the Counter-Reformation) and following bouts of plague and severe flooding in northern Italy, an apothecary named Camilla Erculiani was called before the Paduan Inquisition in response to her published comments on the global environmental destruction of which, she alleged, mankind was all too capable. In *Letters on Natural Philosophy* (1584), Erculiani proposed several natural causes of planetary catastrophe, including the threat posed by human overpopulation to the equilibrium of natural elements that normally make the earth a safe and stable place to live. But she also identified human sin as a related causal factor, arguing, like Francis, that the ruin of global nature was driven in part by humanity's spiritual failings. Nor was this way of thinking about humans and the environment through a religious lens limited to Catholic Europe. Toward the end of the seventeenth century, as Britain's coal industry boomed at the height of the Little

Ice Age and a Catholic monarch was deposed in favor of a Protestant one in the Revolution of 1688–89, the Anglican minister Thomas Burnet would anticipate Francis's description of the damaged planet as "an immense pile of filth" by calling the present planet a "dead heap of Rubbish," degraded by human sin and natural disaster from its original state of Edenic perfection.[8]

These early modern writers relied on a single historical event in order to imagine how humanity could have ruined the global environment: Noah's Flood. The biblical story in which Noah and his family were saved in an ark while the balance of humanity drowned in a flood became the springboard for a scholarly research agenda into the intersecting histories of global nature and global humanity.[9] While nearly all Christians in this period accepted the story of Noah's Flood as an item of faith and as a historical fact, the scriptural text is unclear or silent on many salient details about the causes and effects of this catastrophe, lacunae that natural philosophers, naturalists, and antiquarians—scholars who studied nature's history and human history—eagerly sought to fill.[10] Inspired by new methods for studying nature in the Scientific Revolution and new ways of reading scripture in the Reformations, pious natural philosophers reinterpreted the biblical story of Noah's Flood as a global catastrophe that did just as much damage to the natural world as the human. They came up with various theories about the diluvial transformation of the planet's structure, surface, astronomical orientation, natural elements, and climate. The Flood's drastic alterations, perhaps permanent, to nonhuman nature were then imagined to have caused profound, and perhaps permanent, changes to human bodies, lives, and histories—vitiating human health and longevity, inaugurating new regimes of labor and commerce, and determining the pathways of postdiluvial migration and repopulation. The rearrangement of landforms and waterways; the degradation of the earth's air, water, and soil; and new extremes of weather and seasonality were all seen as well-deserved punishment, given that human sin had sparked this natural disaster in the first place.

In short, the pious philosophers and historians of early modernity sought to understand the spiritual decline of humanity and the physical deterioration of the natural world—including human bodies—as part of the same tragic history of catastrophic and potentially irreversible ruin. The story of Noah's Flood became a means of exploring how humanity's spiritual failings were made physically manifest in the natural world and how nature in turn became the medium through which humanity was punished for their sins. No one disputed God's involvement, but a just God would not unleash a ruinous catastrophe on the world without just cause. In a public lecture delivered in the Swiss city of Neuchâtel in the early eighteenth century, the naturalist Louis Bourguet identified humanity as the unambiguous culprit behind the Flood: "[T]he part of the Human Race that lived before the Catastrophe was the cause of that event."[11]

Nor did the fact that humanity's role in causing this disaster was completely unintentional absolve anyone of moral responsibility. The boomerang effects of humanity's environmental agency on humanity itself became a powerful way for European Christians to imagine the wages of sin working their way through both civil society and the natural world.

This study brings to light a forgotten episode in the history of environmental thought, illuminating several early iterations of the idea that humans have the unwitting power to transform nature on a global scale and are, in turn, vulnerable to the consequences of their unintended alterations to the natural world. This rich synthesis of human and natural history grew out of scholarly debates about Noah's Flood, a biblical disaster that became, in the early modern period, an emblem for the vulnerability of humanity and nature to one another. Long before modern infrastructures for collecting and analyzing long-term scientific data on a global scale came into being, this idea developed out of materialist approaches to sacred history and within the imaginative framework of Christian universalism, which conjured a unified humanity capable of acting as a single moral agent. All living humans were united by their common descent from Adam and Eve, their shared history of near-extinction and postdiluvian diaspora, and by a shared set of spiritual constraints and predispositions inherited from their biblical progenitors; perhaps they also shared a bodily constitution shaped by the postdiluvian environment. Premodern accounts of how this united humanity, acting collectively and unintentionally, could be responsible for planetary transformations was enabled by a deeply pessimistic belief in the pervasiveness of human sin and in its uncontrollable power to wreck the world.

Early modern scholars looked to the past in order to understand their present and discern their future. Protestants and Catholics saw in the Flood a way to explore for themselves and to debate with each other the most urgent and divisive religious questions of the day. Philosophical debates about the causes, consequences, and scale of Noah's Flood engaged questions about the causes, consequences, and scale of human sinfulness. Was sin so ingrained in human nature that it predisposed humans to destroy themselves and the planet they lived on, or had they sinned freely? Was the global catastrophe of the sin-caused Flood a fresh new start or the origin of present misery? Did humanity now possess the capacity to redeem itself and rehabilitate its ruined world? Did human history unfold independently of nature's history, or were they inextricably linked? Were all the world's peoples equal inheritors of original sin? Had they been equally punished in and by the Flood, or had some of Adam's descendants—perhaps the peoples of America—been spared—and if so, why? The idea of humanity as a planetary force emerged as a by-product of debates between and among Protestants and Catholics over questions about sin, salvation, and free will.

These theological questions in turn responded to pressing social and political issues. Early modern Europeans lived in an unfamiliar and broken world, one that was both deeply divided and rapidly expanding. In the aftermath of the Lutheran Reformation, Europe was religiously and politically fractured as never before. Questions about humans' capacity to damn and to save themselves, both individually and collectively, rose to the forefront of religious debate as individuals and polities rearranged themselves along hardening confessional lines and fought bloody wars for territorial control in the states of Europe and in their expanding empires.[12] Colonial conquest and commercial expansion promised wealth, territory, and new converts to the warring factions of Europe. Europeans' growing awareness of the diversity of human cultures across the globe prompted scholarly efforts to gather them all under the aegis of Christian universal history, a research agenda that incidentally lent legitimacy to imperial and evangelical agendas. The long-distance spread of new diseases across the Atlantic coincided with the long-term climactic disturbance known as the Little Ice Age, which caused widespread crop failure, hunger, and social unrest—and in turn exacerbated ongoing religious and political conflict.[13]

During this tumultuous period, Europeans turned to the study of nature in order to answer the religious questions that, they believed, lay at the heart of present-day conflicts and catastrophes. The Renaissance and the Scientific Revolution stimulated broad interest in the study of nature, an interest further sparked by the influx of natural resources and unusual natural specimens from Europe's overseas colonies and trading partners.[14] Many members of Europe's educated elite became convinced that the study of nature and humanity's joint history could reveal what united and divided the human race in the wake of this tragedy, which in turn might help them to reunite Europe's Christians, convert non-Christians outside of Europe, and ultimately bring them all together under a triumphant and universal Christian empire.

Scientific inquiry into Noah's Flood thus formed part of a broader movement in early modern Europe that Ann Blair has called "Mosaic natural philosophy": the effort to unify science and religion and, in so doing, reunify the Christian world in the wake of the Lutheran Reformation.[15] Understanding the Flood as a pivotal event in the intertwined histories of nature and humanity might thus also reveal the hidden unity of scientific knowledge and religious faith. It held out the possibility of discovering where humans had gone wrong and what they could do to make things right again. It promised that everyone, everywhere on earth was part of the same human story, living on the same ruined earth, laboring under the weight of original sin but perhaps also capable of redeeming themselves and the world they had inadvertently destroyed. The multiple unities and reconciliations promised by the story of the Flood might hold the key to

The frontispiece to Georg Horn's 1666 history of Noah's Flood represents the four continents as mythical winged beasts fighting for world domination. Europe, represented by the German imperial eagle, appears to be gaining the upper hand as it makes off with the globe in one of its talons, an imperial scepter clutched firmly in the other. The presence of the ark bobbing on the floodwaters beneath this aerial battle implies a direct historical connection between the settling of the world after the Flood and the European expansion of Horn's early modern present. Newberry Library, Chicago, F 09.417.

bridging the divides and resolving the conflicts that beset the(ir) modern world. The Flood was a way of thinking about unity and division, place and globe, in a world that was both deeply divided and rapidly expanding.

Letters, both published and private, formed one of the main strategies by which early modern Europeans talked with one another about the Flood, a register of their sense that understanding global nature necessitated the long-distance exchange of information. Many of the major works of earth history from this period were published collections of letters, such as Erculiani's *Letters on Natural Philosophy* (1584), Vallisneri's *Critical Letters . . . of Marine Bodies* (1721), and Bourguet's *Philosophical Letters* (1729). The Flood was religiously divisive, philosophically controversial, potentially heterodox, and also potentially irenic, as evidenced by the relationships forged between scholars of different countries and confessions in pursuit of this shared topic of inquiry. I pay particular attention to Protestant Britain and Catholic Italy as the places where histories of the Flood were most numerous and most boldly innovative, and to Protestant Switzerland as a politically neutral but religiously partisan mediator in the intellectual exchanges between northern and southern Europe. In order to contextualize these exchanges within the complex religious and political terrain of early modern Europe and its colonies, I also consider historical figures and texts from France, Germany, the Low Countries, and Sweden, and from colonial New England, New Spain, and Peru. I draw on source material from diverse countries, media, and genres—including published and manuscript correspondence and print matter such as treatises of philosophy, theology, and medicine; natural history catalogues; historical chronicles; meteorological dialogues; commentaries; learned journals; query lists; medical advice manuals; and works of geography, linguistics, and ethnography. The long-distance exchange of letters and specimens formed the material basis for the formation of social relationships, a concrete way the Flood acted as a binding agent across oceans and geopolitical divides.

Ranging widely across western Europe and its American colonies, this study traces the rise and fall of Noah's Flood as a popular though controversial topic of long-distance intellectual exchange from the late sixteenth through the early eighteenth century. Chapter 1 begins in Italy in the era of the Renaissance and the Reformations, when the Paduan apothecary and philosopher Erculiani developed her thoughts on the natural and man-made causes of Noah's Flood in correspondence with male contemporaries in Italy and Europe. Chapter 2 telescopes out to consider the Atlantic world in the long seventeenth century in order to shine a light on the crucial role of European empire and Christian evangelism as key drivers of Flood-based inquiry in early modernity, focusing on the early-seventeenth-century Peruvian Catholic scholar Antonio de la Calancha and the early-eighteenth-century Swiss Protestant scholar Louis Bourguet. Chapter 3 zeroes in on revolutionary England in the late seventeenth century, when Thomas Burnet and John Woodward debated the ruin of humanity and nature after the Flood and their potential for mutual redemption at the dawn of the fossil fuel economy. Chapter 4 uses Woodward's European and transatlantic epistolary network, centered on his correspondence with the Swiss naturalist Johann Jakob Scheuchzer, as a case study of the social forces that drove the Flood's popularity (and the Apocalypse's demise) as a collaborative research project in the Republic of Letters. Chapter 5 returns to Italy and to Padua, where the early-eighteenth-century naturalist Antonio Vallisneri challenged the diluvial theories of his English contemporaries and developed new ones in conversation with Swiss Protestant and Italian Catholic colleagues. Together, the chapters tell the story of how a transnational and interfaith community of scholars interpreted, debated, and ultimately dispensed with the biblical story of Noah's Flood as the centerpiece of a unified history of humanity and nature.

## Inventing the Anthropocene

The term *Anthropocene* is of relatively recent coinage, but the animating idea behind it—that humans as a species are capable of provoking profound and lasting change in the natural world on a planetary scale—is not.[16] Early modern accounts of the causes, crisis, and aftermath of Noah's Flood evince an early version of what Dipesh Chakrabarty has called "geological agency": the human capacity to "have an impact on the planet itself," to instigate geological change on human timescales.[17] Other definitions that have recently been offered for the Anthropocene also find anticipatory expression in early modern discussions of Noah's Flood. Bronislaw Szerszynski suggests that the Anthropocene should be dated, not to the moment when human activity starts to appear in the geologic record—the "golden spike" indicating the onset of a new geological epoch—but to the moment when humans come to consciousness of their own power and agency as a geological force.[18] By this definition, the first inklings of the Anthropocene began in the sixteenth century, if not earlier, and certainly well before humanity was capable of acting as a geological force. Denis Cosgrove's *Apollo's Eye* (2001) shows that humans imagined and represented the "God's-eye view" of the planet Earth for hundreds of years before they were able to actually view the earth from space. So too, I argue, did humans imagine that they possessed the power to transform global nature long before the widespread use of fossil fuels or the dawn of the nuclear age actually gave them such power.

For much of the twentieth century, historians of science and the environment implicitly assumed that premodern people either lacked an awareness of their impact on the environment or could only see themselves as the passive victims of natural disasters and environmental change. In his now-classic *Green Imperialism* (1995), Richard Grove asserts that "until the 1750s, most climactic theorists had, with the notable exception of John Woodward, implicitly envisaged man as a passive actor (or even victim) in the face of monolithic or even global climactic forces over which he had no control. Processes of 'degeneration,' it was implied, could not be forestalled in any way, and humanity could take no overall responsibility for the nature and consequences of climate change."[19]

Historians of science and the environment in the intervening two decades have done much to document the emergence of an environmental consciousness in Europe and Euro-America in the seventeenth and eighteenth centuries, showing how individuals and groups came to believe that they possessed environmental agency on a local or regional scale. Clearing forests, draining swamps and fens, and "improving" arable land were all undertaken with the intention of hastening or forestalling climactic changes that they believed would either promote or interfere with their political, social, and economic goals.[20] The present volume builds on this literature by connecting it to the burgeoning studies on the early modern European imagination of the globe—which also identifies Europe's colonial expansion as a key driver of global imaginaries—in order to illuminate a strain of early modern thought that saw humans as makers of climactic, geologic, and environmental change on a planetary scale. While Grove is correct to note that the British naturalist John Woodward espoused a theory of human responsibility for global climate change (as will be discussed in chapter 3), he was far from the only person before 1750 to think this way.

Another recent vein of scholarship in environmental history and the environmental humanities has begun to document various iterations of the idea of humanity as a planetary force in the eighteenth and nineteenth centuries, long before Paul Crutzen's momentous millennial announcement of our new Anthropocene era.[21] As Christophe Bonneuil and Jean-Baptiste Fressoz argue in *The Shock of the Anthropocene* (2017), even a cursory historical survey of "environmental reflexivity" reveals that "this is not the first time that humans have asked themselves what they are doing to the planet."[22] Bonneuil and Fressoz's critical history of the Anthropocene concept has been furthered by scholars like Fredrik Albritton Jonsson, Vicky Albritton, James Rodger Fleming, and Jesse Oak Taylor, who have documented its emergence across science, literature, and political economy as the social and environmental costs of the Industrial Revolution became ever more apparent.[23] This book joins and extends this scholarship by showing that the history of the Anthropocene concept is both far longer and far more indebted to theology than has yet been acknowledged. Early modern accounts of the human causes and environmental impacts of Noah's Flood demonstrate that the idea of humanity as a planetary force can be traced at least as far back as the sixteenth century, when it was driven by the simultaneous fracturing and expansion of Europe during the Reformations and the post-Columbian launch of overseas evangelism and empire.

The long-standing misperception that premodern people could only see themselves as the passive victims of forces beyond their comprehension or control likely drew some of its force from an implicit assumption that belief in an active, interventionist, and punishing God was incompatible with an account of humanity's power to make or mitigate significant changes in the natural world. Disaster studies scholarship tends to situate case studies of human responses to disaster within a historical trajectory from a premodern and providential world-view to the emergence of a secular "risk society" or a "managerial" approach to natural disasters.[24] The necessity of a secular worldview is also assumed in some of the recent literature on the "Enlightenment Anthropocene," which identifies the geohistorical discovery of secular deep time as a key step toward recognition of humanity's planetary agency.[25] The short biblical timescale of geohistory that preceded the Enlightenment's concept of deep time endowed human history and natural history with congruent timelines, making their intersections in some ways easier to imagine. Moreover, a religious framework in which disasters signified divine punishment for sin provided a necessary role for humanity not only as victims but also as agents, whose moral responsibility for nature's ruin made their recursive subjection to those natural changes moral and just.

The mistaken notion that premodern people could not conceive of themselves as environmental agents may also spring in part from a lack of appreciation for the rich and distinctive conceptual vocabulary of agency, causality, blame, and responsibility that early modern Europeans inherited from Aristotelian philosophy and Christian theology. Craig Martin, a historian of Renaissance Aristotelianism, reminds us that "causation was conceptually far broader for Aristotle and for nearly all of premodern natural philosophy than for contemporary natural sciences."[26] Premodern understandings of causation were capacious enough to accommodate the Christian belief that the material world was shaped by spiritual agents and forces. Alexandra Walsham shows that in the seventeenth-century British Isles, "the physical appearance of the earth" was seen as "a direct consequence of human sinfulness" and also as "a kind of palimpsest upon which the Lord inscribed messages to the people of Britain."[27] Understanding how the material and spiritual worlds interacted became one of the major questions animating philosophy and theology in the early modern period. The controversial efforts of Descartes, Spinoza, and Leibniz toward that end accompanied the controversial efforts of their contemporaries to explain how God, an immaterial deity, and human sin, an immaterial force, could cause the planet to be inundated in water. Nevertheless, the impossibility of articulating an account of how the spiritual world acted on the material that would satisfy Christians of all stripes was not enough to dislodge the fundamental conviction that divine wrath and human sin were jointly responsible, though in very different ways, for the global catastrophe of Noah's Flood. When Bourguet declared that "the part of the Human Race that lived before the Catastrophe was the cause of that event," he was talking first and foremost about humanity's moral culpability. The logic of unintended consequences was central to Christian accounts of the wages of sin. Understanding the distinctive conceptual framework in which pious people in early modern Europe sought the causes of catastrophe is vital in order to fully appreciate the very real sense of responsibility they felt for natural disasters to which they also ascribed providential meaning.

### Global Imaginaries, Parochial Universalism, Fantasies of Loss

Early modern accounts of what it meant for humanity to act as a planetary force were crucially and significantly different from twenty-first-century accounts of anthropogenic climate change. Distinguishing the two requires acknowledging the speculative, imaginative, and fabulist dimensions of this early modern research agenda as well as the very different social, political, and scientific contexts in which it flourished. Denis Cosgrove, Ayesha Ramachandran, and Joyce Chaplin have described the global imagination of early modern Europeans as rooted in colonial expansion, commerce and travel, the Scientific Revolution, and Christian universalism, providing a backstory for histories of the modern global imagination by Alison Bashford, Sebastian Grevsmühl, Matthias Dörries, and others.[28] Imagining global catastrophe, I argue, was a crucial vector through which early modern Europeans imagined the globe.

Part of the Flood's appeal in the early modern period was its use in imagining the destruction of Eden, reimagined as a natural condition of most or all of the earth. The salubrious climate of the Edenic earth was invoked in support of the correlative fantasy of ancient humanity as a race of supermen, whose enormous size and strength were matched by unthinkable longevity and enviable virility. The English bishop Richard Cumberland connected the strength and longevity of antediluvian men to their virility: "The constitution of such longer-liv'd men must needs be much stronger than our's [*sic*] is, and consequently more able and fit to propagate mankind to great numbers than men can now do."[29] *Climate*, a concept from ancient Greek philosophy, was repurposed in early modern European science, medicine, ethnography, and political philosophy in order to explain what made people across the early modern world sick or healthy, tall or short, dark- or fair-skinned, quick- or slow-witted, capable of self-government or slaves to despotism.[30] The Universal Deluge helped early moderns to imagine a *global* climate, indebted equally to the classical Golden Age as to biblical Eden, whose postdiluvian transformation would explain the collapse of human health, size, and longevity following the Flood and the present state of human debility relative to the ancient state of human awesomeness. The sense of inferiority and emasculation conjured by these theories of diluvial climate change—the sense of having lost something as prized in early modern culture as the ability to father numerous, healthy children—is indicative of a larger suite of ideas about humankind's guilt in bringing on its own ruin. It also indicates a distinctly gendered preoccupation with the effects of sin on male health and male bodies, which, according to their logic of the congruence of sin and punishment, also figured men as the main actors in the drama of sacred history. Early modern histories of the Flood evince a palpable sense of loss and sometimes also a perverse pleasure in imagining this vanished and fabulous past in mournful contrast to the degraded present.

Another major dimension of the Flood's appeal was its role in imagining the world just after its occurrence: a depopulated wasteland waiting to be restored and reclaimed. The global migrations of Noah's descendants in the period of human history after the Flood was a staple of early modern European scholarship. This idea of postdiluvian migration and settlement became the framework, in the long seventeenth century, for narrating histories about the origins of race. After the Flood, humanity was subject not only to a changed climate but to a transformed topography that literally directed the course of subsequent human history. Historians of the Flood speculated endlessly about the pathways by which Noah's descendants traveled from the Old World to the New; many of them devoted considerable effort to determine whether the vanished land bridge they surely crossed stretched across the Atlantic or the Pacific. Behind this seemingly arcane antiquarian exercise was a drive to gather all living humans into a single family tree—a key justification for overseas evangelism, ensuring that indigenous Americans were participants in the same spiritual history of sin and salvation as Christian Europeans—while at the same time articulating racial divisions within that monogenetic human family, a key prop to Christian empire.

Fantasies about changing climates and landscapes after the Flood were also fantasies about gender, race, and empire, offered as explanations or justifications for present-day inequalities in Europe and its American colonies. Throughout this study, I highlight the parochialisms and exclusions that lurked beneath the surface of these purportedly universal stories of nature and humanity—those exceptions to the rule that in fact reveal these global histories always applied, intentionally or not, to some people and some parts of the earth more than others. Erculiani's suggestion in private correspondence that her theory of a man-made Flood applied only to men tallied with the public (though unacknowledged) focus of her male contemporaries on the Flood's effects on male bodies. Historians like the Creole Peruvian friar Antonio de la Calancha were explicit about the racial divisions within humanity that emerged in spite of— or perhaps because of—humanity's monogenetic origins from Adam and from Noah. Burnet, the British philosopher who called the postdiluvian planet "a dead heap of Rubbish," may have intended his theory of diluvial climate change to apply only or mainly to the earth's middle latitudes, including his own Britain and excluding its new Caribbean colonies. When the parochial perspective of Christian Europeans on world history was made explicit, it was quickly suppressed and replaced with more subtle mechanisms of exclusion under the guise of universality. When the French Calvinist-turned-Catholic Isaac La Peyrère provincialized biblical history as an account of Jewish history rather than world history, arguing that Noah's Flood only decimated Palestine while leaving the rest of the planet's populace unscathed, he was widely condemned by Protestants and Catholics alike as an atheist and a heretic.

The blowback to Peyrère's polygenist and polycentric vision of world history signals the utility to European Christians of pretending their global histories applied equally to everyone and to all parts of the earth. Their story of humanity and global nature linked through the Universal Deluge was a thoroughly Christian discourse, and it was recognized as such by thinkers outside of Christian Europe. The early-nineteenth-century Shinto nativist scholar Hirata Atsutane, after serious study of the Bible and European natural science, decided that Noah's Flood had never touched Japan because the gods had created Japan higher up on the earth's surface than the Eurasian mainland.[31] Hirata's refusal to allow his country and culture to be incorporated into the Christian European story of the Universal Deluge signals how well non-Christians and non-Europeans were able to perceive the particularist assumptions and agenda, evangelical as well as imperial, behind this allegedly universal story.

The cultural biases that shaped the limits of their horizons on world history also acted as a limit on the community of inquirers devoted to this scholarly agenda. The social composition of the scholars who exchanged ideas, letters, books, specimens, and data about the Flood was relatively homogenous in terms of race, class, gender, religion, and place of origin. The religious homogeneity is reflected in the limits of the present study, which does not consider writers and thinkers from all branches of Christianity, let alone from Judaism or Islam, the other major religious groups of early modern Europe. While I have highlighted several ways in which philo- and anti-Semitism shaped Christian writings on the Flood, especially in chapter 2, Jewish scholars themselves do not appear to have been welcomed into these cross-confessional conversations about sin and global disaster that were carried out in the Republic of Letters. The lone Jewish author I have been able to identify who participated in scholarly conversations about the Flood, Melchior Leydekker, published his 1704 critique of the Protestant Burnet on behalf of "the Republic of the Hebrews."[32] When early modern savants trumpeted the inclusivity of their Republic of Letters, they seem to have been congratulating themselves for their network's bringing together Protestants and Catholics—no small feat—but not its inclusion of Jews, Muslims, or people of non-Abrahamic faiths.

Projection, imagination, and exclusion thus formed key strategies for writing universal histories of humanity and nature. The character of this conversation was influenced by the fact that its participants pretended or aspired to universality but could not actually attain it. At the turn of the eighteenth century, a group of naturalists in Britain and Switzerland attempted to build a global network of travelers and informants who would collect fossils from across the globe in order to prove the universality of the Flood. Their network was nowhere near global, but the pretext of building one proved remarkably effective at bringing together white Christian European and Euro-American men, across divisions of country and confession, in a shared intellectual pursuit laden with political and religious significance.

The Flood thus enabled mediation, social and intellectual, between local, national, transnational, and global scales. In attending to these practices of scaling, I aim to contribute toward what Deborah R. Coen has called a history of "the scalar imagination."[33] Coen suggested such an enterprise in response to the increasingly popular argument that the problem of the Anthropocene is fundamentally a problem of scale. This argument proposes that the long-durational and planetary scale of climate change overwhelms our limited human capacities to act and even to comprehend.[34] Coen points out, however, that ours is hardly the first generation in history to confront the challenge of mediating between vastly different spatial and temporal scales, and she suggests we might begin to write the history of scaling in nineteenth-century Europe, when the expansion of railroads, the plurality of state forms, the birth of new systems of standardized measurement, and the recognition of geological deep time prompted new ways of imagining the relationship between different spatial and temporal scales. I would add to Coen's well-taken point that if we go back further in time, we can see other, equally transformational processes unfolding: expansion of European empires in the Atlantic and Pacific worlds, growth of the transatlantic slave trade, and widening of commercial, religious, diplomatic, and intellectual networks. And just as the rapid technological and political changes of the nineteenth century brought new ideas about the temporal scale of nature, so too did the growth of long-distance networks in the sixteenth, seventeenth, and eighteenth centuries produce new ways of imagining nature on a global scale, as well as new ways of building social networks across space and time.

The global Flood was a machine for transforming local knowledge into global knowledge, offering a narrative and conceptual framework for how this kind of highly desirable scaling might be achieved. One of the most important means of scaling in this period was the exchange of letters across the scholarly network known as the Republic of Letters. From their situated position in the locales of Europe (or European America), most scholars lacked the means to travel widely. Many composers of world histories never left their hometowns and countries. The task of writing the world's history was enormous, and the networks of knowledge to which they had access were not nearly as global as they liked to imagine. Despite their best, optimistic efforts to collect specimens and observations from across the globe, the world-makers (as William Poole astutely calls them) were also aware on some level that their access to parts of the world beyond Europe was extremely limited. The half-acknowledged limits on their universalist and empiricist aspirations helps us to understand one of the key features of the histories of the world from the early modern period: global disaster was always displaced onto the past or into the future. The Flood was seen by many Protestants and Catholics as a harbinger of the Apocalypse, and this temporal doubling formed another major dimension of the Flood's appeal as a subject of scholarly research. Knowledge of this ancient global catastrophe could yield insight into how the world would end and who would be saved. If European scholars felt insecure about their lack of knowledge of the globe in their present moment, conjuring a lost world and a past disaster—as well as a future disaster and a future world yet to be born—gave their imaginations free reign, largely unconstrained by empirical reality or by living people who knew more than they did about other parts of the world.

The Flood was empirically unavailable to early modern naturalists and philosophers in much the same way that the "lost continent" of Lemuria was empirically unavailable to nineteenth- and twentieth-century geographers and geologists. As Sumathi Ramaswamy argues in her brilliant book about Lemuria, this cousin to Atlantis in the Indian Ocean captured the attention of American, European, and South Asian scientists precisely because it could not be approached empirically in the same manner as actually existing continents. Indeed, much of what Ramaswamy argues about Lemuria as an object of study in the nineteenth century applies equally to the antediluvian earth and to the Flood that destroyed it from the perspective of Flood-writers in the long seventeenth century: "Lemuria is by definition not empirically available to its makers for seeing, surveying, and occupying, as places routinely are in the clear and present light of everyday reality. . . . [I]t is a place-world that can only be summoned into existence through imagination and has no existence beyond it. . . . I characterize this place-world as fabulous, to underscore both the primary location of Lemuria in the imagination of its place-makers, and its virtual absence outside their labors."[35] Like Lemuria, the antediluvian Eden, the Flood that destroyed it, and the postapocalyptic planet yet to be born had no existence beyond the labors of imagination undertaken by its "place-makers." Displacing global disaster onto the past or future freed European naturalists from the rigorous standards of empiricism to which they increasingly held themselves. There was no there there anymore (or yet), demanding to be visited and surveyed.[36] The empirical unavailability of their object of study left them free to imagine the world they had lost and the terrifying process by which it was destroyed, to ruminate on the sins of their forefathers, and to speculate about how they might redeem themselves and their ruined planet in the future.

## Science, Religion, and Environment

Historians of science have long been aware of the centrality of Noah's Flood to the early modern earth sciences and of its historical role, down to the present day, as a flashpoint between science and religion. The early modern focus on the Flood has been variously interpreted as an intellectual constraint on free scientific inquiry by pious believers; as an outward show of religious conformity, dutifully embraced by less-than-pious scientists; or even, in recognition of the frequent charges of impiety provoked by philosophical theories of the Flood, as a sign of incipient deism by freethinking philosophers. I believe that early modern conversations about Noah's Flood reveal a different story about science, religion, and secularization. The Flood was actively embraced by many early modern scholars because they found it both good to think with and satisfying to believe in.[37] Among its other virtues, the Flood appealed as a means of crossing the disciplinary and professional boundaries that kept human history and natural history, natural philosophy and theology, distinct from one another. Recent literature on the imbrication of knowledge and faith by Peter Harrison, Alexandra Walsham, Massimo Mazzotti, and others reveals the many ways religious faith spurred and shaped the pursuit of natural knowledge in early modern Europe.[38] The religious motivations behind materialist approaches to sacred history furnish an illuminating counterexample to the notion that philosophical materialism, long associated with Spinozists and Enlightenment *philosophes*, necessarily tended toward deism or atheism.

I also take seriously the religious motivations of those who critiqued materialist approaches to sacred history and rejected the Flood as a subject for philosophy, a diverse group of laymen and men of the cloth, Protestants and Catholics, philosophers and theologians. I show how the waning popularity of the Flood in the eighteenth century cannot be attributed to declining religious piety or belief in the Flood as a matter of historical fact. Instead, I argue that the norms and practices of scholarly sociability that governed the Republic of Letters first encouraged and later discouraged transnational and cross-confessional collaboration on this topic. Where once the Flood promised to build bridges across a divided Europe—and in some cases actually did—it later came to irritate those divisions. The persistent divides of church and confession, combined with incipient forms of cultural nationalism, conspired to produce apparently secular histories of the earth in the Enlightenment that were nevertheless structured by earlier and explicitly religious accounts of the dynamic relationship between nature and humanity.

Scholarship on the early modern earth sciences in the past two decades has revealed their essentially transnational and interdisciplinary character. By situating early modern debates about the Flood within the social space of the Republic of Letters, I build on influential studies by Rhoda Rappaport, Paolo Rossi, and Martin Rudwick, as well as the rich and growing literature on the importance of long-distance exchange and networked sociability to early modern knowledge-making by Anne Goldgar, Dániel Margóscy, Harold J. Cook, and others.[39] Late-seventeenth-century Britain has long played a starring role in accounts of the origins of modern geoscience.[40] In choosing to begin my narrative in late-sixteenth-century Italy and adopting a more expansive geographic scale—neither of which exhausts the spatial and temporal scale on which the history of the early modern earth sciences can and should be written—I firmly situate British science in European and Atlantic networks of knowledge. I also seek to extend Rappaport's, Rossi's, and Rudwick's insights that early modern earth science was fundamentally historical in orientation, liberally borrowing from the toolkit of antiquarians and attending to human history as well as geohistory. Geology was never just about rocks; it was always, equally, about people shaping the planet and climate and about the planet and climate shaping human history. By theorizing how sin was made manifest in nature, the early modern scholars discussed in this volume produced new understandings of humanity's impact on the natural world. By showing how the man-made Flood transformed the structure and surface of the earth, they sought to understand how sin's geophysical effects directed the future pathways of human history. By debating the Flood's effects on air, water, and soil, they sought explanations for human debility and mortality and remedies for redemption through climactic and agricultural improvement. Beyond interdisciplinary borrowing, the fusion of human and natural history into a single grand narrative frequently figured as an explicit goal. Recent calls for a robustly interdisciplinary mode of scholarship suited for the new geological epoch in which we find ourselves might find additional force in an increased appreciation of the vibrant interdisciplinarity of premodern intellectual culture in general and the study of Noah's Flood in particular.[41]

It is not my intention to re-award credit for "discovering" man-made global environmental change, which awaited scientific developments in the nineteenth and twentieth centuries. Nor do I intend in any way to call into question the scientific consensus on anthropogenic global warming by making the historical claim that the Anthropocene was invented, as an idea, prior to its establishment as an empirically well-founded scientific account of geohistory. My intention is to illuminate the ways in which key components of the Anthropocene concept were anticipated long before the Anthropocene itself existed as a physical reality or was apprehended by scientists. This research agenda, which is already well under way, merits further attention for several reasons—not least because these premodern visions of the human impact on the global environment provide a possible point of origin for some of the more durable and problematic myths about climate change in the modern era.

Recent scholarship in the environmental humanities and social sciences by Andreas Malm, Alf Hornberg, and others has shown how responsibility for and suffering from anthropogenic global climate change is unequally distributed across the human population, in both space and time.[42] It has become commonplace to point out the massive injustice that those groups of people who have historically done the least to contribute to greenhouse gas emissions—Pacific Islanders, those in the global South, for example—will be among the first and hardest hit by the devastating effects of sea level rise, changing patterns in rainfall, storms, desertification, and so forth. This has led to calls from the humanities and social sciences to abandon the "species-level" framing that characterized the earliest definitions of the Anthropocene from the natural sciences.[43] All of humanity acting in concert has not led to our present climate crisis, and not all of humanity is suffering from its effects at the same rate or intensity. Early modern histories of Noah's Flood and the Christian vision of universal history on which they depended offer a possible point of origin for the fallacy of monolithic human agency and equal victimhood proposed by the species-level Anthropocene. Early modern stories about a man-made global flood imagined humanity's responsibility for, and subjection to, catastrophic environmental change as collective and indeed universal, a precursor to the problematic *anthropos* of the Anthropocene.

At the same time, premodern histories of the planet and people centered on the biblical story of Noah's Flood were premised on several ideas that fell out of fashion in the nineteenth century but have rightly returned to view in recent decades: the idea that human history must be written in reference to nature's history; that the earth's future should be of equal concern as its past; and that multidisciplinary collaboration is necessary in order to reconstruct the past, understand the present, and discern the future of the human species and the global environment. The influence of religion on the environmental consciousness of early modern Europeans was ambivalent and complex. Just over sixty years ago, the medievalist Lynn White Jr. influentially linked the trajectories of religion and environment in his classic essay "The Historical Roots of Our Ecologic Crisis" (1967) by arguing that Christianity had played a historically malign role in stimulating anthropogenic environmental destruction by fostering an attitude of entitlement and control over the natural world. White's simplistic and entirely negative assessment of the influence of religion on environmental thought and practice deserves reconsideration in light of the historical complexity of that dynamic both within and beyond premodern Christian Europe. Further research will undoubtedly continue to uncover connections, contrasts, and parallels between diverse modes of religious environmental thought across the early modern world. The Flood story is common to all three Abrahamic faiths, and recent scholarship on the global imagination in Islamic art, politics, and science indicates the existence of alternative ways of imagining the globe from a religious perspective in early modernity.[44] Looking beyond Christian Europe would likely yield a distinctive and equally vital picture of how religious stories and concepts fostered global environmental imaginaries in other times and places.

Reconstructing the intellectual history of these ideas beyond the early modern period is beyond the scope of the present study, but further research will undoubtedly enrich our understanding of the complex legacy of these premodern ideas for the modern era.[45] In the meantime, I want to call attention to one particular pattern of representation that seems to link past and present. Popular representations of present and future climate change frequently deploy the language and imagery of flooding. *Laudato Si'* figuratively describes consumerism, identified as a leading cause of man-made environmental degradation, as a "flood."[46] Climate fiction like *The Day after Tomorrow* (2004) visually represents global climate change as a literal flood inundating the city of New York. Darren Aronofsky's *Noah* (2014) uses the biblical story of Noah and his ark as a parable for global warming.[47] Anthropogenic climate change has already provoked and will continue to provoke a host of disruptions in the natural and human worlds: drought, crop failure, abnormally hot (or cold) winters, abnormally cold (or hot) summers, political violence, mass migration, and so on—and yes, catastrophic flooding, as we witnessed in the Caribbean and southern United States in the hurricane seasons of 2017 and 2018. But the images that tend to predominate, at least in popular and public-focused portrayals of anthropogenic climate change, are all of water: of icebergs melting, sea levels rising, waves lapping at the coast, and cities being inundated. This language and imagery, I suspect, derives considerable force from its recollection and reactivation of deep cultural myths about the awesome power of floods to ruin the world as the unintended result of human behavior—myths that are, it must be noted, not universally shared across the diverse human cultures on the planet and thus not equally compelling everywhere. The image and story of Noah's Flood, as reinterpreted by Christian European scholars during the Scientific Revolution, may be the unacknowledged ancestor of much environmental thinking in the European and Euro-American scientific tradition. As we debate how best to counter and communicate the threat of global warming, it might be helpful to better understand the origins of this pattern of representation and the cultural baggage that comes along with it.

# CHAPTER ONE

# Before the Flood

## Gender, Embodied Sin, and Environmental Agency

In 1584, a small but remarkable volume appeared in Kraków. Although published in the Commonwealth of Poland-Lithuania, *Lettere di philosophia naturale* (*Letters on Natural Philosophy*) was written by an Italian woman named Camilla Erculiani, who ran an apothecary shop with her husband adjacent to the renowned University of Padua.[1] Even more significant than its publication in a foreign country several hundred miles from the place where it was written was the author's gender, the book's innovative and controversial contents, and the interplay between the two.[2] Erculiani's *Letters* is the only currently known original work of natural philosophy by a woman in Renaissance Italy. Erculiani was also the only woman brought before the Inquisition in the sixteenth and seventeenth centuries because of her philosophical views.[3] Finally, *Letters on Natural Philosophy* may very well be the first work by an author of any gender or country in early modern Europe to offer a coherent, systematic, and scientific account of the capacity of humans as a species to cause catastrophic harm to the global environment. In her efforts to understand sin as an embodied and possibly gendered human trait with world-changing effects, Erculiani produced a powerful account of man-made environmental change on a global scale.

The combination of these three unique characteristics makes it all the more remarkable that this book and its author have languished in complete obscurity for the past four centuries. The Paduan Inquisition's interest in her book goes a long way toward explaining why it vanished from public view so soon after its publication, which in turn explains why Erculiani has been entirely absent from modern histories of the premodern earth and environmental sciences. But she deserves a place in these histories. While her *Letters* ranged widely over a series of related topics in Renaissance medicine and philosophy, the topic she kept returning to was the natural causes of Noah's Flood, then a little-understood and deeply controversial problem for natural philosophy. Erculiani advanced a theory that humankind's divinely created physical embodiment predisposed them, collectively, to disrupt the planet's natural balance of elements, thus triggering a global inundation of water. She may have meant this theory to apply to all of humanity or specifically to men. In killing off most of the human population, the Flood restored the planet's natural balance of elements to their original, harmonious state. *Letters on Natural Philosophy* thus inaugurated a tradition of recasting the biblical story of Noah's Flood as the pivotal event in the natural history of the earth and using it as a case study for understanding the reciprocally destructive effects of global humanity and global nature on each other, a tradition that received its most visible expression in Thomas Burnet's far-better-known *Sacred Theory of the Earth* (1681–89), published in London almost precisely a century later.

## The Reformation's Anthropocene

We are all familiar with the story of the Enlightenment philosophers who scandalized pious readers across Europe with their radically deistic accounts of the earth's origins and history. This story features men such as George Louis Leclerc, comte de Buffon, whose *Histoire naturelle* (*Natural History*, 1749) suggested that the earth was born millions of years ago from a collision of a comet with the sun, and Benoît de Maillet, whose anonymous 1748 *Telliamed* (the author's name spelled backward, fooling no one) suggested that the planet's continents were created by the slow recession of a universal sea over the course of approximately two billion years.[4]

We have also all learned to be skeptical of this and other triumphalist narratives of secular science prevailing over religious orthodoxy as Europe moved into the modern age. Yet there are strange and subtle ways in which this celebratory account has been resuscitated along with the surge of interest in the Anthropocene among scholars of early modern science and culture. Recent efforts to trace the roots of the Anthropocene back to the eighteenth century—not just the geological age itself, perhaps inaugurated by the first stirrings of the Industrial Revolution at the century's end, but the very idea of a geological age characterized by profound and permanent human impact on the earth—have produced new historical narratives that privilege the Enlightenment's discovery of deep time as a key moment in the modern discovery of the Anthropocene.[5] Implicit in these new narratives is the assumption that the short, biblically based timescale of the earth's history that reigned supreme before the Enlightenment precluded the development of an awareness of humanity's potentially catastrophic impact on global nature. Pious premodern science has once again emerged as a stumbling block to the development of modern, secular science, which this time assumes the guise of an emergent "Anthropocenic" worldview.

Valorizing the Enlightenment's discovery of deep time as part of the intellectual genealogy of the Anthropocene faces problems on several fronts. The first is that deep time was "discovered" long before the eighteenth century, and the adoption of a short biblical timescale happened not so very long prior to its alleged discovery. The earth was assumed to be extremely old, if not eternal, in mainstream Aristotelian natural philosophy from the late Middle Ages through the second half of the sixteenth century, when, as Ivano dal Prete has recently shown, philosophers in Italy and elsewhere in Europe began rather suddenly to adopt short geological timescales and to construct novel accounts of the earth's ancient history based on both nature and scripture where none had existed before.[6]

The second difficulty is that the idea of deep time was not a necessary prerequisite for the idea of the Anthropocene to emerge, as this chapter is concerned to argue. If we adopt Dipesh Chakrabarty's definition of the Anthropocene as the era in which humans begin to possess geological agency—namely, the ability to transform the earth in ways profound enough to leave traces in the geological record—then natural philosophers in Europe had already articulated an account of humanity's geological agency well before the Enlightenment's particular articulation of the concept of deep time.[7] The short timescale of biblical history that structured so many accounts of the earth's history in the sixteenth, seventeenth, and eighteenth centuries was in some ways, I argue, more conducive to the rise of an "Anthropocenic" worldview. Within this framework, human history and natural history existed contemporaneously, beginning with the Creation and ending with the final destruction of the earth by fire. The impingement of natural history and human history on each other during their short, congruent timescales was most vividly demonstrated by the global catastrophe of Noah's Flood, which was productively reimagined in the early modern period as the single most important event in the history of both nature and humanity. The Flood came to be seen by many naturalists and philosophers as an event that had thoroughly transformed the global environment and maybe also the physical structure of the planet itself; an event that had left abundant traces in the geological record still recoverable in the present day, in the form of both strata and fossils; an event that profoundly impacted the human race as a near-extinction episode and as a permanently debilitating climactic disaster; and, most crucially for dating the emergence of the Anthropocene concept, an event for which human behavior was the ultimate cause.

Finally, the short time span and accompanying biblical framework for understanding the earth's history was neither universally adopted, nor was it obviously orthodox, especially in the sixteenth century during the decisive shift that dal Prete has identified. Recognizing the relatively recent vintage of the short biblical timescale prior to the Enlightenment's "discovery" of deep time also forces us to recognize that biblical earth history was both religiously controversial and unevenly embraced across the confessional and political terrain of Europe during the Reformations and beyond.

This book argues that the idea of a global environment susceptible to alteration by human behavior emerged out of transnational and cross-confessional net-works of intellectual exchange in early modern Europe, particularly in natural-philosophical publications and conversations about the Flood, between the late sixteenth and early eighteenth centuries. This chapter recovers an important and neglected early figure in this history from the late sixteenth century in order to show the crucial role of the Reformations' reconfiguration of the rela-tionship between philosophy and faith in this development. In dialogue with Catholic and Protestant interlocutors in France, Italy, and Poland, Erculiani's philosophical letters promoted novel ways of imagining the reciprocal relation-ship between global humanity and global nature via the science of the Flood.

Camilla Erculiani's philosophy of the Flood vividly demonstrates how the idea of anthropogenic global environmental change was able to flourish without postulating deep time and without secular science. While adopting the major events of nature's and humanity's ancient history as outlined in the book of Genesis—Creation, Fall, and Flood—Erculiani makes no pronouncement about how much time had elapsed since the beginning of the world to the Flood to her own day. The six thousand–year biblical chronology later made famous by the seventeenth-century Irish archbishop James Ussher makes no appearance in Erculiani's *Letters*. Indeed, the age of the earth is totally irrelevant to the philosophical discussions about the earth's history contained in the volume. Moreover, citing the words of Moses alongside those of Aristotle and Galen as evidence for her philosophical theory is crucial to her ability to articulate an account of the dynamic and destructive relationship between humans and non-human nature. Erculiani illustrates well how the emerging practice of Mosaic natural philosophy allowed philosophers across Europe in the long seventeenth century to imagine humans, acting collectively in their capacity as moral agents, to enable the physical destruction of global nature.

## Faith, Philosophy, and the Scale of Disaster

Erculiani was among the first natural philosophers in the sixteenth century to fashion biblically sensitive accounts of the earth and its history, often but not al-ways centered on Noah's Flood. They did so in violation of long-standing prohi-bitions against discussing global and supernatural phenomena within Scholastic natural philosophy, and they did so in an attempt to use the methods of natural philosophy in order to address theological questions about sin, suffering, and salvation. Eager to use the methods of natural philosophy as a way of exploring the great religious questions that animated this age of religious reform, Italian Catholic philosophers around the turn of the seventeenth century broke down barriers separating philosophy and faith. By bringing the Flood into natural philosophy, they created new stories about global nature, global humanity, and the reciprocal relationship between the two.

Natural philosophers before the Enlightenment were, as a general rule, less interested in the temporal scale of the earth's history than they were in the spatial scale of geological and environmental phenomena. Much of the discussion of the earth and its history in the late Middle Ages and Renaissance took place under the rubric of meteorology. A popular genre of natural philosophy from the thirteenth century onward, meteorology encompassed the subjects covered in Aristotle's *Meteorologica* (*Meteorology*), from which it derived its name. These subjects included the state of the planet's surface, interior, and atmosphere, its weather and climate, and the periodic disruptions and disasters to which it was subject, including hail, comets, earthquakes, and floods. The sixteenth-century German meteorologist Marcus Frytsche defined his field of study as "the part of physics that is concerned with what comes to be in the regions of the air or in the belly of the earth." Meteorology, in other words, was the earth and atmospheric sciences of its day.[8]

While authors in the meteorological tradition regularly dealt with local natural disasters—the annual flooding of the Nile, for example, was a perennial favorite—they were broadly dismissive of the idea of a universal deluge on the grounds that it was a physical impossibility. For many centuries, Noah's Flood and other catastrophic global phenomena were categorically excluded from natural philosophy owing to their miraculous universality. A local flood could be easily explained according to the regular laws of nature; a global flood, on the other hand, could not.[9] It was in this context that the Venetian scholar Sebastiano Fausto da Longiano claimed in his 1542 vernacular popularization of *Meteorology*: "It is therefore not possible, according to Nature, that the deluge was universal, because it is impossible for water to cover the entire earth."[10] The idea can be traced back to at least the fourteenth century, when the nominalist philosopher Jean Buridan, in a commentary on Aristotle's *Meteorology*, proposed that "by natural means, it would be impossible to produce a Universal Flood, which is to say, for the entire earth to be covered by waters, unless God made it happen by supernatural means."[11] These meteorologists did not deny that Noah's Flood had actually happened, as a matter of historical fact. Rather, the intent behind such statements was to categorize the Flood as a miracle—a supernatural occurrence that was inexplicable according to natural laws and processes. It was precisely the Flood's global scale that pushed it beyond the pale of physical possibility and thus also beyond the boundaries of natural philosophy. The larger goal of excluding the Universal Deluge from their collective consideration was to ensure the integrity of the boundary between meteorology and theology.

There were several reasons why the meteorologists of the sixteenth century were so categorical in their rejection of Noah's Flood and so eager to police the boundary between philosophy and faith. The first was that the Flood had long been a vexed topic of discussion among theologians, biblical commentators, and chronologers. The meteorologists' rejection of the Flood as a major event in earth history may have been motivated in some measure by a desire simply to bypass the difficult questions that consumed the time and energy of scholars in other disciplines. Biblical commentators bravely attempted to resolve the question of the quantity of water needed to flood the earth, where it could have come from, and where it might have gone when the Flood receded.[12] The chronologers were busy debating when the Flood happened and how long it had lasted. The Hebrew, Greek, and Latin Bibles were difficult to reconcile with each other on the question of the Flood's timing, creating a persistent exegetical problem that was compounded by the increasing awareness in Europe, from the sixteenth century onward, of Flood stories from the historical traditions of the Chinese, Egyptians, Incas, and other cultures across the globe.[13] All of this could be safely ignored as somebody else's problem as long as the natural philosophers stayed within the discursive boundaries of their field as they themselves defined it.

The categorical exclusion of the Flood from meteorology was also part of a larger effort to keep philosophy and theology distinct, a tradition dating back to the thirteenth century that was revived in response to the rapid religious transformations of the sixteenth century. The meteorologists' use of phrases such as "according to Nature" and "by natural means" were intended as disciplinary qualifiers, flagging their assertions about the Flood as true only within the limited discursive domain of natural philosophy. In doing so, they drew on the doctrine of "double truth," developed by medieval philosophers such as Jean Buridan in response to the condemnations of Aristotle in Paris in 1270 and 1277.[14] The medieval doctrine of "double truth" as redeployed by Renaissance meteorologists functioned to define meteorology, and natural philosophy more generally, as a discursive, disciplinary, and professional space largely independent of theology.

And yet, countervailing trends in sixteenth-century Europe helped to break down the disciplinary boundaries between philosophy and theology and bring the Flood into natural philosophy. The Protestant and Catholic Reformations, the Wars of Religion, the Scientific Revolution, and the Renaissance revival of ancient philosophies other than Aristotle's all worked to reconfigure the relationship between knowledge and faith in profound and lasting ways. The Reformations in particular exerted a profound, and profoundly ambivalent, effect on the relationship between philosophy and faith. Appealing to the doctrine of "double truth" had long been a strategy designed to ensure maximal *libertas philosophandi*, the freedom to philosophize, and doing so in the wake of the Council of Trent and the newly empowered offices of the Inquisition has to be understood as a strategy to protect philosophical inquiry. But the Catholic Reformation worked through other channels to bring philosophy and faith closer together, not all of them coercive. New ideals of unitary truth, in tandem with increased pressure from religious authorities in both Protestant and Catholic Europe to demonstrate the harmony of philosophy with faith, paved the way for a new generation of philosophers to explore the intersections of these domains. These sixteenth-century developments helped catapult the Flood into the center of natural philosophical debate.

### From Double Truth to Mosaic Natural Philosophy

The century from 1550 to 1650 witnessed a significant turn toward the Bible as a key resource for natural philosophy. The emergence of what Ann Blair has called "Mosaic natural philosophy" dovetailed with the Scientific Revolution's slow decentering of Aristotle's corpus as the foundation of natural knowledge.[15] Proponents of the new science wishing to challenge Aristotle's status as "the Philosopher" turned to the Bible instead, insisting that Moses, the author of the Pentateuch, gave a truer account of nature's history than Aristotle had. A 1578 English translation of the Swiss Calvinist Lambert Daneau praised Moses, "who at the commaundement and appointment of God wrot that historie . . . of the beginninge of the worlde, and creation of all things." Inspired by God, the Mosaic history was thus a true and accurate record of the creation and ancient history of nature. Therefore, "whoso shall deny that the knowledge of Naturall Philosophie may not truely and commodiously bee learned out of holy scripture, gainsaith the sacred woorde of GOD."[16] Proponents of the new science sought to replace Aristotle with Moses both out of a pious desire to make natural philosophy more Christian and out of a philosophical desire to contest some of the fundamental assumptions about nature derived from the Aristotelian tradition. Mosaic natural philosophy helped them do both.

As proponents of the new science sought to replace Aristotle with Moses as the ultimate source of authoritative knowledge, the Protestant and Catholic Reformations worked in different ways, though often along complementary channels, to bring religion and natural philosophy closer together. The birth of Mosaic natural philosophy coincided with the heightened emphasis, in both the Protestant and Catholic Reformations, on scripture as a source of knowledge—not just as a source of religious truth but as a source of *scientia*, or "certain knowledge," the gold standard in medieval and early modern scholarship. Martin Luther's emphasis on scripture as the single source of truth worked to promote pious natural philosophy in Protestant Europe, as seen in the efforts of the German reformer Philip Melanchthon to forge a distinctively Lutheran natural philosophy for the new Lutheran universities.[17] Meanwhile, the Fifth Lateran Council's call for Catholic philosophers to use their skills in order to demonstrate theological truths like the immortality of the soul encouraged, if not outright mandated, the growth of pious natural philosophy in the Catholic world.[18] Although some Catholic philosophers resisted the call for increased commerce between philosophy and faith, others rushed to bring them closer together, supplying philosophical evidence not just for the soul's immortality but for a host of other items of faith.

Even as the meaning and status of scripture was hotly contested between and among Protestants and Catholics, scholars of all faiths increasingly embraced the notion that the Bible could serve as a solid foundation for natural philosophy, and indeed for all forms of truth-seeking and knowledge-making. As Blair argues, Mosaic natural philosophy was an irenic agenda at the height of the Wars of Religion, reflecting a widespread desire for unitary truth during a time of violent religious and political conflict.[19] It made scripture appear as a valuable resource for philosophy, and philosophy as a valuable resource for defending the faith.

The new ideal of Mosaic natural philosophy began to manifest itself in meteorological texts in the late sixteenth century. The preface to Francesco de' Vieri's 1573 vernacular *Meteorology* promised readers that they would learn about "grand and marvelous" things "like the Comet, the Rainbow, thunderbolts, earthquakes, and many other testaments of the marvelous virtue and power of God and of Nature."[20] De' Vieri, a Florentine philosopher, repackaged the typical range of topics covered in meteorology as exhibits of the hand of God at work in the natural world. The generic form authors chose for their meteorologies also shifted. Where once treatises and commentaries were the norm, now dialogues became more popular, which enabled one fictive speaker to voice Aristotle's claims and another to critique it, whether on philosophical or religious grounds.[21] *Discorsi . . . sopra la Metheore di Aristotele* (Discourses on Aristotle's *Meteorology*, 1584), written by a nobleman from Ragusa (now Dubrovnik) named Nicolò Vito di Gozze, turned the traditional topics of Aristotelian meteorology into the subjects of a dialogue between two interlocutors. "G.," the main speaker and stand-in for di Gozze, challenges Aristotle on several points, including the scale of floods. The move to the dialogic form reflected an increasingly critical stance toward Aristotle even within this most Aristotelian of fields.

## Providentializing Natural Disasters, Globalizing Local Disasters

And yet, the Mosaic turn in natural philosophy did not on its own mandate that the Flood become the centerpiece of historical accounts of the earth and its history, as would come to be the case by the end of the seventeenth century. The Pentateuch recorded many historical events; Noah's Flood was only one of them. An equally important way that the Reformations worked to bring the biblical Flood into natural philosophy was by bringing increased attention to local, contemporary floods as signs of God's providence. This in turn helped blur the distinction between local and global floods and between natural and supernatural floods, distinctions upon which meteorology's exclusion of Noah's

Flood had been founded. Earthquakes, plagues, comets, famines, and various forms of extreme weather had long been interpreted as signs of God's wrath at human sinfulness, but Luther's belief that disasters were increasing in frequency as the end times approached directed increased attention their way. The social, religious, and political conflict occasioned by the Reformations seems likewise to have encouraged people to view local catastrophic events as a form of divine intervention in, or commentary on, contemporary confessional strife.[22]

Floods were especially susceptible to providential interpretation, for obvious reasons. As the Reformations progressed, local flood events were increasingly analogized to Noah's Flood, such that local, modern events came to carry some of the religious freight and meaning of the ancient, planetary, biblical deluge. A clear example of this transformation is seen in popular representations of the All Saints' Flood of 1570. The Netherlands had been experiencing increased flooding in recent decades as a result of the Little Ice Age, but the flood of 1570 was particularly devastating, killing an estimated three thousand people.[23] Contemporary representations of the All Saints' Flood invoked parallels with Noah's Flood. A woodcut of the event shows several people and a small house drowning in floodwaters, while an ark bobs safely above water. Over the ark hovers a dove with an olive branch, in case anyone remained unclear about the analogy being drawn. A shining star to the ark's right suggests an astral cause of the flood, while thunder, wind, and hail rain down, further emphasizing the miraculous conjunction of meteorological phenomena.

The woodcut was one among many providential interpretations of the All Saints' Flood in contemporary religious pamphlets, which stressed that this flood, like Noah's, was the result of sin. Whose sin, though? Here contemporary observers differed. The All Saints' Flood occurred in the middle of the Dutch revolt against Spanish Hapsburg rule. While Spanish imperial forces saw the timing of the disaster on a Catholic holiday as a sign of the righteousness of their cause in reimposing Catholicism on the rebellious Protestants, observers sympathetic to the Dutch revolt saw the matter differently. A Protestant observer writing from Strasbourg, for example, noted that the flood destroyed a Spanish fort at Groningen while leaving the town itself relatively unscathed. No doubt this detail was meant to signal to readers that God intended the flood to punish the Catholic Spanish and to aid the Dutch Protestants in their godly fight.[24]

As Elaine Fulton argues, Catholics and Protestants alike were inclined to assign a divine cause to local natural disasters in the sixteenth century, and they were likewise inclined to blame each other.[25] Generic interpretations of disasters as the result of sin were confessionalized by specifying whose sin, which then mobilized these events for use in religious and political polemics. The Reformations and Wars of Religion thus played an important role in encouraging the public to view local natural disasters, particularly floods, through a providential

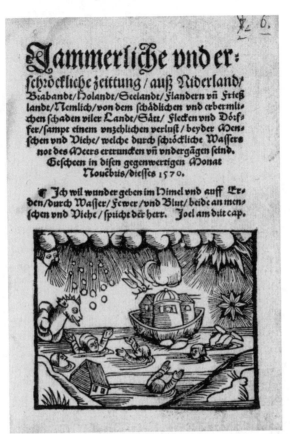

A pamphlet announcing the terrible events of the All Saints' Day Flood, *Jammerliche und erschröckliche Zeittung auß Niderland* (1570), featured a woodcut on its title page visually comparing the event to Noah's Flood. Intriguingly, this often-reproduced woodcut was also used in several pamphlets earlier in the century predicting a global apocalyptic flood in 1524. Bayerische Staatsbibliothek München, Res/4 Belg. 186 c#Beibd.6, fol. 1r, urn:nbn:de:bvb:12-bsb00026861-2.

and often explicitly confessional lens, which in turn prompted an increased association between local floods and the Universal Deluge.

An even more dramatic example of the growing association between contemporary floods and the biblical flood of Noah was the widespread panic over predictions of an apocalyptic global flood in the early decades of the sixteenth century. In 1499, two German astrologers, Johannes Stöffler and Jacob Pflaum, predicted disaster and disruption across the earth in February 1524, occasioned by a rare conjunction of planets in the house of Pisces, a water sign. Shortly thereafter, Stöffler and Pflaum's vague yet dire predictions of unspecified calamities morphed into the specific idea that the year 1524 would witness the return of Noah's Flood: a planetary flood designed to punish humanity for their sins. The Italian astrologer Luca Gaurico, for example, repeatedly published predictions of a global, biblical flood between 1501 and 1522. These astrological predictions of an apocalyptic global flood were widely broadcast via pulpit and broadsheet, causing widespread panic and fear. In Rome, people decamped to the upper stories and rooftops of their houses. The wealthiest citizens fled the city entirely, moving to their country villas in the hills. The Florentines, perhaps remembering the apocalyptic predictions of the popular preacher and condemned heretic Fra Girolamo Savonarola, who preached a sermon in 1494 predicting that a second Noah's Flood would destroy Italy sometime soon, responded to the latest round of diluvial predictions by stockpiling grain. Throughout the peninsula, there were special masses, penitential processions, and other public religious rituals. Many built special boats designed to ride out the flood. In southern France, a wealthy magistrate allegedly ordered the construction of a small ark designed to safely house himself, his family, an experienced captain, a few domestic animals and several months' worth of provisions.[26]

As with the All Saints' Flood of 1570, the predicted global flood of 1524 was drawn into confessional polemics. Pro-Lutheran pamphlets that recast the 1524 prediction as an augur of the imminent success of Lutheranism circulated at the 1521 Diet of Worms.[27] Meanwhile, many Catholics viewed the predicted flood as God's punishment for, and attempt to extinguish, the spread of the Lutheran heresy. Indeed, one Italian ecclesiastical official referred to Luther himself as a "deluge" threatening to engulf the church. Analogizing the spread of Lutheranism to Noah's Flood indicates the seriousness with which Catholic officials were beginning to take the Lutheran movement in the 1520s, as its growth threatened the Roman Church's own claims to universality. Along with the panicked preparations for the global flood of 1524, this comparison of Lutheranism to the deluge was yet another manifestation of the growing tendency, sparked by the Reformation, to assign providential and indeed also global significance to local events by analogizing them to Noah's Flood.[28]

This tendency can also be seen in a subtle but important linguistic shift in early modern Italian in the terminology of flooding. For several decades before the future global flood became the talk of Europe (Machiavelli listed it among those topics of conversation that were so clichéd that they should be avoided), the Italian peninsula had been experiencing a significant increase in the size and number of floods as a result of the accelerating effects of deforestation. Moreover, the same cool weather and heavy rains associated with the onset of the Little Ice Age, which would later play a role in the All Saints Flood, were also at work in southern Europe. Early modern Italian had different terms to distinguish a global, providential, and supernatural flood like Noah's (or perhaps like the one predicted for 1524)—*Diluvio* (the Deluge)—from a local flood with natural causes, called an *alluvione* (flood). However, during these same decades of increased flooding, the distinction between *Diluvio* and an *alluvione* began to blur. Even though the global flood of 1524 failed to materialize, serious floods were reported in Cesena in 1525 and in Rome in 1530, and a number of publications referred to these local events as a *diluvio*. This growing tendency of referring to local floods as "deluges" (*diluvii*) signaled not only the scale and severity of these flood events but also the providential meaning that people attached to them.[29]

## "We Catholics Believe": Toward a Catholic Meteorology

The notion that the Flood's universality put it beyond the pale of philosophical inquiry changed dramatically in the Reformations' wake. The emergence of Mosaic natural philosophy and the providentializing of local floods during the Reformations played a key role in effecting this shift. Treating the Bible as a central source of natural knowledge made it increasingly difficult or undesirable (depending on one's point of view) to maintain a barrier between philosophy and faith. It also made biblical history, and particularly the book of Genesis, appear increasingly relevant to the task of reconstructing the history of nature. Meanwhile, the growing tendency in the public at large to ascribe a supernatural cause and religious significance to local natural disasters, whether by rhetorically comparing them to the global Flood of Noah or by predicting that they would actually be global, helped to break down the distinction between local, natural floods and global, supernatural ones.

In the second half of the sixteenth century, philosophers tentatively began to consider the Flood as a topic for natural philosophy. No one denied the Flood's supernatural origins or its theological significance, but philosophers increasingly began to assert that the Flood also had natural causes and effects that could be considered from within their disciplinary sphere. The first efforts exhibited a great deal of uncertainty about exactly how to bring the meteorological tradition into dialogue with the scattered and often opaque hints in scripture regarding the earth's ancient history. An illustrative example of the ambivalence on display in these early attempts to consider the Universal Deluge from within meteorology is di Gozze's *Discourses*, published in Venice in 1584. The main expository interlocutor "G." declares, "According to Aristotle in this place [*Meteorology*] and . . . according to Plato in the *Timeaus* and in the books *Of the Laws*, the Flood was natural; but we Catholics believe that the Noachian Flood was supernatural."[30]

On the one hand, di Gozze gestured toward the standard philosophical objection to the Universal Deluge, namely, that its supernatural character excludes it from consideration within natural philosophy. Immediately after this statement, "G." launches into a typical discussion of the difference between local floods, which were natural (and hence part of meteorology), and global floods, which were supernatural (and therefore excluded from meteorology). On the other hand, "G."'s statement of belief—"we Catholics believe"—marks a departure from previous meteorological discourse. A supernatural Flood thus formed part of a confessionalized credo, something that di Gozze told his readers they should believe not just as Christians but as good Catholics.

Rather than flatly denying the physical possibility of a universal deluge and then moving on, as earlier meteorologists would have done, di Gozze went on to list possible natural causes for the Flood before reaffirming that these natural causes were merely hypothetical, because the Flood was of course actually supernatural in origin. "G." states that just as "part of the heavens has the power of causing a deluge in just one part of the earth," so too could "all of the heavens" cause a global flood and "occupy the whole earth with water." "G." then quickly pivots to reaffirm that Noah's Flood "was supernatural according to the sheer will of God, because it could never happen naturally," but then he backtracks toward his original position, saying that even though the stars do not currently possess the power to flood the whole earth, "nevertheless by the great virtue of God who so willed it, they could have had it." By suggesting that Noah's Flood could have been caused by the stars as long as God endowed them with the necessary "natural virtue," di Gozze continued to endorse the idea that the Universal Deluge did not belong in philosophy even as he blurred the distinctions between natural and supernatural phenomena, between local and global phenomena, and between philosophy and faith, all the distinctions upon which meteorology's exclusion of the Universal Deluge had for so long depended.[31]

## Medicine, Astrology, and Meteorology: Erculiani's Man-Made Global Flood

The most striking example of the new philosophical openness to global catastrophe is a remarkable short work that has only recently begun to attract the critical attention it deserves. *Letters on Natural Philosophy*, written by the Paduan apothecary Camilla Erculiani, is perhaps the earliest attempt by any early modern European author, of any gender, country, or religion, to forcefully make the case in print that Noah's Flood was both universal *and* natural and thus intelligible by means of natural philosophy. Published in Kraków in the same year that di Gozze's *Discourses* appeared in Venice, Erculiani's *Letters* is a collection of philosophical missives written by Erculiani to two foreign correspondents, the French physician Georges Guarnier and the Hungarian humanist Márton Berzeviczy, as well as one response from Guarnier. The dialogic structure of the new meteorologies thus became, in Erculiani's *Letters*, a true exchange of ideas across political borders and also across confessional lines: Berzeviczy, chancellor to the Polish king Stefan Báthory, was a Lutheran.[32] In the course of her epistolary conversations with these two men, Erculiani showed none of the ambivalence about naturalizing the Flood displayed in the meteorological dialogues of her male contemporaries like di Gozze, offering instead a boldly materialist account of its natural causes. Chief among these are the astrological influences of other planets and the environmental impact of human bodies on the earth's balance of natural elements. Humans, in other words, are among the natural causes that Erculiani believed had triggered the Flood. But the naturalism and materialism of her account of the Flood did not signify a rejection of its religious significance. Rather, her ideas about a man-made Flood flowed from a sense that sinfulness was embodied in humankind in a profound way and was, by virtue of its material instantiation in human bodies, capable of literally destroying the earth. Erculiani's *Letters* reveal her to have been deeply interested in using the idea of a naturally caused and at least partially man-made global deluge in order to answer essentially religious questions about sin, suffering, and salvation. The idea of nature destroyed worldwide by humanity arose in dialogue between men and women, Protestants and Catholics, at the height of the Reformations.

Although she did not come from an elite family, Erculiani was unusually well-educated in natural philosophy for a woman of the sixteenth century. Twice-married to apothecaries, she was also an apothecary in her own right. This is borne out by several pieces of evidence, not least by her self-identification on the title page of the *Letters* as "Camilla Herculiania, Apothecary at the Three Stars in Padua."[33] Like many wives and daughters of artisans, Erculiani almost certainly learned the trade herself and worked actively within the profession as a part of the family business. As the wife of an apothecary and a partner in the business, she would have received a solid education in vernacular botany, pharmacopeia, *materia medica*, and related fields of natural knowledge. Moreover, the close proximity of the Three Stars to the University of Padua makes it highly likely she had the opportunity to converse with scholars of medicine and philosophy who came into her shop.[34] Her first letter to Guarnier, for example, recounts a conversation with an unnamed physician on the topic of man's mortality in relation to his earthly composition, a conversation that could have transpired in the Three Stars or at least been modeled on the kind of intellectual exchanges she had the opportunity to engage in there. Guarnier and Berzeviczy both spent time at Padua's famous university in the 1560s and 1570s. The four letters collected in *Letters on Natural Philosophy*—two from Erculiani to Guarnier, one from Guarnier to Erculiani, and one from Erculiani to Berzeviczy—represent the long-term intellectual relationships Erculiani enjoyed with these two scholars during and after their time in Padua as well as her active engagement in cultures of science and medicine in Padua, Italy, and Europe.[35]

The growing number of vernacular translations and popularizations of Aristotle's *Meteorology* printed in the later sixteenth century gave women access to this key work of natural philosophy.[36] Many of these translations were dedicated to women or contained prefatory material defending women's capacities in the sciences, which strongly suggests that extending the field of meteorology to include elite women, at least as readers, was an intentional goal of producing these vernacular editions.[37] Di Gozze's *Discourses* begins with an introductory letter in praise of learned women written by his wife, Maria Gondola, and addressed to the poet and noblewoman Fiore Zuzori. Whether or not Erculiani read any of these vernacular *Meteorologies* is unknown, but it is certain that she was familiar enough with the field to draw on it while also freely departing from it in several key respects.[38] In her *Letters*, not only does she insist, contrary to meteorological precedent, that the Universal Deluge had natural causes that could be elucidated by a natural philosopher, but she also draws liberally and syncretically on astrology, Galenic medicine, and the text of Genesis in order to prove it.

The subtitle of Erculiani's *Letters on Natural Philosophy*—"in which is treated the natural cause of Floods [*Diluuij*], the natural constitution of man, and the natural formation of the rainbow"—repeats the adjective *natural* three times, emphatically announcing the author's intent to treat the Flood and related topics of biblical history as natural phenomena.[39] Over the course of her letters contained in this slim volume, Erculiani describes two complementary ways that a Universal Deluge could have happened naturally. The first identifies astral influences as the natural cause of the Flood, while the second blames humanity's physical constitution and embodied predisposition to harm the earth.

In the first scenario, outlined in the letter to Berzeviczy and dated April 9, 1581, Erculiani turned to the science of astrology in order to elucidate the role of heavenly bodies in causing the Universal Deluge. The planet Venus was rising, Erculiani proposed, just as Noah was boarding his ark. "Lady Venus" was responsible for "raising [the waters] from the depths to the heights, and this she could do because . . . she has power over the waters." After Venus recessed and another planet became dominant, the floodwaters subsided, and dry land reappeared. "In truth," Erculiani told Berzeviczy, "the natural philosopher and the astrologer cannot assign any other cause to this or to any other universal floods."[40]

Erculiani's theory of a Venereal Flood bore strong similarities to the astrological theories that animated predictions of a global Flood for 1524. Indeed, astrology appears to have played an important role in prompting philosophers to consider the possibility of a naturally caused global flood. Hippocrates said that universal phenomena required a universal cause, and the stars, owing to their location beyond the earth, might be capable of effecting change on all parts of the earth simultaneously. In 1520, the Italian scholar Tiberio Russiliano suggested that Noah's forewarning of the Flood came not from a divine messenger but from his skills as an astrologer; he had read in the stars that the Flood was coming and made all of the necessary preparations.[41] Erculiani's reference to "the great philosopher and astrologer Noah" echoes Russiliano's claim.[42] Di Gozze's suggestion in his 1584 *Discourses* that the stars could have caused Noah's Flood at God's command reflects a similar sense that astrology provided a more promising framework than did meteorology in which to account for a natural, global Deluge. As a branch of natural philosophy with far greater openness to religion, astrology offered Erculiani and her peers a way to consider Noah's Flood—a global catastrophe with tremendous theological significance—as a subject for natural philosophy.[43]

Erculiani's second theory regarding the natural causes of the Flood, by no means incompatible with the first, relies on the influence of people, rather than the stars, on the earth. Human embodiment and overpopulation, specifically, was the culprit, an idea she couched within a more general theory of the relationship between living creatures and the global environment. Drawing on Galenic matter theory, Erculiani asserted that all creatures are made up of a mixture of elements, but one element predominates in each kind of creature.[44] "The ox holds more of the element of earth," she wrote, "and the birds more of the element of air, and the eagle and salamander more of the element of fire, and the fish more of the element of water."[45] Each individual organism thus draws part of its material substance from the elements around it, and this holds true for humans as well. "The elements from which we are formed," she asserted, "join together with our particular substance in our creation."[46] After a creature dies, the element of which it is principally composed returns to the general fund of that element. The water of which a fish is composed, for example, rejoins the general mass of elemental water on Earth after the individual fish dies. Her vision of a process of dynamic exchange between living creatures and their geophysical environment was likewise in line with Galenic thinking about the dynamic relationship between human and nonhuman Nature. Galen theorized that the mixed human body was permeable, and that diet and climate were two major ways the mixed human body interfaced with the mixed world of nature in, around, and beyond it. This cycle of birth and death, composition and decomposition, ensured a natural equilibrium between the four elements.

According to Erculiani, this natural equilibrium between the elements had been out of balance for quite some time prior to the fatal moment when Noah boarded the ark and Venus ascended in the heavens. For many centuries before the Flood, humans had been growing too large and too numerous and thus were taking up too much elemental earth. A land animal like the ox, man was likewise materially composed of more earth than of the other three elements. "The Deluge came," Erculiani wrote in the first letter addressed to Georges Guarnier, dated August 7, 1577, "because men had grown too many in number on the earth, too large in body size, and too great in the length of their lives, so that after the sin, the element of earth was much diminished, earth being that element which made up the major part of their gigantic bodies." The sheer number and size of antediluvian men meant that they were taking away too much of the element of earth from the rest of global nature. Their longevity meant that they were not dying quickly enough for the earth contained in their bodies to return to the general fund of elemental earth. Too many people were being born, growing too large, and living too long; consequently, the element of the earth "was not replenished for many hundreds of years." Overpopulation and overconsumption of elemental earth caused a planetary imbalance between the elements of earth and water. When the disequilibrium reached its tipping point, the earth was "engulfed in water, water being that element which had, for its own part, contributed little to the bodies" of the gigantic antediluvian men. In killing off most of the world's people, the Flood returned the earth in their bodies to the general fund of elemental earth, thus restoring the equilibrium between earth and water on planet Earth and causing the floodwaters to recede.[47]

Erculiani's two theories of the natural causes of the Flood thus combined meteorology with astrology and medicine in unique ways. They also show the influence of the turn toward Mosaic natural philosophy in the world of Catholic learning. She frequently cites Moses along with other philosophers, both ancient and modern, and in some places privileges the testimony of Moses over the words of the pagan philosophers of antiquity. The *Letters'* epistolary structure offered Erculiani the same opportunity to critically distance herself from Aristotle as did the dialogic structure adopted by many of her meteorological contemporaries. When Guarnier objected to Erculiani's theory of humanity's overuse of elemental earth on the grounds that it contradicted Aristotle and Galen, Erculiani responded that many other philosophers agreed with her, including "the great legislator Moses, who says this clearly in the beginning of his account of the creation of the world."[48]

Other philosophers paid lip service to Moses's philosophical prowess without actually demonstrating how the Bible could or should inform philosophical inquiry. Erculiani's second, Galenic theory of the Flood depended crucially on her interpretation of the text of Genesis. Writing to Berzeviczy, Erculiani stated: "At the time of the famous flood, which is said to have been universal, men were living many hundreds of years, and there were still giants during that same time, as one can read very well in the fourth chapter of Genesis. In the sixth chapter, one reads that the earth was submerged by the waters after it was diminished."[49]

The idea that antediluvian men (either all humankind or perhaps only men) were unusually big and tall and extraordinarily long-lived was common enough in early modernity, as will be discussed in greater detail in chapter 3.[50] Erculiani's contention that antediluvian longevity and gigantism were somehow responsible for causing the Flood was, however, quite novel, as was her use of biblical evidence, citing chapter and verse, in a work of natural philosophy. Erculiani immediately followed this citation of Genesis with another assertion: "Speaking in terms of natural philosophy [*naturalmente*], it is clear that the cause of the diminution could proceed because it was transmuted in water, since the dryness of the earth repelled that water."[51] Clearly, Erculiani believed that she was finding natural causes for the Deluge in the text of scripture itself, causes upon which she as a philosopher could expound and elaborate.

Erculiani's use of scriptural evidence is all the more remarkable given that her access to the Bible would have been limited, either by the Inquisition's growing hostility to vernacular editions or by the linguistic challenge presented by the Latin Vulgate, whose language she may not have been able to read.[52] The difficulty she must have encountered in accessing the specific words of Moses reveals how crucial scripture was to her philosophical project. It was scriptural evidence that enabled her to refashion the theories of Aristotle and Galen into something new: an account of the human role in causing global natural disaster. Her impulse to use the words of scripture to elucidate matters of natural philosophy ties Erculiani to the milieu of pious physicians in north-central Italy in the sixteenth century, as recently described by Andrew Berns.[53] Bringing the Flood into natural philosophy was clearly not, for her, a means of desacralizing it, but rather a means of exploring the intersection between philosophy and faith. It was also a means of constructing a biblically based history of the earth where none had existed before.

## "I Have Offered up My Own Ideas": Gender, Authority, and Orthodoxy

In arguing that Noah's Flood had natural causes, Erculiani challenged, perhaps intentionally, the divide between philosophy and faith that previous authors writing about the earth from within natural philosophy had sought to maintain. Moreover, the fact that one of the first people to attempt this was a woman is probably not a coincidence. When Guarnier objected again that many of the opinions expressed in her letters were contrary to Aristotle, Erculiani tried a different tactic, defending her theories by saying: "I know this from reference to Nature [*naturalmente*], without looking at Galen or Aristotle."[54] Erculiani's rather cavalier treatment of Aristotle and other ancient authorities shows the influence of the Scientific Revolution's decentering of classical, textual authority and the emerging rhetorical authority of nature, the close observation of which trumped the words of Aristotle—as did the words of Moses in other contexts. Erculiani's bold self-defense "I know this from reference to Nature" may also have stemmed from her desire, as a woman writing in a male genre, to rhetorically construct her own intellectual authority. She insists several times in the *Letters* that her opinions about the Flood are entirely her own, formed in reference to reading the books of nature and scripture, not from reading the works of other philosophers. In response to a query from Berzeviczy about where she first read about antediluvian longevity and gigantism as causes of the Flood, Erculiani responded, "I tell you that I have not taken this from any author." She went on to say that while "I do not deny that I have read various authors . . . [but] I have offered up my own ideas in writing as well."[55] This assertion of radical intellectual independence interestingly works to efface her deep engagement with key philosophical texts, both ancient and modern. It also reveals how this woman philosopher used the growing rhetorical authority of nature, independent of classical authority, to establish her own authority independent of the philosophical authority of men.

Erculiani knew she was entering a male discursive space, and the structure and language of the *Letters* works to constantly remind the reader that all of her interlocutors are men: not only the two named addressees and respondents, Guarnier and Berzeviczy, but also the unnamed man whose recounted *viva voce* conversation with Erculiani forms the substance of the first letter to Guarnier.[56] Indeed, the fact that this anonymous interlocutor is referred to throughout simply as "the man" or "that man" continually reminds the reader of the gender of her interlocutors.[57] Erculiani was certainly aware of elite women who acted as patrons to male philosophers, such Queen Anna Jagiellon of Poland, to whom *Letters on Natural Philosophy* is dedicated. Erculiani's dedicatory letter to the Polish queen, wife of Stefan Báthory and daughter of the Italian Bona Sforza, demonstrates her awareness of the learned ladies of classical antiquity, and she surely must have known the names of at least a few of the famous learned women of Renaissance Italy. In spite of these illustrious examples, Erculiani had no role model other than herself for women who publicly participated in intellectual exchanges on topics in natural philosophy.[58]

Her self-consciousness of being the lone woman in a male discursive space is clearly shown in the prefatory material, where Erculiani anticipates her readers' surprise at reading a work of natural philosophy authored by a woman. "It will undoubtedly appear as a marvel to some," she wrote, "that I, a woman, have set myself to writing and putting into print things that do not, as custom would have it, pertain to women."[59] Women authors of poetry, belles lettres, and the like were widely known and celebrated in Renaissance Italy, Erculiani knew, as were defenses of women's intellectual capabilities and praises of women's learning written by both men and women.[60] But publishing her philosophical exchanges with learned men aimed to do more than establish women's capacity for intelligent thought and for creative expression in literature and the arts. She wanted her book to demonstrate that women were equally capable in all areas of learning, not just in those that pertained to, or were deemed suitable for, women. "I wanted, through my studies, to let the world know that we women are just as capable in all of the sciences as men," she declared in the dedication to Queen Anna.[61] Erculiani was quite right to draw readers' attention to her singularity as a woman natural philosopher. Her self-authorizing rhetorical strategy ("I have offered up my own ideas in writing") can therefore be read as an assertion of independence from classical authorities like Aristotle and also from the homosocial space of sixteenth-century Italian natural philosophy.

The publication of *Letters on Natural Philosophy* attracted the attention of the Paduan Inquisition, who brought Erculiani in for questioning in 1585 or 1586 on suspicion of heresy and released her with absolution or perhaps a light punishment.[62] During her interrogation, Erculiani's gender became ever more salient, a point to which I will return shortly. First, it must be noted that Erculiani's two theories about the natural causes of the Flood did raise serious theological difficulties independent of her gendered authorial voice. Around the time of her interrogation, Pope Sixtus V released the bull *Coeli et terrae creator Deus* (1586) condemning judicial astrology, that is, astrological predictions of the influence of the stars on human affairs, which he claimed was the sole province of God.[63] Erculiani's ideas about a Venereal Flood was a theory about the past rather than a prediction about the future, but it certainly did give the stars a major role to play in shaping the course of human history. Far more problematic, however, was her theory of the Flood being caused by humanity's taking up too much elemental earth. Her theory regarding "the natural constitution of man" in relation to the surrounding environment was on its own a highly original reading of Galen. But proposing that the physical constitution of human bodies predisposed them to destroy the planet's natural balance of elements and trigger a Universal Deluge carried with it several unsettling, if not outright heretical, implications. Not only did her theory appear to de-emphasize the role of human sinfulness as a cause of the Flood but, taken to its worst extreme, it redirected blame from humans to God.

The theological pitfalls of this theory become even more apparent when considered alongside her theory of the Fall. The earthy composition of Adam's body, she wrote in the first letter to Guarnier, predisposed him to die: a bold departure from the orthodox position that mortality was one of several divine punishments meted out to Adam after his fall from a state of innocence and grace.[64] Erculiani's materialist reading of the Fall transformed death from the contingent outcome of original sin to the preordained outcome of man's physical embodiment. Her theory of the Fall—or rather, her theory of the Fall's irrelevance to human mortality—appeared to strip humans of their free will to sin. It also appeared to fault God for creating human bodies in such a way that their owners were destined to die. In precisely the same way, her theory of the Flood appeared to fault God for having given humans bodies that forced them to destroy both themselves and each other in a global calamity. Erculiani thus proposed a natural and indeed highly materialist basis for the spiritual history of humankind. If taken to its very worst conclusion, it implied that God was a monster who would not only punish people for behavior over which they had no control but would design them to engage in this behavior in the first place, essentially setting humanity up to fail.

While a sixteenth-century reader could have easily drawn such conclusions from Erculiani's cryptic prose, other evidence suggests that it was almost certainly not her intent to advance the notion that sin was irrelevant or that God was a monster. She makes several references in the *Letters* to a "treatise that I am now composing on sin," apparently never published and now lost.[65] The fact that Erculiani was composing a work on sin at the same time that she was composing these letters on the Flood indicates that sin was very much on her mind and was also a topic of considerable importance to her. Her one, tantalizing reference in the *Letters* to sin in connection with the causes of the Flood is maddeningly vague: "after the sin, the element of earth was much diminished."[66] We can only assume that Erculiani's lost treatise on sin would have shed light on her thoughts about the connection between humanity's collective sin and the planetary deluge triggered by their depletion of elemental earth.

As it is, the temporal connection she does make between human sin and human depletion of earth implies a causal connection, even if she refused in the context of *Letters on Natural Philosophy* to spell out precisely how that chain of causes worked. Just as it was presumably not a coincidence that Venus was rising in the heavens and raising the waters of the earth at the same moment Noah and his family were stepping on board the ark, surely it was also providential timing that "the element of the earth was much diminished" and overpowered by water "after the sin." Her *Letters* collectively suggest that all of these events—Venus rising, ark-boarding, earth diminishing past the tipping point, waters overflowing— happened at roughly the same time, and they all happened "after the sin." This temporal conjunction suggests first of all that her two theories of the Flood (astrological and environmental) were in fact one unified conjecture. It also suggests that she did see a causal connection between human sin, on the one hand, and the natural causes of the Flood outlined in the *Letters*, on the other. The movement of the heavens, the materiality of human bodies, and the physical degradation of the earth did not replace human sinfulness as causes of the Flood, but rather all of these factors seem to have worked in tandem in some unarticulated way. What seems certain is that she was looking for a way to relate the spiritual and material causes of the Deluge in a way that linked human history and natural history on a planetary scale.

## Gendered and Embodied Sin

It is highly unlikely, then, that Erculiani wished to argue that sin was irrelevant to the drama of sacred history. It is far more likely that she wished to advance a new understanding of sin, and that she found a way to accomplish this by combining the biblical stories of the Fall and the Flood with the methods of natural philosophy. Her theory that humans' earthy bodies predisposed them to individual mortality and also to collective depletion of earth did indeed divest humanity of free will—but it did not necessarily diminish the importance of sin as a motor of sacred history. Rather, it reflected a deeply pessimistic view of sin as pervasive and inevitable. If humans' bodies were the root cause of both individual mortality and the near-universal genocide of the Flood, then perhaps sinfulness was simply embodied in humankind in a deep, inevitable, and heretofore unacknowledged manner.

There is tantalizing evidence that Erculiani may have viewed sinfulness as not only embodied but gendered. Her treatise on sin is now lost, though two surviving, unpublished letters to Erculiani from the Venetian scholar Sebastiano Erizzo offer intriguing clues to its possible contents. Erculiani's side of the correspondence has also been lost, but the second of the two letters, written the same year as the *Letters* was published, includes Erizzo's critical response to a provocative assertion in Erculiani's preceding letter about gendered differences with respect to sin. Erizzo paraphrased Erculiani as claiming that men unjustly "made laws and decrees pertaining to women", even though women "did not ever want to be subject to [their] theological injunctions." Women should not be subject to these laws because Eve was less sinful than Adam. Erizzo summarized Erculiani's position as stated in her previous letter: "You say that Adam and Eve our first parents . . . had laws that women would not have to abide by, because women are exempt from all other human labors and obligations, except for the generation and production of men."

Presumably, Erculiani was here referring to the fact that scripture records different punishments for Adam and for Eve as a result of their choice to sin: Adam was cursed with toil and mortality, Eve with pain in childbirth. Their differing punishments therefore indicated differences in the severity or at least the kind of their sin, Erculiani seems to have suggested. Erizzo counterargued that God made Adam and Eve alike and that both men and women were bound by the same divine law. "I want to believe you're joking with me," he wrote in reply, "and that this is not your true opinion," perhaps trying to warn her away from espousing such an obviously unorthodox position. Erizzo went on to make an argument about the equality of the sexes as the basis for equal subjection under divine law, "both sexes man and woman being subject to God and to his divine injunctions." As far as can be determined from Erizzo's rebuttal, Erculiani seems to have ventured the opinion that men and women were differently oriented toward sin and thus toward divine law—or at least toward men's interpretations of divine law. This gendered spiritual difference imposed unfair burdens on women, who should not be required to observe the same religious injunctions as men.[67]

The view that men and women had different orientations with respect to sin would harmonize well with the theory about embodiment and sin she advances in the *Letters*. We know she had argued that humankind's embodiment predisposed them to sin. And if she followed Galen in believing that men and women possessed different physical constitutions, even if the composition of elements that made up their bodies was only slightly different, it would follow that men and women were differently orientated with respect to sin. What if the theory of the Flood's causes that Erculiani articulated in the letters to Guarnier and Berzeviczy applied to men in particular and not to humankind in general? The term she used most frequently as the nominal subject of her natural-Mosaic history is *huomini*, which can be rendered either as the gendered "men" or the quasi-genderless "mankind." Erizzo's letter opens up the possibility that Erculiani believed that sinfulness was differently ingrained in the fabric of men's bodies than in the bodies of women. If so, then presumably she also meant that men, not women, were responsible for human mortality after the Fall and for global environmental destruction before, during, and after the Flood.

It is possible that her goal in writing about the Flood as a natural phenomenon was not just to better understand sin but to understand and to critique gender differences in the world she lived in. In the prefatory letter to the reader, which begins by declaring her intention to shock people by writing about topics not pertaining to women, Erculiani provides a peek into her life as a woman, which included "the toil of raising children, the burden of running a household, the obedience I owe to my husband, and my weak constitution [*complessione*]."[68] Although she claimed not to be bothered by these things as much as by the backlash she anticipated from her critics, Erculiani clearly implies that her gendered labor as wife and mother as well as the physical constraints on her body's energy competed for her time with the labor of philosophy. Comparing this passage in *Letters on Natural Philosophy* with her unpublished letters to Erizzo suggests yet another way of understanding Erculiani's position on the embodied and gendered dimensions of sin: that the burden of original sin fell too heavily on women, a burden that they continued to experience in the "toil" of their everyday labors and in the physical weakness of their postdiluvian bodies.

The view that women were less sinful than men was probably a minority position in early modern Italy, but it was not unprecedented. A century earlier, the widely praised Veronese humanist Isotta Nogarola had defended the view that Eve was created less perfect and more ignorant than Adam, which diminished the severity of her sin so much relative to Adam's sin that she and her daughters had barely been punished by God and stood in no need of redemption, as Adam's sons did.[69] The more specific view that women's unique and superior spiritual state was rooted in their differential physical constitution was strongly championed by the Venetian scholar Lucrezia Marinella in her virtuosic misandrist screed *La nobiltà, e l'eccellenza delle donne, co' diffetti, e mancamenti de gli huomini* (*The Nobility and Excellence of Women and the Defects and Vices of Men*, 1600).[70] Published fifteen years after Erculiani's *Letters*, Marinella's *Excellence of Women* relates embodiment to morality, as Erculiani's *Letters* had, while also explicitly linking embodiment and morality to divinely created sex difference.[71] Marinella's treatise thus furnishes context and possible clues to the theory of gendered spiritual conditions that Erculiani might have proposed in her lost treatise on sin.

Written in haste as a furious rebuttal to Giuseppe Passi's misogynist tract *I donneschi difetti* (The defects of women, 1599), Marinella's treatise promises, in its subtitle, to "destroy the opinions" of Passi and other similar authors with "strong arguments" and "infinite examples."[72] Marinella drew on the evidence of both history and natural philosophy in order to prove that "the female sex is, in its actions and operations, more singular and excellent than the male sex."[73] Her historical approach involved compiling examples of great women in history, both ancient and modern, such as the learned Aspasia, teacher of Pericles; Sappho, "the glory of poetry"; and Hildegarde of Bingen, the learned author of "four books about Nature."[74] Instead of treating these notable women as exceptional, as did so many other works in the *querelle des femmes*, Marinella argues that they illustrated the superiority of the female sex more generally, as a group. Famous women were not the exceptions that proved the rule of female inferiority. Nor were they examples of female parity with men; they were evidence of women's "excellence" relative to men.[75]

Marinella reasoned from specific historical examples of high-achieving women to categorical claims about women in general by using the methods of natural philosophy. In order to make clear that the achievements of these particular women are reflective of the capacities of all women, she sought to establish that women are a biological kind who, by virtue of sharing a common body type, share a common set of moral and intellectual capacities. She began, as Erculiani had, by confronting Aristotle. Marinella deployed the language of Aristotelian causality in order to reverse the Aristotelian opinion that men were physically, morally, and intellectually superior to women. God was the efficient cause, the maker, of both men and women, she reasoned, and although one might be tempted to believe that their common origin and shared speciation was a source of equality and similarity between them, that would be a mistake. All people are "engendered by the same cause," but "the Eternal Artisan" nevertheless designed them "with different forms."[76] Sex difference was therefore divinely created. God intentionally created women with bodies and souls that were significantly different from the bodies and souls of men.

Marinella then turned to the Aristotelian language of material causality in order to argue that women's bodies are not just different from, but indeed superior to, the bodies of men. "The female body is more noble and worthy than the male body," she asserted, as demonstrated by "its delicacy, its own complexion or temperate nature, and by its beauty."[77] She conceded Aristotle's claim that women's bodies do not possess as much heat as men's, but she used this fact to argue that the relative coolness of women's bodies allows them to reason more calmly and clearly than do men's elevated body temperatures.[78] Women's cool bodies and cooler heads operate in concert with their more rational and contemplative souls, which are likewise "much more noble than those of men."[79] This is the reason why women are more temperate, merciful, and virtuous and also why they are "able to learn more perfectly the same arts and sciences that men learn."[80] The embodied souls of women enable them to behave more ethically and also to reason more intelligently. Created biological sex difference is therefore the basis for women's superior intellectual and moral capacities.

Marinella drove home her point about women's spiritual superiority in reference to the Age of Reformations in which she and her readers lived. Men, she wrote, are more prone to heresy than women. Calvin and Luther are simply the most recent and famous examples of this dangerous moral weakness to which men are uniquely susceptible. "I am stupefied," she wrote, "how some writers dare to claim that women have invented new religious sects and new heresies." While Marinella acknowledged the existence of plenty of women heretics in the present day, she insisted that the original promoters and "inventors" of heresy were "all men." In her effort to overturn the misogynist allegation that women are more likely to become heretics—or, like their mother Eve, sinners— Marinella incidentally exonerated her Protestant sisters in order to place the blame for the Protestant Reformation squarely on the shoulders of men. Marinella thus furnished a concrete, familiar, and strongly confessionalized example of how women's spiritual superiority and men's spiritual weakness were made manifest in the world of early modern Europe. The gendered differences in body and soul she was concerned to demonstrate were not merely matters of academic disputation; they had real and profound effects on the social fabric and also on everyone's chances of salvation.[81]

Like Erculiani, Marinella used natural philosophy in order to think through the relationship between gender, embodiment, and spirituality. As with Erculiani, Marinella's engagement with the world of medicine may have prompted her to adopt a bodily, materialist perspective on religious questions. The daughter of a prominent physician and medical author, and later the wife of a physician, Marinella came from a higher-class background than Erculiani and had access to a far bigger library. But the concern with embodiment and spirituality was one they both shared, and Marinella's far more explicit consideration of physical sex difference as the root of gendered differences in spiritual states allows us to imagine what Erculiani's lost treatise on sin might have contained.

One final piece of evidence from Erculiani's *Letters* suggests that her theory of the Flood was not merely of historical interest but was indeed a means of meditating on a looming crisis in her own historical moment. A single, cryptic line in the first letter to Guarnier puts forward the tantalizing suggestion that the very same causes behind Noah's Flood were happening again, in her own day. "At the hour of the Flood," she wrote, "and now in the present hour, with respect to the growth and habits of man and animals and plants and clothes and buildings, the earth was much diminished."[82] Humans were again becoming too numerous, as were other living creatures ("animals and plants"). This untenable growth in the population of living things was once again taking up too much elemental earth, she hinted. While she included animals and plants in this destructive pattern of overpopulation and elemental depletion, Erculiani seemed to be saying that humans were primarily to blame by identifying man-made artifacts, "clothes and buildings," as part of the problem, presumably because they were also using up too much earth. Erculiani thereby implied, in this letter of 1577, that a second global flood might be triggered sometime in the near future. As the foregoing discussion of sin and embodied gender suggests, the environmental crisis she warned about had roots in both natural and anthropogenic processes, insofar as humankind's physical embodiment, which disrupted the natural planetary equilibrium, was linked in some way to its sinful nature. Mankind—possibly, men—was once again running up against the limits of the earth's natural capacity to sustain a balance between the elements and thus to support life. The specter of the apocalyptic global Flood of 1524 reared its head once again.

## The Inquisition and the Fate of the Flood

Erculiani was not the only thinker in sixteenth-century Italy to emphasize the depravity, or the predisposition to sin, of humanity after the Fall. The doctrine of total depravity was associated with the Protestant Reformers, especially Calvin, but it traced its roots to the writings of Augustine of Hippo, one of the most eminent of the church fathers. Unfortunately for Erculiani, the decision made by the Council of Trent earlier in the century to emphasize the maximal extent of free will against the Protestant doctrine of predestination made this position difficult for Catholics to maintain, despite its Augustinian lineage. Erculiani could have been influenced by Protestantism, of course. Padua's famous university attracted scholars of many countries and confessions, and her apothecary shop adjacent to the university certainly would attracted the patronage of many such scholars.[83] Luther himself once argued that Adam's "earthy" body foreshadowed his fall into sin and subsequent punishment with mortality, which resonated with Erculiani's theory about humankind's "earthy" bodies predisposing them to destroy their planetary home.[84] Another possible vector of Lutheran influence on Erculiani was her relationship with Márton Berzeviczy. Her correspondence on a religious topic with a Hungarian Lutheran certainly helps to explain her decision to publish *Letters on Natural Philosophy* in tolerant Kraków rather than Padua or Venice, as well as the interest of the Paduan Inquisition in the book after its publication.

But it is equally likely that she was responding to religious and intellectual trends within Catholic Italy that, however unorthodox or even heretical they might be, did not necessarily trace their origins to contemporaneous Lutheran or Reformed theologies. Erculiani's focus on the embodiment of sin rhymed with materialist philosophies of the Italian Renaissance and in particular with the controversial philosophies of her contemporary Cesare Cremonini, who began teaching at Padua a few years after *Letters on Natural Philosophy* was published.[85] It may also have been rooted in neo-Augustinianism, popular across both Protestant and Catholic Europe, which emphasized the pervasive sinfulness of human nature.[86] Augustine, moreover, had defended the reality of Noah's Flood against pagan skeptics and influentially defended the practice of reinterpreting scripture in light of advancing natural knowledge, an exegetical practice that Erculiani adopted and that Galileo would vociferously defend in the next century.[87]

The specific theological influences on Erculiani's thought are difficult to discern in the absence of a larger body of her writings and evidence regarding her life. I would hesitate to assume that she was a crypto-Protestant until new evidence comes to light, given that materialist philosophy and Augustinian theology were fully, if tendentiously, part of Catholic intellectual and religious life. I do not believe that there is any strong evidence to support an interpretation of Erculiani as a libertine, an atheist, or even a deist *avant la lettre*. Instead, the evidence we have now suggests that Erculiani saw natural philosophy as a means of doing theology, not as a means of rejecting or replacing it. In particular, she seemed eager to bring the methods of the new science to bear on the great question of sixteenth-century religion: the relationship between sin and free will. I would argue that Erculiani was trying to advocate a materialist understanding of sin and of its effects on the human body and the natural world. Through her natural-philosophical treatment of key events in biblical history—the Flood and the Fall—she advanced a novel understanding of the dynamic relationship between human beings and their global environment.

It is hardly surprising that the Inquisition did not see things her way, and it is even less surprising that Erculiani's statements to the court during her proceeding did not betray a hint of her desire to advance a new, materialist understanding of sin, if that is indeed what she truly wanted. Instead, her strategy, insofar as it can be discerned from the published account of her interrogation before the Paduan Inquisition by her lawyer, the learned jurist Giacomo Menochio, was to eschew claims to theological innovation based on natural-philosophical inquiry and to embrace instead the old notion of double truth, so frequently employed by generations of meteorologists before her. "I tell you that in philosophy, I do not affirm any of these things as certain truth," she told her inquisitors, according to Menochio. "In theology, I confess, in accordance with the Holy Scriptures, that the Flood and death came into the world because of sin." Her specific confession thus offers strong evidence that the Paduan Inquisition had indeed read her *Letters* as denying the role of original sin in causing both Fall and Flood.[88] Rather than defend her theories by explaining how they rested on an understanding of sin opposed to that which had been officially adopted by the church, Erculiani wisely chose to defend her theories by claiming that they had nothing to do with faith at all. She reiterated at least half a dozen times over the course of the interrogation that she had spoken philosophically and not theologically in *Letters on Natural Philosophy*, a strategy affirmed and probably encouraged by Menochio, who summarized his defense: "Lady Camilla spoke not as a theologian, but as a philosopher." Suspect and defense lawyer both invoked the privilege and protection of *libertas philosophandi* that the meteorologist's strict separation between philosophy and faith was supposed to afford.[89]

Erculiani's tactical deployment of the double truth defense when facing the inquisitors may have been undercut, however, by statements in the *Letters* in favor of unitary truth. For example, Erculiani's invocation of Moses in her conversation with the unnamed physician earned her the praise of Guarnier, who wrote, "You then responded to him very well in theology and in philosophy."[90] Moreover, as Eleonora Carinci has argued, the double truth defense was not always a winner with the Inquisition—in fact, it usually was not—and so it seems likely that the argument that actually got Erculiani off the hook had nothing to do with double truth and everything to do with gender.[91] Menochio reinforced Erculiani's self-defense that she had been speaking in the *Letters* as a philosopher and not as a theologian, but he also advanced a second argument that Erculiani did not make: that women's public speech should be taken less seriously than that of men. Women who made heretical statements in public should therefore, Menochio implied, be held to a lower standard than men who did the same, because their readers or auditors would take them less seriously and therefore be less inclined to adopt their heretical opinions.

Erculiani's uniqueness lay not only in her authorship, as a woman, of a work of natural philosophy, but also in having that work taken seriously enough by church officials to bring charges of heresy against her. Most women condemned by the Inquisition were brought in on charges of witchcraft, whereas most men were charged with public heresy, whether in speech or in print.[92] By charging her with heresy, the church treated her as a man, but by declining to punish her harshly or to place her book on the *Index of Prohibited Books*, it treated her as a woman. Her absolution, welcome though it must have been, was rooted in the Inquisition's refusal to recognize her authority—her right to speak publicly as a philosopher, to enter a male discursive space on an equal footing with her male peers.

In sum, Erculiani's gender seems to have worked both for and against her. On the one hand, her escape from punishment shows how her gender worked to deauthorize her, which may explain why her pathbreaking book is now so little known and appears to have been quickly forgotten. On the other hand, as Carinci argues, her gender afforded her greater freedom to philosophize than was typically granted to her male peers, who presumably would have been punished more harshly and seen their books placed on the *Index* for making the same assertions in print. Thus Erculiani's interrogation has significance far beyond whatever limited impact the ideas expressed in her book may have had.

The case of Camilla Erculiani illuminates the significant challenges faced by both women and men attempting to publicly engage in philosophical discourse about the Flood in sixteenth-century Italy. The most obvious victim of Erculiani's interrogation was Erculiani herself; it certainly worked to silence any further philosophical speech. Even if she had not died a short while later, it is inconceivable that she could have published her treatise on sin, or her treatise on the soul, which she likewise mentioned several times in the *Letters*, in the aftermath of her interrogation by the Inquisition. Insofar as news of her troubles traveled across Padua, Italy, and the Republic of Letters, it would have encouraged women eager to philosophize to do so in private, whether in *conversazione* in private homes or in private correspondence. No woman author that we know of dared follow in Erculiani's footsteps. She was both the first and last named woman author of a philosophical work on the Flood, and indeed on earth history, in early modern Europe. If Erculiani had indeed been effectively granted a degree of *libertas philosophandi* on account of her gender, it likely appeared to her contemporaries and to women in the next generation that such a defense would not be made routine.[93]

Erculiani's case also indirectly reveals the challenges faced by male philosophers, especially when their discourse touched on religious matters, as it inevitably did when the subject of conversation was the Flood. Menochio's argument that women's heretical speech was less dangerous and hence more excusable than men's was not original to him. The frequent use of the *imbecilitas* defense of women in legal settings and the relative dearth of women authors on the *Index* would have suggested to any keen observer that men were generally held to a more stringent standard when it came to public speech—especially philosophical speech.[94] This could not have been encouraging to Catholic men who likewise wished to consider the Flood as a topic in natural philosophy.

By the end of the sixteenth century, a dialogic popularization of Aristotle's *Meteorology* by the Paduan theologian and monk Vitale Zuccolo would contain the declaration that "by means of the Universal Flood . . . the entire face of the earth was transformed." Zuccolo joined with his fellow Paduan Erculiani, whose book he certainly could have read, in placing the blame for this terrestrial transformation on humanity. "The cause of this universal evil," he wrote, "was the conjunction of all human evils, which induced the all-powerful God to produce these tremendous effects." Zuccolo flirted with the idea of blaming humanity for the Universal Deluge by saying that sin caused the Flood that transformed global nature. And yet, Zuccolo also insisted ("since I must speak the truth") that the Flood was, first and foremost, "the work of the divine. . . . I say that everything depended on the force of God's omnipotence." Unable to say with any precision how human "evil" caused the Flood naturally, and unwilling to follow Erculiani in postulating various other (bodily, astrological) causes of the deluge, Zuccolo insisted that God, not humans, was the ultimate cause of the Flood. In arguing that the Flood was universal and produced natural effects, Zuccolo acknowledged the rising dominance of Mosaic natural philosophy. At the same time, by refusing to attribute the Flood to humankind, he successfully sidestepped the questions about sin, its material instantiation, and its catastrophic effects that Erculiani had so forcefully raised. Zuccolo thus found a way to have the cake of unitary truth while eating it too, avoiding theological controversy and the attention of the Inquisition.[95]

The next chapter explores the growing popularity of the Flood in natural philosophy, in Italy and across Europe, in the seventeenth century. The tricky business of constructing an orthodox account of Noah's Flood as a global and natural phenomenon did not become any less difficult. Rather, new and ever more heterodox ways of telling the story of the Flood arose, which made the sixteenth century's halting attempts look tame by comparison. Proving the Flood's universality became an increasingly orthodox way of studying earth history in the wake of Isaac La Peyrère's widely reviled *Prae-Adamitae* (*Men before Adam*, 1655), which proposed that the Flood had covered only Palestine and that the Mosaic history was a record of the Jewish people alone. La Peyrère's dismissal of a global flood and global history led him to polygenist speculations that the different peoples of the world had been separately created by God and had since enjoyed their own distinct histories and spiritual trajectories. The Flood came to seem ever more appealing and far less controversial as scholars realized that proving the universality of the Flood was the single best way to decimate La Peyrère's proposal to provincialize biblical history. A Universal Deluge entailed the monogenetic origin of all humans alive on earth in the present day—because global genocide proved universal human descent from Noah and his kin—which in turn re-universalized biblical history as world history, a key component of the ideological justification for Christian imperialism and evangelism beyond Europe. This is how the Flood's global scale, once a sticking point for natural philosophers, became the single most important and valuable thing about it.

# CHAPTER TWO

# After the Flood

## Biblical Monogenism, Global Migrations, and the Origins of Scientific Racism

In his 1606 history of the New World, the German physician and cosmographer Heinrich Martin launched his discussion of Native American origins by invoking the biblical history of Noah's Flood:[1]

> Holy Scripture tells us, in the sixth chapter of Genesis, that all the creatures of the earth perished in the General Deluge, excepting those that were saved by divine Providence in Noah's ark, so that the world could be repopulated, and from them it was once again populated. The survivors grew in number and spread from one land to another, starting in Asia, until they occupied all of Europe and Africa as well. They could occupy these lands without much difficulty because these three parts of the world are connected, each one to the other, and the arms of the sea that reach into these lands are narrow and easily traversed. But the land of the West Indies is trapped, on both the western and the eastern coasts, by broad and spacious seas that divide and separate it from other habitable lands, so there is not the same ease of migration. Therefore we must inquire how the first people arrived there.[2]

Writing from his new home in Mexico City, the German-born Martin, who published his history of New Spain under the name Henrico Martínez, felt that the best place to begin his discussion of the mystery of Indian origins was not with Adam and Eve and the first chapter of Genesis, but with Noah and his descendants, in chapters 6 through 10. Why Noah and not Adam? As Martin/Martínez's Spanish counterpart Gregorio García explained in his own book on Indian origins published in Valencia the following year, "all men and women" are descended from "our first parents Adam and Eve, *and subsequently from Noah and from his sons*, the only ones who remained alive after the general deluge."[3] García, a Dominican friar, saw the Flood as a near-extinction event that turned Noah into a second Adam, the more proximate patriarch of the entire human race. After the Columbian voyages revealed the existence of lands and peoples previously unknown to Europeans, defending the reality and the universality of Noah's Flood became central to European efforts to prove the consanguinity of

Old and New World peoples. The Flood, if it indeed killed off the entire human race save Noah and his family, guaranteed that every human alive on earth in the present was descended from a single common ancestor. The planetary scale of Noah's Flood became an essential means of defending biblical monogenism in the first age of globalization.

Noah's Flood—specifically, the notion that a global Flood with religious significance could be discussed within the disciplinary space of natural philosophy—remained tendentious well into the seventeenth century. As the previous chapter has shown, pious natural philosophers in the age of the Reformations remained skeptical about the Flood as a legitimate topic for philosophy on the twin grounds of its global scale, regarded by many as a physical impossibility, and its supernatural origin, which was the means of reconciling its physical impossibility with its historical reality. The Inquisitorial interrogation of Camilla Erculiani in the late sixteenth century pushed her pioneering attempt to naturalize the Flood into obscurity, helping to keep this line of philosophical inquiry dormant until its reemergence in the second half of the seventeenth century.[4] In the interim, historians like Martínez and García embraced the Universal Deluge as a key explanatory resource. While the previous chapter explored the Flood's role in the emergence of a global environmental consciousness—particularly in stimulating the idea that humans, through sin and through their bodies, could collectively harm the global environment—this chapter highlights the Flood's role in the development of a world-historical consciousness. As European soldiers, colonists, and missionaries claimed more and more territory in the Americas, European scholars crafted speculative histories of Native origins that relied on the Flood in order to prove that colonizer and colonized were part of the same human story. The biblical story of Noah's Flood was pressed into service to show that all peoples of the world shared a common ancestor and an ancient history.

The world-historical consciousness enabled by the Universal Deluge was, however, intimately tied to conceptions about global nature, which only became more prominent as the seventeenth century wore on. Martínez, García, and hundreds of other European and Euro-American writers across the long seventeenth century converged on the same basic answer to the puzzle they set themselves of how Adam's and Noah's descendants came to live in this New World, separated from the Old World by two massive oceans. The English writer Matthew Hale, in his 1677 defense of biblical monogenism, postulated that "there might have been in former times Necks of land, whereby communication between the parts of the Earth, and mutual passage and re-passage for Men and Animals might have been, which in long process of time . . . may have been since altered: That those parts of Asia and America which are now disjoyned by the interluency of the Sea, might have been formerly in some Age of the World contiguous to each other."[5]

A vanished land bridge seemed the easiest solution to a problem that was religious and political as much as it was historical. Humanity was unified first by Noah's Flood, as its common point of origin, and, second, by the land bridge that connected the Old and New Worlds after the Flood and enabled humanity's global diaspora from the site of the ark's landing place. Biblical monogenism thus fostered a world-historical consciousness that viewed global nature as the literal ground on which human history transpired, shaping and directing its course in profound ways.

Early modern European scholars embraced the Flood because it became vital to defending biblical monogenism, which in turn became a central plank of the ideological justification for Christian evangelism and empire in the New World. This chapter sketches the contours of the scholarly debate about transoceanic land bridges, postdiluvian migrations, and Native American origins among Protestant and Catholic scholars across the Atlantic world in the long seventeenth century. The first part of the chapter considers the European assumption that the ancestors of the Native Americans must have immigrated by land rather than sea, before turning to the controversy surrounding Isaac La Peyrère's scandalous *Prae-Adamitae* (*Men before Adam*, 1655), whose polygenetic account of human origins threatened the universality of the Bible's account of human history and thus the very basis for Christian evangelism beyond Europe. As a means of re-universalizing biblical history and demonstrating the monogenetic origins of humankind, the Universal Deluge became an ever more popular and orthodox part of natural philosophy in the second half of the seventeenth century. The second part of the chapter turns to European debates on Native American origins that were staged within the framework of a global Bible and a global Flood. In the context of these debates, the idea of a land bridge linking the world's continents after the Flood emerged as a means of proving monogenetic origins while at the same time accounting for contemporary human difference. This second section focuses on two writers who lived a century apart in different worlds: the Spanish Creole Mendicant scholar Antonio de la Calancha (1584–1654) and the French-born Swiss Protestant naturalist and linguist Louis Bourguet (1678–1742). Despite their many differences, both men believed that the indigenous peoples of the Americas were descended from Tartars (and ultimately, of course, from Adam and Noah) who immigrated to the Americas after the Flood via a Pacific Ocean land bridge. As Calancha and Bourguet so vividly demonstrate, scholars who sought to establish the universal kinship of all people were nevertheless deeply attentive to questions of human difference. The particular theory that the Native Americans were Tartars was a way for Europeans to gather them into the Adamic human family while, at the same time, demonstrating their inferiority to Europeans, thereby lending legitimacy to overseas evangelism and imperialism.

Exploring the religious and political contexts in which this idea was elaborated —Protestant and Catholic, European and Euro-American—allows us to see how narratives of Noah's Flood spoke to contemporary questions of race, religion, migration, and empire. The Flood became central to European science and scholarship in the seventeenth century because of the ways it could prop up particular visions of world history and of contemporary geopolitics. Chapter 1 highlighted the complex ways in which the Renaissance, Scientific Revolution, Little Ice Age, and Reformations intersected to make the biblical deluge an appealing, if tendentiously orthodox, subject for natural philosophy in the sixteenth century. This chapter argues for the crucial role of European imperialism and evangelism in making the Flood ever more popular and orthodox in the seventeenth.

Scientific theories about the changing configurations of land and water during and after Noah's Flood played a crucial role in early modern European theories about the origins of human unity and diversity. However, historians of the earth sciences have not delved very deeply into the ways that debates about the Flood's scale and aftermath were implicated in contemporaneous debates about Native American origins.[6] Meanwhile, historians of race and empire in the early modern Atlantic world have largely overlooked the crucial role played by geohistory in elaborating racial taxonomies and supporting colonial hierarchies.[7] By attending to the racial and imperial dimensions of early modern theories of the Flood, historians of science can better appreciate the ways in which the early modern earth and environmental sciences were shaped by European colonialism, as has already been demonstrated so well for natural history in this period.[8] In paying attention to the geological underpinnings of early modern monogenism, historians of race can better appreciate the ways in which biblical monogenism fostered the construction of racial taxonomies in the context of early modern settler colonialism.[9]

These oversights may be a function of the misleading anachronism *geology* that has been retroactively applied to so many early modern accounts of the earth's history, which implies a field of study focused on long-durational geophysical processes and structures and devoid of human actors. These oversights may also be a function, more broadly, of the modern disciplinary boundaries between the natural sciences, on the one hand, and the humanities and human sciences, on the other, which can make it difficult for modern scholars to appreciate the synthetic richness of these early modern attempts to unify natural and human history on a planetary scale.[10] This misapprehension may be intensified by the rhetoric surrounding the "discovery" of the Anthropocene, which can make it seem as if ours is the first generation in history to recognize the deep imbrication of human and natural history.[11] Whereas the previous chapter treated early iterations of the idea that humans could provoke catastrophic changes in nature worldwide, this chapter surveys a variety of ideas about how those natural changes attendant on the Flood in turn shaped the course of human history. The idea that the Flood killed off most of humanity and then changed the face of the earth, scrambling the planet's oceans and continents and creating post-diluvian land bridges for Noah's descendants to traverse from one landmass to another, was a materialist way of reading biblical history—and it was similarly pursued with the goal of better understanding sin and salvation on a global scale. The centrality of geology to early modern debates about human origins indicates broadly how the Flood functioned to unite human and natural history on a planetary scale. At the same time, the geological determinism inherent in European accounts of land bridge–based migration also worked to articulate hierarchies of difference within humanity. Imagined histories of land and sea were critical to early modern European accounts of the historical unity and contemporary divisions of the human race. Early modern theories about the Flood as an agent of global near-genocide and land bridges as an enabler of global diaspora exhibit a form of geological determinism that perceived human history—and especially Native American history—as profoundly determined by geological structures and events.

## By Land or by Sea: Geological Determinism in the European Search for Native American Origins

The rapid expansion of long-distance networks of trade, religion, and empire in the early modern period generated an increasing awareness in Europe of the sheer scope of human diversity, challenging European scholars to figure out how these diverse populations were related to one another and to themselves. The European discovery of the Americas at the turn of the sixteenth century posed the greatest challenge of all. Without any clear and obvious path of migration from the Old World to the New, and with Moses and Aristotle both frustratingly silent on who these people were, Europeans wondered how to square the Pentateuch's supposedly universal account of ancient human history with the reality of a heretofore unknown continent full of unknown peoples. Were the peoples of the Americas descended from Adam and from Noah, whom Genesis indicated were the joint patriarchs of the entire human race? If so, how did their ancestors travel from the ark's landing site somewhere in Asia to this new continent separated from the Old World by two vast, and only recently navigable, oceans? If they were not descended from Adam and Noah, had they escaped the punishment of the Flood or the stain of original sin? Were they an entirely different race of humans with a distinct orientation to sin, salvation, and divine law? And if that were the case, how could European Christians justify the evangelical missions currently under way throughout the Americas?[12]

Biblical polygenism, or the idea that Native American and possibly other peoples had been independently created by God and descended from patriarchs other than Adam and Noah, was so unacceptable to most of Europe's Christians that the mere thought it might be true fueled centuries of scholarship whose explicit aim was to prove the opposite. Martin/Martínez's question "we must inquire how the first people arrived there" was posed in the imperative because of the range of politically and religiously acceptable answers to the bigger question lurking behind it. The only acceptable answer, generally speaking, was that all the peoples of the Americas were descended from Adam and Noah and migrated to the Americas at some point after the Flood.

Perhaps because this general answer had been decided in advance, Christian scholars in Europe and Euro-America invented other questions within this predetermined framework and debated them for hundreds of years: Had the Indians' ancestors traveled to the Americas by sea or by land? If by land, where was the land bridge they had crossed? Did it stretch across the Atlantic Ocean or the Pacific Ocean? What part of the Old World had they departed from—Europe, Asia, or Africa? Which subgroup or groups of the Noachian family tree were the ones to make this transoceanic voyage? Did they leave for the New World immediately after the Flood as a single group, or was it a long and drawn-out process consisting of multiple waves of migration? Did they all come from the same place or from multiple points of origin? The answers that European and Euro-American scholars gave to these questions were varied and, at times, dizzyingly complex. Opinion differed greatly on the question of the Indians' most proximate progenitors, for example. Proposals included the Jews, Chinese, Japanese, Phoenicians, Scythians, Greeks, Carthaginians, Egyptians, Ethiopians, Britons, French, Spanish, Frisians, Norwegians, and even the folks of the lost island of Atlantis.

Amid this welter of possibilities, two main points of consensus emerged over the course of the long seventeenth century. European scholars first came to agree that the Indians' ancestors almost certainly arrived by land, not by sea. Later in the century, they converged on the idea that this land bridge must have stretched across the Pacific Ocean, not the Atlantic. The preference for a Pacific land bridge, which will be discussed in the second half of the chapter, reflected the concern to articulate racial hierarchies within a monogenetic framework. So too did the preference for migration across a land bridge rather than across open ocean, which reflected in particular the mistaken assumption that long ocean voyages were a novelty in human history and the unique achievement of Europeans.

Martínez's injunction about the necessity of discovering how Noah's descendants first came to the New World is immediately followed by this observation about the emerging scholarly debate: "To some it seems they came by sea, others say they came by land, and they all try to save the appearances of their opinion as best they can." The possibility of migration by sea was widely considered but almost universally rejected. Many centuries before Thor Heyerdahl sailed across the Pacific Ocean in a handmade wooden raft in order to demonstrate the possibility of ancient human migration by boat, few Europeans thought that an ocean crossing was the most likely explanation for the peopling of the New World. Martínez voiced what was then a common sentiment when he argued that the state of navigation in biblical times was far too primitive to have enabled a successful ocean crossing. "In ancient times," he declared, "navigation had not yet reached the point it has today, nor was there sufficient dexterity in its practice as we now have, to be able to make long voyages and to settle remote and distant lands." His invocation of "we" aligned the German-born Martínez/Martin with his Spanish co-colonists and implicitly grouped together all European nations currently expanding into the Atlantic world, making long voyages, and populating distant lands. This "we" was defined by possession of navigational technology superior to that of any ancient civilization. The contrast Martínez drew was not just between ancient and modern technology, but also between Indian and European technology. Martínez expressed a European presumption that Native Americans, not only in antiquity but also in their early modern present, lacked the requisite knowledge of shipbuilding and navigation that would allow them to cross the Pacific or the Atlantic Ocean. Implicit in this statement were several interrelated assumptions about European superiority that inclined Martínez and the majority of Europeans writing on the Indian origins question against the theory of migration by sea.[13]

Early modern Europeans, reflecting on the globalizing world in which they lived, regarded themselves as a uniquely mobile and seafaring people. From their perspective, America could only be reached from Afro-Eurasia by boat, and this only recently. As far as they were concerned, no other culture had reached the advanced level of technology necessary to cross the major oceans until Portuguese and later other European sailors first did so in the fifteenth and sixteenth centuries. The English courtier and explorer Walter Raleigh spoke from experience when he cited the current difficulty of "long voyages to Sea" as an argument against ancient migration by boat. In his *History of the World* (1614), Raleigh wrote: "[E]ven now when this Art is come to her perfection, such voyages are very troublesome and dangerous."[14] The uniqueness and novelty of the European "perfection" of the "Art" of navigation could explain both the Old World's historical ignorance of America's existence as well as contemporary European supremacy there. It was the reason no other Old World group had joined the competition to claim pieces of the New World. For Europeans, their unique ability to travel to the New World from the Old by sea was a constitutive fact of their globalizing early modern world.

The assumptions about civilizational differences between Europeans and Indians that undergirded the skepticism around migration by sea was also bolstered by a kind of geological determinism. If one were to concede that the Indians' ancestors could have made it to the New World by boat, across an open ocean, then they could have come from anywhere and landed anywhere. That version of ancient history would have afforded the Indians' ancestors a degree of freedom of movement across the globe that Europeans were, in this early modern age, trying to claim as their unique prerogative. Insisting on a crossing by land not only denied oceanic mobility to the Native Americans; it also meant that the possible points of origin and pathways of migration were tightly restricted, indeed entirely determined, by what types of landforms existed on the earth in the ages after Noah's Flood.

Further evidence of the geological determinism guiding the preference for migration by land comes from discussions of the postdiluvian migration of non-human animals. The Noah story was generally read as an account of the mono-genetic origins of nonhuman animals as well, but as many writers noted, animal migration to the New World raised a separate and thorny set of issues if one assumed that humans had arrived by boat. Matthew Hale echoed José de Acosta in pointing out that there were numerous animals who could neither swim nor fly across vast oceans like the Pacific and Atlantic, nor could they be safely transported by boat. "Tame Animals for use, delight, or food," such as geese, cattle, deer, and monkeys, could easily have been transported by ship, Hale argued, but not so "ferine, noxious, and untamable Beasts, as Lions, Tigers, Wolves, Bears, and Foxes." Beasts of prey were unlikely candidates for transport in a small craft, at close quarters with humans, on a long sea voyage. It seemed equally unlikely that large, land-dwelling quadrupeds could have made it across the ocean on their own, "especially for Tigers and Lions, which are not so apt to take the Water." Hale declared himself favorably inclined toward the theory that the Americans' ancestors migrated by sea, but the difficulty of imag-ining lions, tigers, and bears making a transoceanic crossing moved him to concede that a land bridge was the more likely means of postdiluvian migration, for human and nonhuman animals alike.[15]

The denial of seafaring mobility to both indigenous Americans and nonhuman animals worked to align them and to mark them both as inferior to seafaring Europeans. The scholarly consensus in favor of migration by land thus appears to have sprung in part from a desire to defend their world-historical status as the first people ever to achieve truly global mobility by virtue of their skill in crossing oceans. Conversely, the postdiluvian land bridge they imagined must have enabled Native American mobility across the globe in biblical times also strictly determined the pathways of that mobility; these early settlers of the Americas did not have much of a choice about where they went, unlike the current ones. The convergence around the theory of an ancient land bridge was not only a means of justifying their sense of civilizational superiority, however.

It was equally, if not more so, designed to justify evangelism among indigenous Americans.

One of the first people to propose the existence of a postdiluvian land bridge in print was the Jesuit scholar and missionary José de Acosta. His time spent in Mexico and Peru formed the basis for his celebrated *Historia natural y moral de las Indias* (*Natural and Moral History of the Indies*, 1590), which was reprinted in several European languages throughout the seventeenth century and widely cited as an authoritative text on American people and American natural history. Acosta argued, as Martínez and Raleigh would soon do, that it was unlikely the Indians' ancestors had made an ocean crossing, given the state of navigational technology in the time after the Flood. More likely, they traveled to the New World over land, crossing at some unknown point where the world's two major continents intersected or were not far distant from one another. Acosta believed the land bridge likely stretched between Asia and America, in the Pacific, but he could not be sure. More important was that a land bridge existed somewhere, so that universal Adamic origin was guaranteed. "Holy Scripture clearly teaches that all men are descended from the first man" Adam, and therefore "we are forced to confess that men first came here from Europe, Asia, or Africa."[16]

It should not surprise us in the least that one of the earliest writers to seriously discuss the possibility of an Old World–New World land bridge as a subject for scholarship was also a member of an evangelical religious order. Acosta's mission among the Indians would have been exposed as absurd and illegitimate if the Indians turned out not to be descendants of Adam and inheritors of his sin. The land bridge, wherever it had been, legitimized the evangelical undertakings of his fellow Jesuits in the New World. When *Natural and Moral History of the Indies* first appeared in Latin in 1589, it was accompanied by a missionary manual, *De procuranda Indorum salute* (Procuring the Indians' salvation, 1588), which discussed at length the challenges and necessity of evangelism among the indigenous peoples of the New World.[17] The search for the land bridge was fueled by a feeling on the part of scholars and missionaries alike that it was necessary to prove the Indians' descent from Adam and Noah, which in turn was necessary in order to justify missionary work among them.

## Provincializing Biblical History: La Peyrère and the Threat of Polygenism

Understanding exactly how and why polygenism was perceived as a threat to the project of overseas evangelism furnishes crucial context for the fluorescence of monogenist scholarship in the long seventeenth century. The specter of polygenism that had haunted Christian Europe ever since the Columbian expeditions most often took the form of "pre-Adamism," or the idea that the descendants of Adam and Eve formed one of several distinct human lineages currently alive on

earth and that one or more of these other lineages may have predated the creation of Adam's line. In the 1530s, not so very long after the European discovery
of the Caribbean, the German alchemical philosopher Paracelsus wondered how
"Adam's children came to these secret islands." It was so difficult for Paracelsus to
imagine how they could have gotten there from the Old World that he was willing to entertain the idea that they constituted a distinct, non-Adamic branch
of the human race. "Just as there are many kinds of animals on the earth," he
wrote, "maybe there are also many different kinds of men." Paracelsus worked
out the consequence of this observation in the subjunctive, speculating about
whether "there was perhaps another Adam" who was the progenitor of these
island peoples. If they were in fact descendants of a different Adam, they would
not "share our flesh and blood" and perhaps were not even "made in the image
of God." In another work from the 1530s, Paracelsus openly wondered whether
the peoples of the New World even possessed souls. Non-Adamic origin would
almost certainly entail that indigenous Americans were a physically and spiritually distinct branch of humanity from Christian Europeans, with their own
distinct relationship to God, sin, and salvation.[18] In 1512, Pope Julius II officially
declared that the Indians were descended from Adam in response to just this
type of speculation. The church's declaration on this issue was intended not only
to counter widespread depictions of Native Americans as monstrous, bestial,
diabolical, or otherwise less than fully human, but also to shut down polygenist
speculations about the existence of separate human lineages descended from
different "Adams."[19]

Indigenous interventions in the European debate on human origins reveals
their awareness of the ways biblical monogenism was being used to prop up
empire and evangelism. The Christian Andean author Felipe Guaman Poma
de Ayala reworked biblical monogenism and polygenism in order to argue for
Andean sovereignty. Guaman Poma, a descendant of Inca nobility who worked
as a Quechua interpreter for Spanish missionaries, argued in *El primer nueva
corónica y buen gobierno* (*The First New Chronicle and Good Government*, 1612),
that "the Indians are the natural owners of this kingdom" based on the historical
allotment of different parts of the world to different groups of people by God. In
this thousand-page illustrated manuscript history presented to the Spanish King
Phillip III, Guaman Poma argued that God "made the world with all its lands,
and He planted each seed in them: the Spaniard in Castile, the Indian in the
Indies, and the black in Guinea."[20] This polygenetic statement was tempered by
a monogenetic appeal to Adam as a common ancestor. His narration of Noah's
Flood in an earlier section of *The First New Chronicle* suggests that the divine
seed-planting took place after the Flood, which killed all humans save Noah's
family. "Of these sons of Noah, God took one to the Indies; others say that he
came from Adam himself," wrote Guaman Poma, tracing Indian origins back

to Noah or perhaps to Adam (which would, of course, be incompatible with a universally genocidal deluge). This partial adoption of biblical monogenism was in keeping with *The First New Chronicle*'s larger framing of Inca history within Christian universal history and with Guaman Poma's own Catholic faith. However, the polygenist elements of Guaman Poma's history of the peopling of the Americas—the possibility that Indians had not descended from Noah or traveled across a land bridge; the separate "seeds" that gave rise to the Spanish, Indians, and Africans—undercut the potential imperial legitimation of the monogenist elements. Guaman Poma argued for the Incas' divine right to rule their land based on their divine emplacement in the Indies following the Flood: "God knows all; and with his powers, he can put these Indians in a separate place."[21] God had placed the Incas in their present location a long time ago, intending them to occupy that land in perpetuity and to be its "natural owners." Moreover, Guaman Poma argued that his people had long ago been converted to Christianity by one of Christ's apostles, so they had no need of Spanish evangelizers. Guaman Poma's history of human origins shows how biblical polygenism, in the hands of an indigenous scholar, could be used to contest Spanish evangelism and imperialism in the Andes and to insist on indigenous rule.

Guaman Poma's quasi-polygenist history of indigenous origins was never published in his lifetime, and when compared to the fully polygenist history of Isaac La Peyrère forty years later, it is not difficult to understand why.[22] La Peyrère, a French Calvinist of possible *converso* heritage, proposed in *Men before Adam* that multiple races of human beings had been created by God in different parts of the world at different moments in time. Most outrageously in the eyes of European Christian readers, La Peyrère claimed that the inhabitants of the Americas had been created by God in the Americas and formed an entirely distinct lineage of human beings from Adam and Eve's descendants. The backlash against his book was swift and severe.[23] La Peyrère's polygenist history of the world provincialized biblical history and restricted the scale of the inheritance of original sin, both of which posed a direct challenge to Christian evangelism.

The existence of the titular "men before Adam" was the key plank in La Peyrère's more general case for interpreting the Pentateuch not as a universal history of the entire world and all its peoples but as a particular history of the Jewish people and of their historical homeland. Adam and Eve were not the parents of all mankind, but the progenitors of the Jewish and Christian peoples only. Their progeny were, in turn, the only ones affected by Noah's Flood, which had not "overflow'd the kingdoms of the earth with most mighty overflowings" but instead "overflow'd only Palestine, and the Land of the Jews." Noah's descendants had not migrated to the Americas after the Flood: "None of Adams posterity ever arrived" on those shores. The Americans were not Adam and Noah's descendants but were instead simply Americans, which obviated the need to explain how they could have originated in the Old World. Polygenism, in other words, was a means of provincializing the Bible. La Peyrère advanced this polygenetic theory not to overturn or discredit the Bible, but to restrict the scale on which its historical drama had played out and within which its moral laws applied in the present day.[24]

In retrospect, it is easy to regard as inevitable that the book and its author would be condemned as wildly heretical. La Peyrère published *Men before Adam* anonymously, in tolerant Amsterdam, and it was banned before it was even printed by Cardinal Richelieu, to whom it was dedicated.[25] La Peyrère claimed, perhaps sincerely, that he had taken the dramatic step of provincializing biblical history for essentially pious reasons. In the introduction to his treatise, La Peyrère related his childhood misgivings about various aspects of biblical history that had appeared to him irrational and incredible, which had almost caused him to lose his faith at a young and impressionable age. These included apparent inconsistencies in the book of Genesis, such as who Cain had married after killing Abel if Adam and Eve were the first humans with only two offspring, and philosophical problems that compounded the exegetical ones, like the physical universality of the Flood. La Peyrère argued that these philosophical and exegetical problems could be easily solved if one were to simply assume that the Mosaic history only applied to one corner of the planet. The Bible could retain its status as a true and accurate account of ancient history, as long as it was read as a history of the Adamic branch of humanity alone.

La Peyrère's proposal to provincialize the Bible in order to save it did not, however, meet with an especially warm welcome. *Men before Adam* was immediately and widely denounced as heretical by Protestants and Catholics alike. Using language that was not in the least hyperbolic in the context of the pre-Adamic controversy, the Anglican minister Edward Stillingfleet labeled La Peyrère's speculative polygenetic history an "impious, absurd, and rude" attempt to "prostitute the Scriptures to his opinion."[26] La Peyrère was imprisoned shortly after the book's appearance in print in 1655 and was released only after agreeing to convert to Catholicism and to abjure the heresies contained in his book in the presence of Pope Alexander VII.[27] The extremity of the reaction to *Men before Adam* across western Europe signals the depth of commitment to the Bible as a *global* document: a document recording a history of the entire world and expounding a theology that held sway the world over. Tempting as it might have been to abandon the task of determining the dimensions of an ark that could fit two of all the world's animals or calculating the volume of water necessary to inundate all of the planet's dry land, most Christian scholars in Europe decided that it hardly seemed worth it if they would be forced in exchange to concede that there were people who lived outside of biblical history and beyond the scope of the Christian dispensation. A provincial Bible and a provincial Flood reduced Christianity to just one among the world's many religions.

Even worse, polygenism radically undermined the theological justification for Christian evangelism. The existence of human lineages who did not trace their ancestry back to Adam and Eve raised the possibility that those races of people were not stained by original sin. Not having inherited original sin from their first ancestors, such peoples would not need to be saved. Stillingfleet argued that the real problem with La Peyrère's polygenism was that it provincialized original sin. "It is hard to conceive how the effects of mans [*sic*] fall should extend to all mankind," he observed, "unless all mankind were propagated from Adam."[28] La Peyrère's distinction between what he called "natural sin" and "legal sin" made clear that polygenism implied an unequal distribution of sinfulness across the world. Natural sin is "innate in every person by reason of his peccant nature," he wrote. This type of sin is universal and physically instantiated, "deeply and naturally rooted in their very bowels, *Which proceeds from the fat of them,* as is elegantly expressed in the 73rd Psalm." Sinfulness was built into the very fabric of the human body, La Peyrère suggested, much as the Italian natural philosopher Camilla Erculiani had proposed in her short work on the Flood seventy years earlier (see chapter 1) and as her Paduan successor Antonio Vallisneri would suggest again seventy years later (see chapter 5). Legal sin, on the other hand, was the specific inheritance of Adam's descendants. In other words, any creature possessing a human body would labor under the burden of natural sin, while only those humans descended from Adam would carry the additional burden of legal sin. Polygenism exploded the imagined community of all humans bearing the same burden of original sin and the same necessity of salvation.[29]

Maintaining the Indians' full inclusion in Adam's family tree thus became key to constructing the global validity of the Gospels in the age of the European expansion. Anyone who was not descended from Adam and Eve would not share in the Old Testament's grim assurance of the universal inheritance of original sin, nor would they stand in need of the New Testament's rosier promise of salvation.[30] The urgency of maintaining universal sinfulness helps to explain why it became nearly as vital to insist on universal descent from Noah as well as from Adam, as García affirmed in his 1606 history of Indian origins. If the Indians were Adam's ancestors but not Noah's—if, hypothetically, their ancestors migrated to the New World prior to the Flood, remaining there safely while the Flood wreaked havoc on their distant cousins in the biblical lands—they would have been stained by original sin, but their exemption from the divine punishment represented by the Flood would have implied a lesser degree of sinfulness than their Old World kin. Proving that the Indians' ancestors migrated to America *after* the Flood, making them descendants of both Adam *and* Noah, was a crucial means of upholding the truth of the Christian religion and the various political projects which rested on that truth for their legitimacy.

La Peyrère's polygenetic account of world history also undercut justifications for European territorial claims and forced conversions among both Indians and Jews. Asserting that the Indians were created by God in the Americas chimed with indigenous assertions of autochthony that formed the basis to rightful claim to their lands. (The role of monogenism in rhetorically propping up settler colonialism by figuring all people as migrants will be discussed later.) And while provincializing the inheritance of original sin cast serious doubt on the project of evangelism among the indigenous peoples of the Americas, La Peyrère's messianic vision of the world's future challenged the persecution and forced conversion of Jews in Europe. In *Du rappel de Juifs* (On the recall of the Jews, 1643), he chastised Christians for their poor treatment of Jews, a position that could very well reflect his possible descent from Iberian Jews forced to convert to Christianity. His messianic vision of Christ's return in *Recall* involved voluntary conversion to a reformed Christianity (reformed along the lines he would later propose in *Men before Adam*) by Jews and Gentiles alike. La Peyrère's polycentric vision of human history thus seems to have been intended to encourage religious toleration and to discourage settler colonialism in the present. His rejection of Christian universalism enabled an implicit anticolonial and anti-evangelical critique.[31]

*Men before Adam* shocked Christian Europe so profoundly that the natural-philosophical Flood came to seem orthodox by comparison—and maybe even necessary in order to defeat it. Many of the published critiques of La Peyrère appealed to the global Flood as a means of proving monogenism and a global Bible. Likewise, many of the publications defending the all-inclusiveness of the deluge included discourses attacking polygenism. The German scholar Georg Horn, for example, published several tracts in the 1650s and '60s that collectively sought to defeat pre-Adamism, prove the Flood's universality, and settle the question of Indian origins. When the Dutch Anglican canon Isaak Voss tried to salvage La Peyrère's idea that the Flood may have only covered part of the earth in *Dissertatio de vera aetate mundi* (Dissertation on the true age of the earth, 1659), on the grounds that "the human race [*genus*] held only a small part of Asia" in the age of Noah, Horn published several short works attacking Voss and vociferously insisting that "the deluge was universal, the entire globe of the earth was covered with waters and absolutely no parts of it were visible."[32] Natural philosophers increasingly saw it as their task to explain how a global Flood could be physically possible. The Italian philosopher Jacopo Grandi began *De veritate diluvii universalis* (On the truth of the universal deluge, 1676) with the following declaration: "I will show that the Universal Deluge was possible, according to natural causes . . . and finally I will demonstrate, according to physical reasons, that the whole earth, up to the peaks of the highest mountains, was at one time covered by water."[33] The German Jesuit Athansius Kircher's *Arca Noë* (Noah's ark, 1675) contains philosophical demonstrations that an ark big enough to house two of all the world's animals and a flood big enough to cover the entire surface of the earth were, in fact, physically possible. Kircher also discussed how the Flood altered the earth's topography, rearranging landforms and waterways in such a way as to ensure that humans could pass between them. The Flood thus destroyed most human life at the same time it created global pathways of mobility for the saved to repopulate the earth.

This campaign to prove the formerly unprovable derived a considerable degree of its energy from a sense that the Flood, for all its baggage, could be an effective means of re-universalizing biblical history—and therefore bolstering Christian evangelism—in the wake of La Peyrère's provincialized Bible. Universality was precisely what the Universal Deluge had to offer. By washing over the entire planet, the New World as well as the Old, it guaranteed that biblical history was world history. It proved that historical events recorded in scripture had genuinely occurred across the whole world, not just in the biblical lands. It proved that everyone alive on earth was a descendant of Adam and Noah and a participant in the same drama of sacred history. No place, no people on earth were exempt from its transformative effects. A global flood guaranteed the universality of sinfulness and the need for salvation by certifying that God's judgment and God's providence had already reached into the far corners of the world. Lands and peoples whom these scholars' nations were busy conquering, and coreligionists busy converting, were, in effect, preapproved for assimilation into the Christian magisterium by the Universal Deluge.

Polygenism's challenge to the theological basis for Christian evangelism helps to explain why monogenism was the only viable option for Christian scholars considering the "problem" of Indian origins in the early modern period. The religious stakes of this question also help us to understand why scholar-missionaries like Acosta were so well-represented in the debates about Indian origins and why those following in Acosta's footsteps eagerly embraced the idea of an ancient land bridge connecting the Old World to the New.

## Globalizing Biblical History: Land Bridges and the Monogenetic Origins of Race

Polygenism and a provincialized Bible threatened not only the theological basis of Christian evangelism but also the ideological justifications for European imperialism. The depth and breadth of opposition to *Men before Adam* is a telling index of the way that a global Bible propped up not only Christian evangelism but a particular form of Christian imperialism in the New World. While the

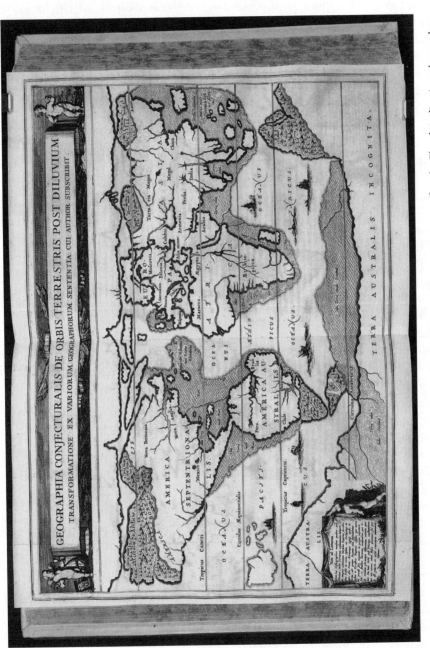

A map from Kircher's *Arca Noë* (1675) depicts the state of the world's continents and oceans after the Flood, indicating places where the Flood changed water to land and vice versa. Linda Hall Library of Science, Engineering & Technology.

Universal Deluge established the truth of biblical monogenism and thus legitimated global evangelism, postdiluvian land bridges established racial divisions within monogenism and thus legitimated both evangelism and empire. In the second half of this chapter, I briefly sketch the contours of the land bridge debate as it unfolded in Europe and its American colonies in the long seventeenth century. In particular, I want to uncover the assumptions behind the evolving consensus that the Native Americans' ancestors migrated to the Americas across a Pacific Ocean land bridge. The theory that Native Americans were originally Asian Americans vividly demonstrates the latent racist potential of biblical monogenism, which in turn helps to illuminate how this body of scholarship furthered the interests of both imperialism and evangelism.

The famous mid-seventeenth-century debate between the Dutch scholars Hugo Grotius and Johannes de Laet over the ancient pathways of migration to the New World has been the subject of several excellent studies, but their high-profile dispute was only one episode in a much bigger conversation that unfolded across countries and centuries.[34] I've chosen to focus on two historical figures who are barely known at all today in order to illuminate underappreciated aspects of this long-running debate. The Creole Mendicant scholar Antonio de la Calancha and the Swiss Huguenot scholar Louis Bourguet lived a century apart on opposite sides of the Atlantic, but both men converged on the belief that America had been populated by Tartars crossing an Asian-American land bridge. Each one saw his scholarly quest to demonstrate this historical fact as crucial to his advocacy for the global supremacy of his own faith in the present day. Calancha was directly involved in missionary activity and imperial expansion as a friar in colonial America, while Bourguet supported several Protestant missionary groups as a donor and propagandist. Out of the hundreds of authors who weighed in on the land bridge debate from the sixteenth through the eighteenth centuries, Calancha and Bourguet demonstrate especially well the imbrication of this field of historical scholarship in evangelical and imperial agendas. Their differing milieus—Catholic Peru and Protestant Switzerland—show how the same idea of postdiluvian migration across a Pacific Ocean land bridge could be made to serve the interests of either Catholic or Protestant empire. Moreover, their positions at the beginning and end of the long seventeenth century allow us to trace the changing racial politics of the Indian origin debate as well as its increasingly geological character.

Once the proposition became commonly accepted that the migration of the Indians' ancestors from the Old World to the New after Noah's Flood occurred over land and not across open ocean, the next question was: Did the land bridge stretch across the Pacific or the Atlantic Ocean? This debate, which began in the Spanish Atlantic world but grew to encompass writers in northern, Protestant Europe over the course of the seventeenth century, played out along broadly similar lines, as had the earlier debate about migration by land or by sea. While the theory of an Atlantic land bridge was widely considered, it held considerably less appeal than the Pacific land bridge theory, which became increasingly dominant over the course of the long seventeenth century. Also like the theory of migration by sea, migration across an Atlantic land bridge seems to have steadily lost ground to its main alternative for reasons primarily having to do with empire, evangelism, and race. Its rise to prominence in the later seventeenth century signaled the beginning of a new stage in the Indian origins debate, in which the land bridge's location took on vital importance because of its implications for the racial genealogy of the Native Americans.

Defenders of an Atlantic land bridge generally did so in order to demonstrate European claims to American territory. The Spanish colonizer Pedro Sarmiento de Gamboa argued in 1572 that long ago, America had been geologically conjoined to the Iberian peninsula before breaking away and drifting westward across the Atlantic.[35] America was, in fact, the lost island of Atlantis, which had been populated by ancient Spaniards. Sarmiento put forward this idea in a history of the Incas commissioned by the viceroy of Peru in order to demonstrate Spain's dominion over the Americas. Perhaps their onetime geological connection appeared to him an even better way than the papal bull *Inter caetera* (1493) to secure Spanish claims to the New World.

An Atlantic land bridge (or transatlantic continental drift, in the case of Sarmiento de Gamboa's theory) created a short, direct route of migration from the Old World to the New. In that sense, it was a natural choice for those looking to defend biblical monogenism. The problem, however, was its racial politics: it turned Americans into the direct descendants of Europeans. Although Sarmiento de Gamboa's speculative geohistory was intended to establish the legitimacy of the modern Spanish Empire, it nevertheless troubled the legal basis of Spanish claims to its colonies as well as pretensions to civilizational superiority by indicating that the Spanish were in the act of conquering people who were likewise already of Spanish descent. This is perhaps part of the reason why the theory of Atlantic migration, while widely considered, was far less frequently championed than the alternative theory of migration across a Pacific Ocean land bridge.

The Atlantic land bridge theory, when coupled with a globally genocidal deluge, guaranteed biblical monogenism, but it implied a degree of parity and kinship that left many Europeans uncomfortable. Placing the land bridge in the Pacific Ocean, on the other hand, exaggerated the geographical, historical, and thus ethnic differences between western Europeans and American Indians to the greatest possible extent while remaining within a monogenetic framework. The farther the Indians' ancestors had to travel from Ararat to reach the New World, the more time would have elapsed since the split between the ancestors of present-day Indians and Europeans, allowing for the greatest possible divergence between them. The Spanish courtier and royal chronicler Antonio de Herrera y Tordesillas decisively rejected any European ancestry of the Native Americans by attributing racial distinctions to the spatial and historical distance between the two groups. "It is known that the People of the New World are in Colour like the Eastern," Herrera asserted in his widely translated history of the Spanish conquest, "and there is not the least Appearance of any having pass'd thither from the politer parts of Europe, before the Spaniards."[36] Their skin color as well as their lack of polished manners were joint indicators to Herrera that the Indians could not be the immediate descendants of Europeans but must instead be descended from East Asians, whom they more closely resembled in both body and behavior. Notably, Herrera justified his racial theory not only in reference to contemporary bodily and cultural differences but also in reference to geography and history. Asserting the impossibility that the Indians' ancestors passed through "the politer parts of Europe" on the way to the West Indies was key to explaining racial differences between Europeans and Indians in the present.

## Calancha: The Asian-American Land Bridge in Service of Catholic Empire

The seventeenth century witnessed a growing number of attempts to demonstrate the previous existence of an Asian-American land bridge, and, by extension, the Asian origins of the Native Americans. Theories of ancient human migrations across a Pacific Ocean land bridge were a powerful means of elaborating racial typologies within the framework of biblical monogenism. As such, this theory was popular with clerics and men in religious orders, who deployed them in order to justify evangelism within a colonial context.

An illuminating early example comes from the Spanish Creole writer Antonio de la Calancha's *Corónica moralizada del orden de San Augustín en el Perú* (Moral chronicle of the order of Saint Augustine in Peru, 1638).[37] Calancha was an Augustinian friar born in Charcas (now in Bolivia) who spent most of his adult life in Peru. Early on in his two-volume history of the Augustinian order in the New World, Calancha turned to the crucial question of Indian origins. Rejecting the then-common theories that they were of Hamitic or Semitic (and specifically Jewish) origin, Calancha argued instead that they were a Japhetic people, descended from Tartars who crossed a Pacific Ocean land bridge in the years after Noah's Flood. Members of colonial religious orders like Calancha cared deeply about questions of biblical history and geohistory because they were seen as directly relevant to their chances of success among the indigenous Americans.[38] In the Tartar theory of Indian origins, Calancha found a way to argue for better treatment of the Native Americans, for the legitimacy and necessity of converting them, and for their continued subordination to Christian Europeans.

The idea that the Indians were descended from Tartars, which would grow in popularity throughout the seventeenth and eighteenth centuries, made sense to Calancha and to many others for reasons both geographic and ethnographic. Geographically, it was widely believed that Europe, Asia, and America intersected, or came close to intersecting, near the North Pole. "As the navigation maps show," Martínez had written thirty years earlier, "in the high latitudes of the pole, there is little distance between the parts of this land [America] and those of Asia and Europe."[39] Even if these major landmasses did not currently touch one another, it was reasonable to assume that they had been closer together in the years following the Flood, during the global diaspora of humanity. Noting that Greenland, Norway, Tartary, and Labrador were, in modern times, close to one another in the high latitudes, Calancha declared: "I hold as certain truth that the deluge happened, and that after the sea returned to its natural confines and the water to its subterranean vaults, there was a continuous stretch of land from Tartary or the northern lands all the way to Chile."[40] These intercontinental conjunctions in the Arctic would have enabled the Indians' ancestors to travel eastward from northeast Asia, crossing the "frozen sea, which comes up against this New World."[41] The idea that frozen waterways could serve as a kind of land bridge may have found additional inspiration in the lived experience of the Little Ice Age, in which waterways froze during unusually cold winters, enabling Europeans to walk (or skate!) across rivers, ponds, and canals that were typically only navigable by boat.

In the eyes of western Europeans, the Tartars were an obvious choice to be the Indians' most immediate Old World ancestors, not only because of their present-day location in northeast Asia close to where the land bridge must have been, but also because of their evident proclivities for migration and conquest. Calancha described the Tartars as fierce and nomadic, "naturally inclined to settle or to vanquish distant and diverse Kingdoms." Moreover, the modern Tartars bore evident physiological and cultural similarities to the modern inhabitants of the Americas. According to Calancha's own observations of the indigenous Americans as well as those of other colonists and missionaries, the Indians "do not have cities or fixed habitations, instead they wander from one part to another, like true Tartars." They also "bear the same color, the same traditions, [and] similar Religion" as the Tartars. Ethnographic comparisons could supplement geographic knowledge in order to reveal the origins of the American Indians.[42]

The idea that the Indians were Tartars was not unique to Calancha or to the Spanish Atlantic world. The English scholar Edward Brerewood, in his widely and posthumously reprinted *Enquiries Touching the Diversity of Languages and*

*Religions through the Chief Parts of the World* (1614), adopted similar reasoning and conclusions in declaring that the Indians had most definitely migrated from northeastern Asia. Using cultural and bodily markers of difference as his metric, Brerewood employed a process of elimination in order to pinpoint the Indians' route of migration and nearest kin. They could not have migrated westward through Europe, because "they have no rellish nor resemblance at all of the Artes, or learning, or civility of Europe." Nor could they have migrated south and west through Africa, since "their colour testifieth, they are not of the Africans progeny." The Indians' "incivility, and many barbarous properties" ruled out China, India, "or any other civill region" in Asia, according to Brerewood, leading him to conclude that "they resemble the old and rude Tartars, above all the nations of the earth."[43] The Tartars and the North American Indians, though presently far distant from one another in space, were seen by Europeans to exhibit striking similarities: primitive, nonagricultural, and seminomadic, yet also fierce and hardy owing to the cold climates in which they lived. Though it is probably fair to say that none of these folks had ever met a "Tartar" before, much less had a strong grasp of the cultural or physiological characteristics of the group, the eagerness of people in both Spanish Catholic and British Protestant contexts to align Tartars and Indians demonstrates the widespread appeal of a theory establishing both the Indians' Old World (monogenetic) origins and their roots in a group regarded by Europeans as primitive and barbaric.

The deeper reasons for the attraction of the Tartar theory becomes clearer in relation to Calancha's arguments against alternative theories of Indian origins, namely, that they were descended from Hamitic Africans or from Semitic Jews. Calancha's stated interest in determining the ancestry of the Indians was to disprove the popular theory that the Indians were a Hebraic people, descendants of Noah's son Shem, in all likelihood one of the ten lost tribes of Israel.[44] The biblical history of Noah's sons as recorded in Genesis 9 and 10 had been used for centuries as a narrative framework for explaining the postdiluvian global diaspora and, by extension, the present-day diversity of the human race. In the early modern period, the story of Noah's sons was becoming increasingly racialized, as the association of the three brothers with the three Old World continents solidified.[45] Once they had granted the reality of biblical monogenism, scholars' next question naturally became: Which one of Noah's three sons—Ham, Shem, or Japheth—was the patriarch of the American settler colonies? Calancha intended to prove that "the populators of these Indies were the sons and descendants of Japhet, third son of Noah," who was likewise the progenitor of Europeans and several other Eurasian peoples.[46]

The Peruvian friar's determination to prove that the Indians were a Japhetic and not a Semitic race was rooted in large measure in his concern to secure the basis for Catholic evangelism among the Indians of the New World. The

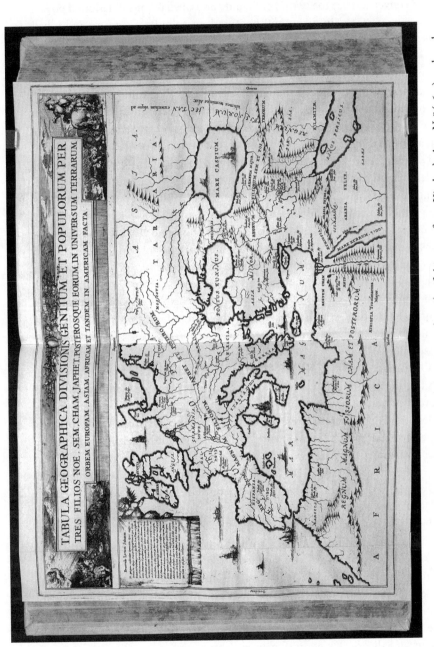

Although it only shows Europe, North Africa, and the Levant, the title of this map from Kircher's *Arca Noë* (1675) makes clear that Noah's descendants divided and occupied the "entire globe of the earth," including the Americas. Linda Hall Library of Science, Engineering & Technology.

Jewish Iberian population had been offered the terrible choice of expulsion or forced conversion in 1492, with Ferdinand and Isabella's consolidation of power. Suspicions persisted in Spain and its American colonies a century and a half later about the sincerity of the *conversos* who chose to remain among them.[47] If, as many alleged, the indigenous Americans were descendants of Jews, then the notorious resistance of Jews to Christian conversion in recent history did not augur well for the missions being undertaken by the Augustinians or any one of the other Catholic religious orders operating in the Americas. Calancha shared the view that the Jewish reluctance to convert sincerely and lastingly to Christianity was a terrible thing. He labeled it a form of "contumacy"—willful disobedience to the orders of an ecclesiastical court—and likewise a heritable "vice [that] cannot be left behind, because of an obstinacy so enduring (in the words of Isaiah and Saint Steven) that some believe it is in their blood." But this heritable obstinacy does not infect the Americans, Calancha argued, for the simple reason that they are not Jews. Against the backdrop of the recent history of Jewish-Christian relations in early modern Spain, Calancha disputed the ancient connection between Jews and indigenous Americans in order to improve the missionaries' chances of bringing the latter "into the house and the tabernacle of the Catholic Church, from which the Jews, descendants of Shem, have fled." For Calancha, working to bring fellow descendants of Japheth to the light of Christ opened up the possibility of more authentic and lasting conversions to the Catholic faith than the prospect of dealing with recalcitrant New World Jews.[48]

On the face of it, Calancha's effort to prove the Japhetic origins of the American Indians may not appear so different from Sarmiento de Gamboa's effort to prove they were originally Spaniards. But the implications of Calancha's theory, which rested on migration across the Pacific rather than the Atlantic, Asian origins rather than European, Tartar ancestry rather than Spanish, were quite different. Like many other European writers who espoused the theory of the Tartar origins of the Native Americans, Calancha was concerned to legitimate both evangelism and empire, to bring the Native Americans into the fold of the Noachian family tree while also distancing them from Christian Europeans.

First, Calancha's Tartar land bridge theory allowed him to articulate a more biblically based justification for Spanish claims to the New World than Gamboa's theory of Spanish-Atlantic origins. It suggestively aligned the present-day Spanish Empire with the ancient Tartars as migratory conquerors, whose joint descent from Noah's favored son Japheth bestowed on each group a divinely mandated right to whatever lands they discovered in their global wanderings. The characterization of the Tartars as "naturally inclined to settle or to vanquish distant and diverse Kingdoms" may well have applied equally, in Calancha's mind, to the modern Spanish.[49]

Calancha may also have regarded the migratory habits of the ancient Tartars as a figure for his own order of Augustinian friars, who first arrived in the New World in the 1530s in the wake of Cortés's bloody Mexican campaign and spread throughout South America along with the expanding Spanish Empire in the sixteenth and seventeenth centuries. The Mendicant orders, which included the Augustinians as well as the better-known Franciscans and Dominicans, were founded on a commitment to engaging the wider world through preaching and ministering to the laity, especially to the poor. Mendicants like Calancha were thus distinguished from other Catholic orders by their mobile and active, rather than cloistered or contemplative, religious life. Calancha's declaration that "the Gentiles from Japhet" were destined to "spread throughout the world with the coming of Christ" simultaneously comprehended the biblical Tartars and modern-day European Christians, and perhaps was also meant to refer specifically to Mendicants like himself.[50] He cast the project of empire as a holy enterprise, in biblical and modern history alike. Determining the location of the land bridge and the ethnic ancestry of the Native Americans was implicated in larger questions about the nature of the Spanish Empire and the subject position of the indigenous Americans within it.

Biblical monogenism and even common Japhetic descent could be mobilized to justify Spanish imperialism. It could also be used to draw race-based distinctions between Europeans and Indians. Even after drawing suggestive parallels between the Tartar-Indians and the Spanish-Americans, Calancha went on to construct a geohistorical account of the inherent differences between Indians, Europeans, Africans, and Euro-American Creoles. This dimension of Calancha's racial thinking becomes even clearer when exploring his arguments against the Hamitic theory of Indian origins.

Many Creole writers in the early seventeenth century preferred the theory of Hamitic origins, rather than Semitic or Japhetic, as a means of drawing racial contrasts between themselves and the other peoples living in the New World.[51] Spanish attitudes toward Indians and Africans grew increasingly racialized in the seventeenth century, as they despaired over the incomplete conversion of the Indians and the proliferation of mixed-race people in Spain's American colonies.[52] First, the biblical curse on Noah's son Ham, which was newly emerging in the early modern period as a justification for enslaving Black Africans, if applicable to the Indians, implied that they, too, were ripe for enslavement or were at least naturally servile.[53] The famous debate in Valladolid between Bartolomé de las Casas and Juan Ginés de Sepúlveda on whether the Indians fit Aristotle's definition of "natural slaves" was part of a larger debate in the Spanish Atlantic world about the validity of the various legal, theological, and philosophical justifications for slavery.[54] Hamitic origins could certainly play a role in justifying the enslavement of indigenous peoples, or at least their inferiority to Japhetic peoples like the Europeans.

Second, appealing to the Hamitic theory of Indian origins helped to explain away a troublesome fact from two fields of natural philosophy—astrology and climatology—that seemed to undermine European claims to racial superiority. Astrology dictated that the moral, physical, and intellectual traits of individuals were influenced by the stars: the configuration they were born under and the ones they currently lived under. Climatology said that those same traits were also shaped by the climate—the unique combination of local air, water, and soil—in which one was born and lived. The problem, which was of great concern to medical writers in seventeenth-century New Spain, was how to understand the persistence of health disparities and bodily differences between Indians and Euro-Americans, despite living under the same skies, breathing the same air, drinking the same water, and eating food grown in the same soil. Astrology and climatology dictated that Europeans now living under American stars in the American climate should have reached parity with the Indians by now. Moreover, Aristotle's theory that people in hotter climates were intellectually superior to those from colder ones dictated, even more unbearably to European writers, that the Indians of New Spain should be more intelligent than Spaniards. If, however, the Indians labored under the curse of Ham, that could explain why Indian bodies and minds remained distinct from, and inferior to, Creole bodies and minds despite being subject to the same astral and climactic influences.[55]

Calancha rejected the theory of Hamitic origins and opposed the idea that Indians were natural slaves. Like Acosta before him, Calancha styled himself as a champion and a defender of the downtrodden Indians.[56] Giuliano Gliozzi has shown how Calancha's hostility to the Jewish theory of Indian origins was motivated in part by his opposition to the forced, privatized labor of Indians in the *encomienda* system.[57] At the same time, Calancha could not accept the notion that the American climate made its inhabitants superior to the Spanish or to the Creoles. So while insisting on the Indians' Japhetic origins, Calancha also espoused a tripartite racial taxonomy that clarified their inferiority to (Japhetic, Christian) Europeans.[58] Calancha cited the German Creole physician Martínez, who had influentially theorized that inherent differences between Native Americans, white Europeans, and Black Africans were so profound that neither the stars nor climate could fundamentally alter them.[59] "The complexions of the *negro*, the Indian, and the Spaniard are very different," Calancha observed. Modifying the Hippocratic dictum that universal causes produce universal effects, he went on to argue that "even though general causes operate across this Kingdom, they cannot produce equal effects in all people." Astrological and climatological influences were the same across Peru, but they affected individuals differently according to the individual's "complexion," that is, their physical constitution and particular embodiment.[60] Much like the curse of Ham—and perhaps also like Calancha's theory that Jewish "contumacy" was a heritable quality that was transmitted by blood—"complexion" was an ingrained trait in human bodies, passed down within specific human lineages and acting as a natural limit to astral and climactic influences. Calancha's *Moral Chronicle* demonstrates well how the Tartar theory of Indian origins, joined to the new Creole philosophy of race, worked simultaneously as a justification for conversion and an argument against unfree labor, while at the same time articulating a hierarchy of racial difference.[61]

Calancha's *Moral Chronicle* gives us a valuable window onto debates about Indian origins in the Spanish Atlantic world in the first half of the seventeenth century. These debates in turn reveal much about the racial politics and political theology of biblical monogenism. The case of Calancha in particular helps us to understand why the Tartar theory began to emerge as a popular contender amid the welter of possibilities offered to explain the mystery of Indian origins. While the Hamitic theory of Indian origins worked well as a justification for forced labor, it did not appeal to missionaries who wished to see the Indians treated more humanely. The notion that they were of Jewish origin was likewise fraught from the standpoint of evangelism, especially in Iberian and Hispanic contexts. The theory of European origins worked very well as a justification for evangelism, but did not easily work to buttress European supremacy in the Americas. The Tartar theory of Indian origins, on the other hand, poised the Indians for conversion while also maintaining their racial inferiority to Europeans. It offered a geohistorical basis for advancing a position that was simultaneously monogenetic and racist, pro-evangelical and antislavery.

## Bourguet: The Asian-American Land Bridge in Service of Protestant Empire

Determining the location of the land bridge and the racial ancestry of the Native Americans was crucial to Calancha's understanding of what he and his brethren —Spanish and Augustinian—were doing in the New World. As European Protestant states launched serious attempts to challenge Iberian supremacy in the Atlantic world in the second half of the seventeenth century, the scholarly debate about Indian origins spread beyond Spain and the Spanish Empire to engage scholars in England and the Low Countries in ever greater numbers. The precise location of the land bridge became a question of even greater urgency as the Dutch and English struggled to establish commercial and imperial claims vis-à-vis each other as well as Spain, Portugal, and France. Moreover, as various Protestant groups and sects began to launch more concerted campaigns to evangelize in Protestant-controlled American territories at the turn of the eighteenth century, new iterations of the Asian-American land bridge theory emerged that reflected distinctively Protestant visions of Christian imperialism and racial hierarchy.[62]

In 1735, the Huguenot scholar Louis Bourguet published a letter in the learned journal *Mercure Suisse* (*Swiss Mercury*, later the *Journal Helvétique*, 1732–69), announcing his discovery of a land bridge connecting northeast Asia and northwest America, somewhere off the coast of Kamchatka. A founding editor of the *Swiss Mercury* as well as a frequent contributor, Bourguet presented his finding as the resolution of more than two centuries of debate about the origin of the Native Americans. "Of all the discoveries made in the sixteenth century," Bourguet declared in the letter's opening lines, "the discovery of America, or the New World as it is justly called, was undoubtedly the most important" and was understandably "a subject of shock for all of Europe."[63] Unfortunately, in Bourguet's view, the ensuing debate over Native American origins had so far been purely speculative and therefore utterly fruitless. "The Unbelievers, delighted to find a plausible pretext for promoting their unbelief," exploited the weak empirical grounding of monogenetic theories of Native origins in order "to combat the truth of the Mosaic History, for which they had little respect."[64] All the efforts of European scholars for the past two hundred years had accomplished nothing except to promote skepticism about the validity of biblical history as a true and accurate record of the history of all lands and peoples.

What made this Swiss scholar of the early eighteenth century so convinced that all previous efforts to prove biblical monogenism had failed and that proving it remained as urgent as when Columbus first returned from the Americas in 1493? Bourguet's stated fears about the spread of unbelief may have been in reference to the resurgent threat of polygenism, which received a fresh new jolt of energy in 1734. In *Traité de métaphysique* (*Treatise on Metaphysics*), the young French philosopher Voltaire argued that Europeans, Indians, Asians, and Africans constituted four distinct races of people with no common ancestry. "The bearded whites, the woolly-haired negroes, the horse-haired yellow men, and the men without beards," by which he meant the American Indians, "are not descended from the same man."[65] In a foretaste of the provocation for which he would later become famous, Voltaire framed his polygenism as a statement of fact rather than a hypothesis, as La Peyrère had more cautiously done nearly a century previously. Moreover, Voltaire's theory is explicitly racialized, as La Peyrère's polygenist theory had not been, using bodily markers of difference— hair color and texture as well as skin color—as evidence of the multiple and distinct origins of the major human groups.[66] Louis Bourguet's first article in the *Swiss Mercury* defending the existence of an Asian-American land bridge was published in July 1735, the year after Voltaire's *Treatise on Metaphysics* appeared. Bourguet had been pursuing the land bridge question since at least 1716, so it is very probable that the resurgent threat of polygenism represented by Voltaire's work formed the impetus for Bourguet to finally publish his thoughts on the subject.[67]

Bourguet's quest to locate the land bridge reflected a continued sense that monogenism remained vital in order to support Christian evangelical goals. It also signaled an emerging desire to make Protestant evangelism competitive with the longer- and better-established Catholic missionary operations in the Americas and beyond. Bourguet, a Huguenot refugee, was eager to assist the Protestant evangelical missions that were gearing up in the early decades of the eighteenth century. His pursuit of the Asian-American land bridge also reflected the changing geopolitics of Russia, China, and northern Eurasia in the early eighteenth century. The growth of organized Protestant evangelism beyond Europe and the changing geopolitics of northern Eurasia provide crucial context for understanding Bourguet's variations on the long-standing Tartar theory of Indian origins. He made three important changes. First, the Tartar migrations to America happened in modern, not ancient, history. Second, the land bridge connecting the Old and New Worlds might still exist. Third, the Tartar migrations to the New World had been spurred by violence in Asia and led to further violence in America, as different Tartar groups battled each other for supremacy. With all of these innovations, Bourguet drew a series of implicit and explicit connections between the pre-Columbian Pacific world and the post-Columbian Atlantic world. Ultimately, his decades of scholarship on Indian origins pressed geohistory into the service of Protestant empire.

Bourguet was well aware that the question he was taking up was an old one, as evidenced by the 1735 article and a subsequent one on the same topic published in the *Swiss Mercury* the following year. They also convey his sense that, even after more than two centuries of research and debate, very little progress had been made in answering it. Bourguet singled out Grotius's theory about the Scandinavian origins of the North Americans for critique. He cited Georg Horn's theory of Tartar origins approvingly, though he complained that Horn had advanced this notion without solid evidence. His multidisciplinary approach to evidence reflects his sense that new approaches were needed if the question of Indian origins were ever to be satisfactorily resolved. Bourguet drew on ethnographic and linguistic evidence just as scholars before him had done, but he also relied heavily on geographic evidence, including the most current maps he could locate and the latest news from geographic expeditions.[68]

Bourguet's reliance on geographic and geohistorical evidence evinces not only the growing power of geological determinism, as will be discussed shortly, but also the very different situation in which European Protestants found themselves relative to their Catholic, and especially Spanish, counterparts when trying to access information that would help them to answer the question of Indian origins. Calancha was a Creole, born in the New World, whose work with the Augustinians gave him firsthand knowledge of the Americas, its natural features, and its indigenous inhabitants. Bourguet, for all his mobility as a Huguenot refugee, never left Europe and could only access secondhand knowledge of the New World, as was the case for most European naturalists.[69] He was doubly disadvantaged by not living in Britain or the Netherlands, Europe's Protestant imperial powers (though he did have a brother in London who helped connect him to British networks of information). Switzerland, Bourguet's adopted homeland, was not an imperial power or really even a country, meaning that he did not have access to bureaucratic, academic, and religious networks of information-gathering about the people of the New World that existed in the Spanish Empire. This may be part of the reason why Bourguet, while ecumenical in his methodological approach, relied far more heavily on geographic and cartographic evidence than did Calancha and other Spanish writers on the question of Indian origins. He was limited to seeking material proof of pre-Columbian Tartar migration that did not depend on actually talking to living Native Americans or knowing very much about them at all.

Bourguet's two articles in the *Swiss Mercury* gesture toward many of the same ethnic analogies between modern-day Tartars and Native Americans that scholars had been drawing throughout the previous century. Just like the Tartars, the Americans "ignore entirely the Art of laboring the Earth." They "resemble very strongly the most uncivilized Tartars" not only in their ignorance of agriculture but in "their morals, their customs [*coutumes*], [and] their Religion." Bourguet exclaimed, "It would take a whole Volume to describe in detail all the similarities" between the two groups. But racial and cultural homologies were not enough to prove common ancestry and defeat polygenism. More solid proof was needed.[70]

An accomplished linguist—he allegedly taught himself over fifty ancient and modern languages through daily reading of foreign-language Bibles—Bourguet was keen to find linguistic evidence that could help him definitively establish the Tartar origins of the American Indians.[71] A mysterious text engraved on a boulder in New England was a major breakthrough. In 1714, the *Philosophical Transactions* published a summary of a series of letters addressed to the Royal Society by the Boston minister Cotton Mather. One of Mather's letters briefly described a boulder, located on a riverbank in Taunton, Massachusetts, with indecipherable letters carved into one side. Mather had no luck deciphering the inscription and even doubted it was writing at all. However, he included a "transcription" of the first two "lines" of writing on the rock's face, which was published in the *Philosophical Transactions* along with Mather's verbal description of the rock and its carvings. Bourguet did not learn about this astonishing discovery until 1722, and he did not get his hands on a copy of the *Philosophical*

*Transactions* until several months after that. Before he had even seen Mather's transcription, Bourguet informed his friend, frequent correspondent, and fellow Swiss Protestant, the naturalist and physician Johann Jakob Scheuchzer, that the "writing is assuredly a type of Tartar character."[72] After finally committing his thoughts to print, Bourguet informed the readers of his journal that here at last, from across the Atlantic Ocean, in the form of a boulder in Massachusetts, was "incontestable proof that some little Colony of Tartars came, several centuries ago, to the country that one calls New England."[73]

While excited about the linguistic evidence offered by the Taunton stone, which to him proved the historical migration of Tartars to America, Bourguet was equally animated about geographic evidence that might prove the land bridge they migrated across still existed in the present day. Focusing his search on the icy northern waters between northeastern Asia and northwestern America, Bourguet actively lobbied his contacts in St. Petersburg for news of the Bering expedition that was plying the cold seas off the northeastern coast of Siberia.[74] He even got the French astronomer and geographer Joseph-Nicolas Delisle to ask Peter the Great about the far extremities of his empire when the Russian tsar was visiting Paris in 1721–22.[75] More helpful yet was Philipp Johann von Strahlenberg's detailed map of Tartary and the northeastern coast of Asia, published in his *Historie der Reisen in Rußland, Siberien und der Grossen Tatarey* (*History of Travels in Russia, Siberia, and Greater Tartary*, 1730). Strahlenberg, a Swedish officer who was taken captive by Russian forces in the Great Northern War and held as a prisoner of war for thirteen years, declared in the introduction that he hoped his new map of Tartary would contribute to the ongoing project of reconstructing migration history: with the creation of "exact Maps, especially of Countries far remote," he wrote, "the Origin and Migration of Nations may in time be set in a truer Light."[76]

Based on the information in Strahlenberg's map and travel accounts of Dutch and Japanese expeditions, Bourguet believed he had determined the precise co-ordinates of the land bridge's current location: between latitudes 48°50' and 51°N.[77] This placed it just beyond the coast of Kamchatka, a peninsula hanging off the coast of Siberia, north of Korea and Japan. Bourguet opened his second *Swiss Mercury* letter on the land bridge with the following declaration: "We are on the eve of the entire discovery of Kamschatka, and as a result, of the existence or nonexistence of the isthmus, across which I believe that Asia is attached to America."[78] He hoped that Captain Bering's expedition would prove him right and would further determine whether the "bridge" was a strait or an isthmus— he apparently believed either one would count as the "junction" of Asia and America he had long been seeking.[79]

Bourguet's fervent hopes for the actual existence of the Asian-American land bridge in his eighteenth-century present can be more readily understood when

Joseph-Nicolas Delisle's 1752 map of the North Pacific (*Carte des nouvelles découvertes au nord de la Mer de Sud, tant à l'est de la Siberie et du Kamt-chatcka*) features the peninsula of Kamchatka, where Bourguet believed the land bridge was still located, and visually hints at the close proximity of the far edges of northeastern Asia and northwestern North America. The two human figures in the top right and left corners, identified as an "Inhabitant of Kamchatka" and a "*Sauvage* of northern Louisiana," reinforce the map's suggestion of a land bridge by implying a geographic and historical connection between the two indigenous groups. Newberry Library, Chicago, Novacco 4F 60 (PrCt).

placed alongside his conviction that migration across the land bridge was a relatively recent phenomenon. In a departure from precedent, Bourguet did not think the Asian-American migrations had happened during the biblical age or even in postbiblical, but still ancient, history. He supposed these migrations had happened in the relatively recent past—as recently as the Middle Ages, in fact. Writing in the *Swiss Mercury*, Bourguet posited three major waves of migration from Asia to America: first the ancestors of the "Peruvians," then the ancestors of the "Mexicans," and finally the Tartars, who mostly settled North America.[80] The Peruvians and Mexicans, he speculated, were so much more civilized compared to their North American neighbors that they likely traced their ancestry to the Chinese or Japanese, who crossed from Japan to California and then to Mexico—possibly even by boat—about a thousand years previously.[81] The third major migration comprised multiple waves of Tartars crossing the land bridge, which Bourguet thought began in the Middle Ages and had ended just a few hundred years before, perhaps right around the time that Columbus first stumbled across the island of Hispaniola.

Both of these attempts to bring ancient history into the present seem to have been motivated by a powerful, if not fully articulated, sense of the deep connections between past and present, the Atlantic world and the Pacific world, and pre- and post-Columbian world history. Bourguet's imagined history of the medieval Pacific world closely resembles the reality of his early modern Atlantic world. In reconstructing the history of Asian-American medieval migration, Bourguet envisioned successive waves of migration to the New World and conflict among different groups as they battled for territory in their newfound continent. The Tartar migration to America was in fact composed of multiple migrant groups, he wrote, who were constantly in conflict with one another. Bourguet speculated that the people who had carved the Taunton stone had been victims of genocide, killed off by a subsequent wave of bloodthirsty immigrants seeking to displace them and claim their land. The stone-carvers came from Asia and settled in Massachusetts only to be "massacred by the Iroquois or some other Barbarians in the area."[82] If the characters could be deciphered, he told Scheuchzer, it could help us to understand "what type [*espece*] of people were first in America, who perhaps perished by the ferocity of other Savages who made war against them."[83] The first settlement of the Americas had been characterized by bloody warfare, displacement, and even genocide—not unlike the early modern settlement of the Americas by Europeans from across the Atlantic.

The similarities continued, as Bourguet outlined his theory about how the migration of Tartars to America began and why it unfolded in such a violent fashion. Strahlenberg's *History*, probably the most authoritative source on Tartary available to European readers in the early eighteenth century, claimed there were six distinct "nations" of Tartars, though it was difficult to determine where exactly the territory of one nation ended and another began, given that "the Tartars have not been very exact in fixing the Boundaries among themselves."[84] Bourguet speculated that migration across the land bridge was initially spurred by war in Asia between these six Tartar nations. "The Tahuglauk"—the Tartar nation Bourguet suspected of being the ancestors of the Taunton rock-carvers— "must have established themselves in Jesso [Hokkaido] in the eleventh or twelfth centuries, when there were many wars in Tartary."[85] In other words, the turbulent era of warfare prior to the unifying reign of Genghis Khan was the likely motor of the migration out of Asia and across the northern Pacific. Moreover, Bourguet imagined that this intra-Tartar conflict spilled over into their settlements in the New World, as his theory of the Tahuglauk rock-carvers massacred by the Iroquois was intended to demonstrate.

One major difference, however, distinguished pre-Columbian Tartar settler colonialism from its early modern European post-Columbian counterpart. Bourguet believed that the different Tartar nations held members of each other's nations as slaves in their new American homeland. He cited a travel account by the Baron de la Hontan of his encounter with a people named the Gnasitares living west of the Mississippi River, who had among them four slaves who were "Mosemlek," that is, members of another Tartar nation.[86] While the idea of warfare practiced by the nations of Tartary upon each other in Tartary and in America echoed the actual intra-European conflicts of the past several centuries in both Europe and America, the enslavement of Tartar groups by other Tartars would have appeared as a bright line distinguishing Tartars from Europeans to all of Bourguet's European readers, given the long-standing norm that forbade Christians from enslaving each other. Of course, the injunction against Christian slaveholding was routinely violated by the practice of selling Black Africans into slavery and continuing to hold them as slaves even after their Christian conversion. Nevertheless, Bourguet's account of Tartars enslaving each other in the New World reflected the self-serving definitions by which Christian Europeans distinguished themselves from peoples and cultures they considered inferior.

The Tartary theory of American origins changed as the politics of the region changed, another link between past and present. Writing in the late seventeenth century, Matthew Hale was probably thinking of the ongoing wars between the Manchu and the remnants of the Ming dynasty when he predicted that the Tartars (who he also believed had settled America) were on the verge of ruining Chinese civilization. As a general rule, Hale declared, civilized countries overrun by less-civilized foreigners "degenerated into the Ignorance and Barbarism of their Conquerers. . . . [T]his possibly may be the condition of China in a few years after the great Irruption and Devastation by the Tartars."[87] Bourguet was also concerned about the shifting geopolitics of northern Eurasia around the turn of the eighteenth century as he was working on his history of Tartar immigration to the Americas. The part of the globe that Bourguet identified as the pathway of Asian-American migration was, as he well knew, a geopolitical hotspot in the early eighteenth century. In a 1727 letter to Scheuchzer, Bourguet expressed worry that the ambitions of the new czar, Peter II, signaled that "the Russians have a great desire to return to their primitive barbarism." If the Russians succeeded in consolidating their power across northern Eurasia, Bourguet worried, geographers and missionaries from western Europe would lose access to that part of the world, harming both the pursuit of knowledge and the pursuit of souls. Geographic knowledge that could help scholars reconstruct the pathways of Asian-American migration would slow to a standstill. "We will lose the entire discovery of the more than 1200 places . . . from west to east which surely there are between Jesso and California and the Mississippi." Moreover, if these same lands were to fall into Russian hands, Protestant missionaries would not be able to enter, and "these numerous Nations will never hope to be enlightened by the light of evangelism."[88]

Bourguet's concerns about the eastward expansion of the Russian Empire reflected not only his commitment to locating the Asian-American land bridge but also his concomitant—and related—commitment to global Protestant evangelism. His own experience of religious migration helps to explain the depth of his commitment to Protestant evangelism and to scholarship that would further evangelical ends. Born in the French city of Nîmes in 1678 to a family of Huguenot merchants, Bourguet was forced to flee the country with his family in 1685 when Louis XIV revoked the Edict of Nantes. The Bourguets emigrated to Switzerland along with thousands of other Huguenot refugees, where they moved back and forth between Geneva, Lausanne, and Zurich, the last of which expelled its Huguenot population in 1700. Bourguet spent most of his young adulthood traveling back and forth between Switzerland and Italy on business trips (his family was in the textile business) before finally settling in the Swiss city of Neuchâtel. His itinerant life and early experience of religious persecution had a strong effect on his self-concept as a stateless Protestant in exile. In a notebook he kept as a young man, Bourguet styled himself an "amateur scholar of antiquities from Nîmes in Languedoc, now living in Zurich as a refugee for the Reformed Religion."[89] As Bourguet grew older, his sense of being a "refugee for the Reformed Religion" grew into a sense of belonging and obligation to the communion of Protestants, not just in the Atlantic world but across the globe. Prevented by poor health from becoming a missionary himself, Bourguet funneled as much money and energy as he could into supporting Protestant evangelical groups in Germany and Britain, including the Moravians and the Society for the Propagation of the Gospel in Foreign Parts (SPFG).[90]

It is not clear whether Bourguet was consciously aware of how closely his speculative history of Tartar migration in the medieval Pacific world mirrored the actual history of European migration in the early modern Atlantic world. He was, however, explicit about his desire to instrumentalize his geohistorical scholarship on Asian-American migration to advance contemporary evangelical goals. In 1730, Bourguet wrote to the London naturalist Hans Sloane asking for an array of Indian-language Bibles, catechisms, and grammars from the British colonies in North America, which he said would enrich his knowledge of "the History of the Travails of Protestants for the Propagation of the Faith and will also be useful to me in revealing the diverse languages that are spoken in the four parts of the world."[91] A fuller knowledge of the Indian languages of New England would help him to reconstruct their affinities with the languages of Tartary, which would bolster his case for the Asian-American land bridge, which would in turn contribute to the missionary efforts of Protestants in the Americas. In fact, Bourguet seems to have had a special interest in converting the Indians of colonial New England. In the same letter to Sloane, Bourguet specifically requested titles by Increase and Cotton Mather on Protestant missions in New England, and he told Sloane that he was especially anxious for the SPFG to succeed in the "complete conversion" of the Iroquois—the same group he believed was responsible for the massacre of the Tartars who carved the Taunton stone.[92] When Bourguet wrote in the *Swiss Mercury* of wanting to prove the land bridge's existence in order to vanquish atheism, perhaps it was not atheists in Europe he was thinking of, but rather the "heathens" of colonial North America.

The political and evangelical commitments that undergirded Bourguet's scholarship sometimes zeroed in on specific places, like New England, and at other times zoomed back out to comprehend the global context of the emerging battle between Protestant and Catholic states. Bourguet's fears about Russian power in Eurasia and the Pacific Rim reflected his broader concern to see European Protestant powers dominate over both Catholic and Orthodox powers on a global stage.[93] In a letter to Scheuchzer in 1728, Bourguet declared himself "stunned at the blindness of the Dutch and the English" for letting Brazil and the Mississippi River valley fall into the hands of the Portuguese and the French, respectively. He predicted the Spanish would advance all the way up the west coast of North America until they reached Japan and, from there, march into East Asia and attempt to "attract the Emperor of China to their Religion." Bourguet explained that his wish to see the Protestant powers prevail was purely instrumental: "The desire I have to see the advance of the English and Dutch comes principally from the love I have for Religion; barring that, I would care very little which Europeans became Masters of all those countries." Bourguet was remarkably candid, writing privately to a Swiss Protestant, about the conditional nature of his support for the English and the Dutch in their battles with other European powers for access to ports and control of territory. One imagines he would not have voiced this sentiment to any of his English or

Dutch correspondents, of which he had several. This passage reveals that Bourguet was willing to support Europe's Protestant powers only because he believed that Protestant imperial supremacy was the best and perhaps the only means of spreading the Protestant faith across the globe.[94]

## Biblical Monogenism and the Origins of Scientific Racism

Bourguet's primary concern was not with building racial taxonomies but rather with weaponizing his scholarship in the fight for global Protestant supremacy. Yet the particular way in which he chose to pursue that goal—locating the Asian-American land bridge and reconstructing the history of Tartar immigration across it—was rife with unspoken assumptions about the superiority of Protestants over other Christians and of Europeans over non-Europeans. The emergence of racial hierarchies from the imbrication of geohistory and biblical monogenism forged in the long seventeenth century is an underappreciated but nevertheless important step in the development of racial thinking in European and American history. The obvious complicity of nineteenth-century polygenism in the development of modern scientific racism and the maintenance of legalized slavery, coupled with the resurgence of monogenism in the postwar period in response to the horrors of the Holocaust, has perhaps led modern scholars to underestimate the racist potential of monogenism. In the early modern period, it was certainly the case that polygenism was widely rejected in favor of monogenism. But monogenetic theories of human origins, especially as exemplified in early modern theories of ancient land bridges and American migration, also had a role to play in the development of modern racial categories.

There are several reasons to doubt the claim that the biblical form of monogenism that dominated European thinking about global human diversity in the sixteenth through eighteenth centuries was antiracist, or even particularly benign. Rhetorics of family affinity and declarations of the universal brotherhood of humankind from this era are frequently cited as evidence for the racial toleration of biblical monogenism. But these statements need to be understood first of all in the context of early modern European conceptions of the family, household, and kinship. As is well documented, early modern kinship was not egalitarian.[95] In early modern European culture, membership in a family by no means guaranteed equal status with other family members; in fact, it usually guaranteed the opposite. Differences of age, gender, race, place of origin, citizenship, marital status, socioeconomic status, and religion sorted individuals within a single family unit into an unequal hierarchy. The same was even more true for households (and the distinction between a family and a household were often blurry). Households comprised not only parents, children, and kin such as grandparents and cousins, but also servants, slaves, apprentices, and wards, none of whom stood on an equal footing with the master of the household, his wife, and his heirs. Most modern observers have difficulty relating to the (utterly typical) family life of the famous English diarist Samuel Pepys, who accepted his unmarried sister as a servant in his household on the condition that she not be allowed to eat at the same table as Pepys and his wife.[96] The lived experience and legal reinforcement of families characterized by stark inequalities among its constituent members provides illuminating context for understanding what it meant to early modern Europeans and Euro-Americans to say that they were part of the same "family" as Indians, Jews, Muslims, sub-Saharan Africans, and so on. Families in early modern Europe were hierarchal and exploitative; so too then, in all likelihood, were early modern European visions of the global human family of Noah's descendants. Declarations that all living people were members of the Noachian lineage cannot, therefore, be interpreted as a necessarily egalitarian gesture, or even a benign one.

A striking example of the elaboration of racial taxonomy within biblical monogenism appears in the widely reprinted *Purchas His Pilgrimage* (1613) by the Anglican divine Samuel Purchas. The passage in which Purchas likens the divinely created racial diversity within the human race to the "incomprehensible unitie" of the Holy Trinity is worth quoting in full:

> [God] hath pleased in this varietie to diversifie his workes, all serving one humane nature, infinitely multiplyed in persons, exceedingly varied in accidents, that wee also might serve that one-most God; the tawney Moore, black Negro, duskie Libyan, Ash-coloured Indian, olive-coloured American, should with the whiter European become one sheepe-fold, under one great shepheard, till this mortalitie being swallowed up of life, wee may all be one, as he and the father are one . . . and their long robes being made white in the bloud of the Lambe . . . without any more distinction of colour, Nation, language, sexe, condition, all may bee One in him that is One.[97]

The multiplicity and variety of human groups of diverse appearance and places of origin is presented by Purchas as, first, evidence of God's power as a Creator. At the same time, he emphasized that the diversity of the human race is contained within a single, unified "humane nature," which he suggested is a holy mystery akin to that of the Trinity: the unity of the Father, Son, and Holy Ghost in one God. Moreover, Purchas envisioned death and union with the divine as the final erasure of human diversity, a figure of mass conversion to Christianity and indeed as a kind of whitewashing. "Being made white in the bloud of the Lambe" links with the earlier reference to "the whiter European" and contrasts with the list of skin colors of non-Europeans. Purchas's notion of "one humane nature, infinitely multiplied" is a vivid demonstration of the core tension within biblical monogenism: the articulation of a racial taxonomy within a single human family tree.

Rhetorics of universal kinship from advocates of biblical monogenism also need to be understood in reference to the politics of migration, indigeneity, and settler colonialism in early modernity. It is true that polygenism was adopted by slaveholders and their political allies as a justification for slaveholding as early as the seventeenth century.[98] But La Peyrère was not an apologist for slavery, and his account of humanity's multiple origins in some respects granted Native Americans the indigenous status that biblical monogenism denied them. La Peyrère's was one of the lone voices in early modern Europe whose historical account of Indian origins tallied with native accounts of themselves as autochthonous peoples who had always lived in the place they called home. Biblical monogenism, on the other hand, worked against native claims to autochthony. The sixteenth-century French humanist Jean Bodin explicitly paired a declaration of universal humanity with a denigration of the idea of indigeneity in *Methodus ad facilem historiarum cognitionem* (*Method for the Easy Comprehension of History*, 1583). According to Bodin, the Mosaic history establishes that "all men . . . share the same blood and the same origins." Belief in these common origins promoted "good will and friendship" among men, whereas believing otherwise threatened religious truth and the social order. "Those who declare themselves to be indigenous, do they not break the bonds of human fellowship?"[99]

A century later, Stillingfleet echoed Bodin in lambasting the "the great pretence of several Nations that they were self-originated" and who "make themselves Aborigines."[100] Claims to indigeneity were viewed as illegitimate and indeed offensive from within the framework of biblical monogenism's uncompromising universalism. Significantly, though unsurprisingly in light of anti-indigenous sentiments like these, the common term *Americans* in the early modern European and Creole literature on human origins was rarely paired with modifiers like *native* or *indigenous*. Biblical monogenism could act as a powerful counterweight to native claims of indigeneity and autochthony, which not coincidentally weakened Native Americans' territorial claims. Biblical monogenism figured everyone on earth as a migrant and a settler. By adopting this ideology, early modern Europeans found a powerful rhetorical tool that could undermine native claims to autochthony and at the same time legitimate their own "migratory" behavior.

This chapter has sought to demonstrate the links between earth history, the Indian origin debate, and the evolution of racial thinking in early modern Europe and the colonial Atlantic world. These links, I argue, were forged in efforts to legitimize evangelism and empire in the New World. Collectively, they show how modern scientific racism emerged out of early modern biblical monogenism. The assumption that biblical monogenism was a form of antiracism or pre-racism, holding in check the more virulent forms of racist exploitation that would later be justified in reference to polygenism in the nineteenth century, is common in the literature on race in early modernity.[101] So is the related notion that "race" is a modern idea, a category of analysis dating to the nineteenth century or perhaps the Enlightenment.[102] These assumptions have been challenged by the an emerging body of evidence that Europeans and Euro-Americans in the sixteenth and seventeenth centuries did articulate theories regarding the inherent mental, moral, and bodily differences between Europeans, Indians, and Africans.[103] This chapter contributes to that trend by showing how biblical monogenism, bolstered by the biblical story of Noah's Flood, provided a flexible and unfortunately generative framework in which racial hierarchies could be elaborated.

# CHAPTER THREE

# Protestant Climate Change
## From Edenocene to Fallocene

'Tis we that have left the tract of Nature, that are wrought and screw'd up into
artifices, that have disguis'd our selves. . . .

—Thomas Burnet, *Sacred Theory of the Earth* (1684)

### The Earth as Artifact

Long before the environmental writer and activist Bill McKibben lamented "the
end of nature" in the age of anthropogenic global warming, the seventeenth-
century natural philosopher Thomas Burnet declared that humanity had "left
the tract of Nature" and was now living in a world that was fundamentally,
though unintentionally, of our own making.[1] Burnet's influential and controversial *Telluris theoria sacra* (*Sacred Theory of the Earth*, 1681–89) located this
profound transition in the ancient past, not in the near future, as McKibben
projected in the late twentieth century. According to Burnet, Noah's Flood was
the bright, sharp dividing line between the ancient, "natural" earth and the
modern, "artificial" one. The structure of *Sacred Theory* reflects this twofold
division of world history. Volume 1, published in Latin in 1681 and English in
1684, first discusses the Flood and the creation of the modern earth, then jumps
backward in time to survey the original earth, from its creation by God through
its destruction by the Flood. The antediluvian world was natural, according to
Burnet, because it was designed by God; the postdiluvian world was artificial
because it had been ruined by people. The Flood was sent by God, of course, but
it was inadvertently provoked by human sinfulness, meaning that its destructive
effects were ultimately humanity's fault. "We have . . . [gotten] screw'd up into
artifices," Burnet wrote, reflecting gloomily on the world he knew in 1684. What
he meant was: We have no one to blame but ourselves.

Like many natural philosophers across Europe in the second half of the seventeenth century, the Cambridge-educated philosopher and Anglican minister
was keen to prove that Noah's Flood was universal. Unlike the scholars surveyed
in the preceding chapter, however, Burnet's efforts to demonstrate the Flood's

global scale did not spring from a desire to furnish additional proof for biblical monogenism or to provide additional legitimation for overseas evangelism. Instead, by cataloging the Flood's devastating effects on the physical earth and climate, Burnet sought to demonstrate that sin was a powerful, world-historical force capable of ruining the global environment. He also wanted to argue that these catastrophic changes in nature the world over, the unanticipated yet richly deserved result of human action, afterward placed limits on humans' capacity to better themselves and their world. In so doing, he actually undermined arguments for global evangelism. Burnet's version of the Universal Deluge was undergirded by a deeply pessimistic view of the pervasiveness of sin, across time and space, and of the extreme selectivity of salvation. He crafted a past, present, and future history of the world that demonstrated that humanity's power to ruin the world and, consequently, their own health, was not matched by an ability, in the present, to rehabilitate either the planet or themselves. In working to articulate a materialist history of sin and salvation, he incidentally formulated one of the first theories of man-made global climate change.

Burnet's history of Noah's Flood evinced a very early version of what Dipesh Chakrabarty and others have called "geological agency"—the human ability to provoke lasting change in the global environment, as recorded in the geological record—and of what Fabien Locher and Jean-Baptiste Fressoz have called "environmental reflexivity"—an awareness of the human capacity to cause environmental change.[2] Chakrabarty is one of many recent scholars who speak of geological agency as a historically recent (natural) phenomenon and an even more recent (human) discovery. Locher and Fressoz have countered this emerging consensus by arguing that humans have been conscious of their power to degrade the environment, perhaps even on a planetary scale, since at least the mid-eighteenth century, predating the most environmentally destructive forms of industrial capitalism and the fossil-fuel economy. In this chapter, I extend Locher and Fressoz's argument further back into the past, to an era before "faith in progress and . . . belief in the regenerative capacity of nature" mitigated any potential feelings of culpability, caution, or regret.[3] Burnet and his English contemporaries like John Woodward and William Whiston demonstrate that environmental reflexivity dates at least as far back as the late seventeenth century, when a declensionist view of human and natural history led to the notion that humans, in their capacity as sinners and collective moral agents, provoked devastating harm to the natural world and thereby also to themselves. Above all, they sought to expose the reciprocal (and mostly negative) influence of humans and the environment on each other, using the biblical story of Noah's Flood as a means of imagining the devastating effects of humanity's geological agency on humanity itself.

My aim is not to reassign priority in the history of climate science, much less to argue that Burnet and his contemporaries "discovered" anthropogenic global climate change in its modern sense. Instead, I want to insist that the concepts—if not the realities—of geological agency and environmental reflexivity are much older than is typically acknowledged and that they owe an unacknowledged debt to religious thought. It is my contention that (at least some early modern) humans imagined themselves having the power to transform the global environment long before the Industrial Revolution and the atomic bomb actually gave them that power. Chronology and context are vital here. Tracing these ideas to the early modern period reveals the key role of religion in fostering the idea that humans could destroy the global environment and be destroyed by it in turn. In particular, the religious, political, and intellectual context of late-seventeenth-century Britain proved to be fertile ground for this particular type of environmental reflexivity. Alexandra Walsham's excellent survey of British histories of the earth highlights the influence of Protestant beliefs that "the physical appearance of the earth . . . [was] a direct consequence of human sinfulness" as well as "a kind of palimpsest upon which the Lord inscribed messages to the people of Britain."[4] Reading spiritual meaning into the natural world was a habit of mind that encouraged early modern Protestants to see the natural environment as a material record of past sins.

At the same time, the turbulent years around the Revolution of 1688–89 stoked religious controversy, and these new histories of the world were nothing if not controversial. Much of the recent scholarship on Burnet and his cohort of "world makers," as William Poole felicitously calls them, focuses on the maelstrom of religious objections they provoked in the 1680s and '90s.[5] The goal of reconciling philosophy and faith under the rubric of Mosaic natural philosophy continued to inspire new histories of the Universal Deluge, as it had for the past century. But there were nearly as many opinions on how this reconciliation of philosophy and faith could be achieved as there were people trying to achieve it, and the challenge of writing an orthodox history of the earth based on the Mosaic text was just as challenging in Anglican Britain as it was in post-Tridentine Italy. It was no easy task to know in advance what would be received as orthodox in the Church of England, which was notoriously established without designating a foundational reformer's writings as canon, as most other Protestant churches had done. Meanwhile, Burnet's proposal that the parallel decline of humanity and nature attested in scripture could be explained not only by God's providence but by a postdiluvian change in climate provoked howls of outrage from ministers and mathematicians alike, who regarded this admittedly materialist history of sin's effects on the world as promoting deism, atheism, and irreligion.

Highlighting the centrality of early modern climate theory to Burnet's *Sacred Theory of the Earth* leads us to a deeper appreciation of why it was both appealing and controversial to readers at the close of the seventeenth century. A major and heretofore unexplored aspect of its appeal, I argue, was its resonance with medical and colonial discourses of climate and with the lived experience of environmental and climactic change in seventeenth-century England, which coincided with some of the coldest periods of the Little Ice Age and with worsening air quality in London. The charges of atheism that *Sacred Theory* provoked, meanwhile, can be traced in part to its environmental determinism and its correlative pessimism about the scope and scale of salvation. John Woodward's equally influential and controversial *Essay toward a Natural History of the Earth* (1695) softened Burnet's environmental determinism, offered a rosier picture of the possibility of salvation, and provided a religious justification for agricultural improvement and resource extraction into the bargain.

Burnet's *Sacred Theory* is frequently cited but little understood. In contrast to the ubiquity of references to Burnet and his famous book in nearly every survey of the earth sciences, natural history, and natural theology in early modernity, there has been only one monograph and a handful of articles devoted solely to him.[6] Part of the reason for this lack of sustained scholarly engagement is surely the paucity of relevant archival sources.[7] More important, perhaps, is the symbolic value that has accrued to Burnet as an important figure in the historical relationship between geology and religion. Ever since the eminent Victorian geologist Charles Lyell denigrated Burnet in his foundational *Principles of Geology* (1830–33), it became common, for most of the nineteenth and twentieth centuries, to hold Burnet up as a poster child for the stultifying influence of religious superstition on the development of the secular and empirical discipline of geology. From the late twentieth century onward, Burnet's reputation has enjoyed a near-complete reversal among historians of science.[8] Finally acknowledging the widespread charges of atheism and deism that *Sacred Theory* provoked, scholars like Poole now regard Burnet as a rather heterodox thinker if not an outright participant in the so-called Radical Enlightenment. Others, like Walsham, see widespread agreement as well as polemical disagreement in late-seventeenth-century Protestant Britain about the spiritual significance of nature, all unfolding within a fractious political and religious culture actively debating what that significance was and who was licensed to discern it.[9] I am inclined to see Burnet as a sincerely pious practitioner of Mosaic natural philosophy whose efforts to sacralize nature's history and to materialize sacred history might not have caused such alarm in a different time and place.

Burnet was many things, which appear as a bundle of contradictions only in retrospect. He was a rationalist philosopher in the Cartesian tradition whose deep commitment to naturalistic explanation guided his inquiries into the natural causes and effects of the Universal Deluge.[10] He was also an ordained minister in the Church of England who harbored a deep pessimism regarding human depravity and the selectivity of salvation. His religious and philosophical commitments combined to produce a remarkably appealing yet also wildly controversial theory about the power of sin to wreck the world and the powerlessness of humans afterward to do very much about it. By focusing on Burnet's interventions in the early modern discourse of "climate" as the interface between humans and nature, the reasons for his book's popularity and controversy become more apparent. In particular, looking at why some readers found his idea of the earth as an accidental ruin so compelling while others found it so repugnant helps to explain why so many of his contemporaries accused this pious minister of atheism and why he has been such an equivocal figure in the historiography of science and religion ever since.

## Paradise Ruined: The Synchrony of Human and Natural History

Like the pious scholars of the previous chapter attempting to disprove the heresy of polygenism, Burnet insisted on the universality of biblical history in general and of the Flood in particular. "Some modern Authors," he wrote disparagingly, no doubt alluding to the widely condemned midcentury French author Isaac La Peyrère, dare to assert that "Noahs Flood was not Universal, but a National Inundation, confin'd to Judaea, and those Countries thereabouts."[11] On the contrary, Burnet argued, the Flood was "a *mundane* change, that extended to the whole Earth, and both to the heavens and the Earth."[12] Burnet was one of dozens of natural philosophers in the second half of the seventeenth century who began to insist not only on the Flood's universality across the entire face of the earth but also on its profoundly transformative effects on global nature. Burnet called the Flood "the greatest thing that ever yet hapned in the world, the greatest revolution and the greatest change in Nature."[13] Noah's Flood was responsible both for the punishment of a sinful humanity and for virtually every observable feature of the earth's surface at the time of publication. His rival John Woodward echoed the same sentiment in his alternative history of the Flood, which he described as "the most horrible and portentous Catastrophe that Nature ever yet saw."[14] The move from the Flood's universality to its globally transformative effects was a significant innovation of the late seventeenth century, directing attention toward the gravity of sin in causing the Flood (as Erculiani had done) but also toward the Flood's permanent effects on nature worldwide (as she had not).

Burnet's first key innovation in the study of the Flood was to argue that it transformed every aspect of the natural world: the geophysical structure and topography of planet Earth, its position in space, its weather and climate, its air and soil, and its living creatures, including the bodily constitution of its human inhabitants. His second major innovation was to argue that the global transformation of nature attendant on the Flood demonstrated the synchrony, even unity, of human and natural history, whereby changes in humanity's spiritual state would be continually reflected in the natural environment. Insisting on the Flood's universality was a way for Burnet to turn the Flood into a "golden spike," or geochronological marker, between two distinct geological epochs.[15] From the time of its first creation by God up until the Deluge, the planet was perfectly spherical. No mountains, valleys, or oceans marred its smooth, flat, unbroken surface.[16] This topographical oddity produced a global regime of climactic and meteorological stasis in which human life flourished to an astonishing degree. Before the Flood, Burnet contended, humans grew to enormous size and lived for several hundred years, just as Erculiani had argued in her account of the Flood nearly a century earlier. In keeping with the global approach of *Sacred Theory*, Burnet insisted that "the whole Earth was, in some sence, Paradisiacal in the first Ages of the World."[17] Just as it was absurd to believe that the Flood was a "National Inundation, confin'd to Judaea," so was it equally incredible that Paradise was "confin'd to a little spot of ground in Mesopotamia, or some other Country of Asia."[18] The Mosaic history was not Asian history or Jewish history, but *global* history. The spot where Adam and Eve lived until the Fall was unusually rich and fertile, of course, but the natural characteristics of Paradise extended across the entire earth and—another major innovation—lasted right up until the Deluge.

After the Deluge, everything that was beautiful and comfortable about the earth, from a human point of view, was destroyed. Much of the popular appeal of *Sacred Theory* lay in the dramatic language Burnet used to describe the present state of the earth, which he likened to a desolate ruin.[19] Mountains were "wild, vast, and indigested heaps of Stones and Earth."[20] Islands lay scattered across the ocean "like limbs torn from the rest of the body."[21] The ocean was a horrible gash on the face of the earth, "the most ghastly thing in Nature . . . as deform'd and irregular as it is great."[22] The ocean floor, if it were to be uncovered by water, would resemble "an open Hell, or a wide bottomless pit."[23] Every aspect of the earth's topography, Burnet asserted, was characterized by disorder, irregularity, inequality, "wild and multifarious confusion . . . Pits within Pits, and Rocks under Rocks, broken Mountains and ragged Islands, that look as if they had been Countries pull'd up by the roots, and planted in the Sea."[24] The placid air of the antediluvian earth was also destroyed, inaugurating new regimes of weather and climate that radically curtailed human flourishing. All of this damage and desolation in global nature was the work of the Flood, and very little had changed since then. We in the modern era, Burnet told his readers, are living in the world the Flood has made.

In keeping with his belief in the synchrony of natural and human history, Burnet insisted that the world's two geo-climatological epochs separated by the Flood were also distinct epochs in the spiritual history of humankind. The natural properties of the earth characteristic of the Edenic period—perhaps we might call it the "Edenocene"—reflected and reinforced the original spiritual purity of the first humans. The natural and spiritual purity of the Edenocene stood in sharp contrast to the present state of things. "We do not seem to inhabit the same World that our first fore-fathers did," Burnet lamented, "nor scarce to be the same race of Men."[25] The Flood ruined the earth and ruined humanity in the process. But crucially, humanity played a key role in bringing about this ruin in the first place. Burnet's Flood was in some respects the true Fall, the time when global Eden was destroyed in order to match the disgraced and degraded state of its human inhabitants. Indeed, Burnet several times used the word "fall" to describe the planet's transition from antediluvian perfection to postdiluvian ruin, as when he declared that "Nature doth not fall into disorder till Mankind be first degenerate and leads the way."[26] The Flood, and thus nature's ruin, was indirectly caused by the Fall of mankind from a state of innocence to a state of sinful depravity. Thus we might label the second of Burnet's geohistorical epochs the "Fallocene," the period in which humankind's spiritual depravity is reflected in the ruin of the natural landscape. Even while blaming human sin for the planet's ruin, Burnet was careful to insist that God's providence was a necessary binding agent between the parallel histories of nature and humanity. "This seems to me to be the great Art of Divine Providence, so to adjust the two Worlds, Humane and Natural," he wrote, so that "they should all along correspond and fit one another, and especially in their great Crises and Periods."[27] The Flood was reimagined as an event that touched human history and natural history in equal measure and, even more crucially, demonstrated the interdependence between the two.

While Burnet's theory of the earth was certainly innovative, it was hardly unprecedented. The idea of the present earth as a ruin, for example, seems to have met with broad approval precisely because it was already commonplace across literature, philosophy, and everyday life. A frequently cited phrase from *Sacred Theory*, "what a rude Lump our World is which we are so apt to dote upon," mimics a line from the early-seventeenth-century poem "Upon Appleton House" by Andrew Marvell, which described "the World" as "a rude heap together hurl'd."[28] When the Anglican divine Gilbert Burnet (later bishop of Salisbury, no relation to Thomas Burnet) traveled through the Alps on a Grand Tour, he recalled *Sacred Theory*'s description of the same mountains and found himself agreeing that these jagged peaks were "vast Ruines of the first World."[29] Part of the appeal of *Sacred Theory* surely derived from the literary and indeed poetic qualities of Burnet's philosophical prose. But even natural philosophers aligned with the Royal Society and striving to imitate its new "plain style" employed the language of a ruined earth left behind by a profoundly transformative and universally destructive Flood.[30] In a paper read before the Royal Society in

1694 on "The Causes of the Universal Deluge," the astronomer Edmond Halley declared that "the Earth seems as if it were new made out of the Ruins of an old world."[31] In his *Essay toward a Natural History of the Earth* (1695), John Woodward described the Flood as "a mighty Revolution" that left the earth "quite unhinged, shattered all to pieces, and turned into an heap of ruins."[32] The idea of a ruined earth, then, seemed to be a popular literary trope, a common philosophical claim, and even something that resonated with people's lived experience of traveling across the earth's variegated terrain.

Burnet's proposed unity of human and natural history was embraced by his admirers and critics alike, signaling that it was neither especially novel nor controversial. The Scottish apothecary Matthew Mackaile's *Terrae prodromus theoricus* (1691), written "by way of Animadversions upon Mr. Thomas Burnet's Theory of His Imaginary Earth," is in perfect agreement with Burnet that there is a strong relationship "betwixt the Spirituall condition of Man, and the naturall condition of the Earth."[33] The astronomer William Whiston, who offered his *New Theory of the Earth* (1696) as a replacement for Burnet's *Sacred Theory of the Earth*, echoes this point as well, writing that "[t]he State of Mankind . . . was before the Fall vastly different from the present; and consequently requir'd a proportionably different State of external Nature."[34] On a theological level, the interdependence of human and natural history made a great deal of sense. A ruined earth seemed to many a more suitable habitation for humankind in its fallen state than the perfect planet God had originally fashioned at the Creation.

Burnet's story of a ruined earth was broadly and intuitively appealing because of its resonance with the vague yet powerful notion of the world's decay, which was widespread across Europe from the Renaissance through the Enlightenment.[35] In contradistinction to the Enlightenment vision of universal progress that would take hold later in the eighteenth century, it was a common belief in the sixteenth, seventeenth, and early eighteenth centuries that the world was in a state of progressive decline. In 1627, the English clergyman George Hakewill complained that this declensionist belief was so widespread across all social groups as to be almost fashionable: "The opinion of the Worlds decay is so generally received, not onely among the Vulgar, but of the Learned both Divines and others, that the very commonnes of it, makes it currant with many."[36] The appeal of this idea to so many—religious and lay, scholarly and uneducated— was due in part to the fact that multiple sources of ancient history indicated that the ancient world was in many ways superior to the modern one. The depiction of Eden in the book of Genesis painted a tantalizing picture of a young, green, and pristine earth. The biblical account of Eden, moreover, resonated with stories of a lost Golden Age found in several writers of pagan antiquity such as Virgil, Ovid, and Hesiod. While John Milton's *Paradise Lost* surely fired Burnet's imagination about the pleasures and perfections of the antediluvian world, so too did Virgil's pastoral epic *Georgics*, which recounts a time when nature was so abundant that people prospered without needing to work, living easily off the fruits of the land. The classical Golden Age so closely resembled the biblical account of prelapsarian paradise that early modern commentators frequently identified the two. Both religious and classical authorities seemed to indicate the superiority of the ancient, golden past over the modern, degraded present.

The general conviction that nature used to be more verdant and fruitful was frequently accompanied by the belief that humans used to be taller and stronger and live for much longer.[37] Genesis records the presence of giants on the earth before the Flood, which formed the basis for the early modern idea that all antediluvian humans were of gigantic size. The sixteenth-century Italian philosopher Camilla Erculiani blamed the Flood on human gigantism, as discussed in chapter 1. Walter Raleigh's *History of the World* (1614) vividly describes the "men of huge bodies" who roamed the earth "in the first flourishing youth and newnesse of the world," whose enormous size "exceeded the bulkes and bodies of men which are now borne in the withered quarter and Winter of the world."[38] The genealogies of the biblical patriarchs recorded in the Old Testament also indicated that people used to live for much, much longer in the ages before the Flood.[39] Life in early modern Europe was often, in Thomas Hobbes's famous phrase, nasty, brutish, and short: average life expectancy hovered around thirty-five years, whereas Methuselah lived to be eight hundred years old. Longevity was appealing not only for extending the lifetime of a single individual but for increasing the number of children that person would be able to have. The English bishop Richard Cumberland connected the strength and longevity of antediluvian men to their virility: "The constitution of such longer-liv'd men must needs be much stronger than our's [*sic*] is, and consequently more able and fit to propagate mankind to great numbers than men can now do."[40]

The idea of antediluvian gigantism and longevity gave natural philosophers a metric, derived from biblical evidence, for gauging the transformation of human bodies and, correspondingly, of the global environment, from ancient to modern times. The physical characteristics of human and especially male bodies that were thought to have changed most dramatically since ancient times—size, virility, and longevity—were also thought to be strongly influenced by climactic conditions. Raleigh's claims about human gigantism in the "flourishing youth and newnesse of the world" and reduced size in "the withered quarter and Winter of the world" hinted that the decline of humans and the decline of nature were not simply parallel processes, but causally interrelated ones. Later in the seventeenth century, Burnet sought to specify those causal links between human and natural decline by invoking not only God's providence but also a change in climate.

Modern interpreters of Burnet have read him primarily through the lens of the history of geology and thus have paid far more attention to his theories of geological transformation attendant on the Flood. But Burnet's repeated use of the phrase "Heavens and Earth" when referring to the changes wrought by the Flood indicates that his interests extended beyond the range of topics that would come to be gathered under the rubric of "geology" in the eighteenth and nineteenth centuries. Climate, a classical concept that got new life in early modern philosophy and medicine, provided Burnet with the means to connect the geophysical state of the earth to the physical state of the human body before and after the Flood. In radically changing the earth's topography and position in space, he argued, the Flood instituted a new regime of weather and climate that punished humanity for their sins by ruining the health and vigor of Edenic human bodies.

## From the Edenocene to the Flood to the Fallocene

Weather and climate are central to Burnet's definition of the distinction between the Edenocene and the Fallocene. Before the Flood, the uniformity of the earth's surface was linked to the calmness of its atmosphere, which produced perfect meteorological placidity. "The smoothness of the Earth made the face of the Heavens so too," Burnet declared. "The Air was calm and serene; none of those tumultuary motions and conflicts of vapours, which the Mountains and the Winds cause in ours."[41] Reworking the Aristotelian meteorological tradition that linked weather events to perturbations in the air and the circulation of watery vapors, Burnet speculated that a planet with no mountains or valleys, continents or oceans, would essentially have no winds or weather, or at least, no meteorological variability.[42] "The Meteorology of that World was of another sort from that of the present," he wrote.[43] Moreover, the antediluvian planet's position relative to the sun—perfectly straight and perpendicular to the plane of the ecliptic—meant that the world before its ruin experienced no change in seasons either.[44]

The climate of the Edenocene was not perfectly uniform across the globe; certain parts of the antediluvian planet were more "Paradisiacal" than others. The earth's smooth surface established a global water cycle that produced different effects on different parts of the earth. Burnet identified these with the ancient Greek climactic zones: the torrid zone circling the equator, the two frigid zones around the poles, and the two temperate zones in the midlatitudes between the equatorial and polar regions. Water from the planet's subterranean ocean (what Burnet called "the Abyss") evaporated from the equatorial and temperate zones and fell back to earth over the polar regions.[45] The global circulation of water above and below ground produced continually dry weather in the warmer climates and ceaseless precipitation at the poles. Meanwhile, the earth's position relative to the sun endowed each of these climactic zones with

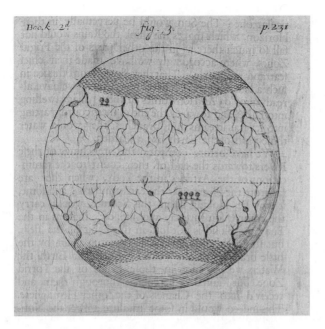

The antediluvian earth "with its Zones or greater Climates," in Burnet, *Theory of the Earth* (1684). Newberry Library, Chicago, Case folio C 257.13.

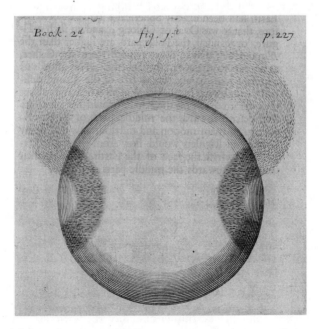

The static global weather pattern that Burnet imagined to have prevailed across the five climactic zones before the Flood. Newberry Library, Chicago, Case folio C 257.13.

a markedly different temperature, producing cold polar regions with unremitting precipitation; dry and scorching desert heat that precluded all life around the equator; and in the midlatitudes, a "perpetual Spring" where it was always warm, sunny, and clear, every hour of every day, every day of the year. The unchanging springlike climate of the midlatitudes produced vibrant plant life, "the Fields always green, the Flowers always fresh, and the Trees always cover'd with Leaves and Fruit."[46] So while the Edenic earth had drastically different climates, the habitable ones were uniquely conducive to the flourishing of human, animal, and plant life, and the entire planet experienced remarkable stability. Within the boundaries of each climactic zone, the temperature remained stable and the seasons and weather never changed. Perhaps one way to sum up Burnet's vision of the Edenocene was to say that he imagined a world with climates but no weather.

The Edenocene came to an abrupt end following the Universal Deluge, Burnet argued, which caused the planet to tilt on its axis, inaugurating seasonality. Meanwhile, the Flood's eruption from the Abyss broke up the smooth surface of the planet, creating mountains and oceans and so forth and thus destroying the perfect calm of the air and the global circuit of water. The meteorological stasis that characterized the Edenocene gave way to variability of weather and season in the Fallocene. As the topography and inclination of the planet underwent a violent change, so too did the climate, bringing with it new temperatures, weather patterns, and regimes of seasonality. At the same time, the five climactic zones of the antediluvian earth dissolved into subtler patterns of global climactic variance. "This distinction of the Globe into five Zones, I think, did properly belong to that Original Earth," he wrote of the classical climatological division of the world, "and improperly, and by translation only into the present. For all the Zones of our Earth are habitable, and their distinctions are in a manner but imaginary, not fixt by Nature."[47] If Burnet's Edenocene was a world with climates but no weather, his Fallocene was the opposite, a world with weather but no climates—or at least, no climates in the classical sense. The idea of a global Flood that caused planetary changes to the earth's topography and astronomical orientation became a means for him to think through the relationship between topography and climate and between climate and weather on a global scale.

In narrating the world's ruin, Burnet focused on the climate of the temperate midlatitudes, which he sometimes conflated with the climactic condition of the entire earth. The earth's position relative to the sun "made a perpetual Æquinox or Spring to all the World," he wrote. "There was no Winter or Summer, Seed-time or Harvest, but a continual temperature of the Air and Verdure of the Earth."[48] In some places, Burnet described the warm air and fertile soil that nourished human life as "universal," writing that the "the long Lives of the Ante-diluvians was an universal Effect, and must have had an Universal Cause," even though he elsewhere specified that the warm air and fertile soil of the mid-latitudes did not exist in the polar and equatorial regions.[49] Burnet contended that the midlatitudes were the only inhabited parts of the earth before the Flood, so perhaps he felt comfortable speaking on occasion as if the Edenic climate of the temperate midlatitudes covered the entire planet. This slippage reveals his true concern with the theory of diluvial climate change: to explain how the climate produced awesome human bodies before the Flood and degenerate bodies afterward, and thus to explain more generally why human and natural history proceeded along the same declensionist pathway from perfection to ruin.

The meteorological stasis resulting from the topographical uniformity of the antediluvian earth was key to Burnet's account of how the Edenic climate created Edenic people. In ancient Greek philosophy, *klima* referred to a zone of latitude on the earth, a natural territory defined by its location relative to the sun, which influenced its typical weather. Burnet's discussion of the Edenocene's climactic zones reflected this classical definition of climate. But his theory of how the temperate climate produced gigantic and long-lived people reflected more recent understandings of "climate" from the Renaissance, when its classical, astronomical definition merged with the Hippocratic tradition of airs, waters, and places. This Hippocratic tradition considered human health to be shaped, individually and collectively, by local environmental factors, notably, air, water, and soil. As European empires and commercial networks spread across the globe from the sixteenth century onward, this novel understanding of climate became a popular means of explaining the diversity of plants, animals, and human bodies and cultures within the terrain of Europe and across the surface of the planet.[50] This early modern merging of Aristotle and Hippocrates, natural philosophy and medicine, in the face of globalization produced new understandings of local climates and their effects on human health and well-being, which can be loosely grouped under the rubric of "environmental medicine."[51] If the air, water, and soil of a particular climate were healthy, so too would be the human inhabitants of that climate. If these elements were out of balance with one another—as in a damp or humid climate, for example, caused by too much water interspersed with the air and soil—then human health in the area would also be poor. The key to "a long and healthy life," proclaimed the English physician Tobias Venner in 1620, was to live in a place that was "free from muddie and waterish impurities: for it is impossible, that a man should live long and healthily in a place, where the spiritis are with impure aire daily affected."[52] "Climate" in early modern medicine became a crucial means of understanding how atmospheric and environmental conditions shaped human bodies and affected human health.

Burnet attributed antediluvian longevity and gigantism to the superior climate of the Edenocene, specifically the climate of the temperate midlatitudes. His description of the "Primaeval Earth" as possessing "temperate Climates, a

clear and constant Air, a fruitful Soil, pleasant Waters, and all the general characters of Paradise" names precisely those geophysical properties—air, water, and
soil—that were understood to define a local climate.[53] The soil, he wrote, was
primarily responsible for the enormous size of antediluvian humans. He wrote
glowingly on the "Spontaneous and Vital fertility" of the Edenic soil, which
produced fruit and other human nourishment without the need for human
labor, just as in Virgil's Golden Age.[54] Human longevity, on the other hand,
was primarily due to the salubrious air, and specifically to the lack of changes
in season and weather. "That vicissitude of Seasons, inconstancy of the Air, and
unequal course of nature which came in at the Deluge, do shorten Life," Burnet
declared.[55] The impoverished soil of the Fallocene could no longer support the
explosive human growth of the Edenocene, while the new variability of seasons
and weather conspired to cut short human lifespans. As a result of this climactic change, "[o]ur fore-fathers . . . liv'd seven, eight, nine hundred years and
upwards, and 'tis a wonder now if a man live to one hundred."[56] Burnet's mar-
riage of the idea of a Universal Deluge to early modern understandings of local
climates produced a novel understanding of large-scale (if not quite global) and
irreversible climactic change, which in turn provoked universal and irreversible
changes in the fabric of the human body.

Burnet's projection of the temperate climate of the midlatitudes onto the
entire planet signals his commitment to narrating the history of natural and
human decline on a global scale. His frequent lamentations for the loss of Eden
("we might have enjoy'd the comfort of a perpetual Spring, which we have lost
by its dislocation ever since the Deluge") drew much of their force from the
universality of the "we" who destroyed their former home.[57] But his projection
of one part of the earth onto the whole also highlights the limits of his ability to
offer a physical description of a truly global climate, something that would not
be achieved until the modern era. The ambiguous globality of Burnet's theory
of diluvial climate change was also a symptom of the partial perspective, Protestant and Eurocentric, from which he aspired to tell a universal story about all
of nature and all humankind.

In addition to the frequent slippages regarding the scale of Eden, *Sacred Theory*
also contains contradictory remarks about the scale of nature's Fall. The text is not
at all clear about whether the polar and tropical regions experienced the same drastic change in weather and climate as did the temperate midlatitudes. Burnet seems
to have thought that the climate changed most drastically at the midlatitudes—
perhaps even Britain in particular—where static, weatherless spring gave way to
hot summers, cold winters, winds, rain, and other weather events. In some places,
he suggested that diluvial climate change affected the nontemperate parts of the
earth too, claiming, for example, that the equatorial regions were too hot to
support human life prior to the flood but became habitable afterward, thereby

implying that the Flood caused the equatorial climate to cool. *Sacred Theory* contains several references to "hurricanos," indicating awareness of the extreme weather event to which the tropics were presently subject. Elsewhere, however, Burnet downplayed the variability and even the existence of weather and seasons in that part of the world, observing that Bermuda, at the time an informal part of the British Empire, experienced "extraordinary steddiness of the Weather, and of the temper of the Air throughout the whole Year, so as there is scare any considerable difference of Seasons."[58] His inconsistent characterization of present-day tropical climates reflects opposing colonial discourses about North America in general and the Caribbean in particular. The first saw the extremes of American weather, epitomized by hurricanes, as signs of the wild, deadly, and dangerous features of nature in the New World; the second saw the warm temperatures and lush plant life as signs of the Edenic qualities of the New World.[59] Burnet's inability to decide if the weather of the postdiluvian tropics was static (hence Edenic) or wildly variable (hence fallen) is influenced by the colonial lens through which he regarded the natural world beyond Britain and Europe. It also suggests a lack of sustained interest in parts of the world other than his own, which he may have simply considered less important to the drama of sacred history than his own midlatitudes.[60] Burnet's ability to use climate to link natural and human history on a global scale was therefore limited both by the state of climate science at the time and by his own cultural biases.

## Acclimatization and the Little Ice Age

The influence of colonial discourses on nature is also evident in Burnet's invocation of "acclimatization," another way in which he mobilized climate theory in order to tell a story about the universal decline of nature and humanity. A theory that first appeared in European natural philosophy and medicine during the age of the European expansion, acclimatization referred to the process of biological change undergone by a living thing as it moved from one climate to another. This process could be beneficial or harmful, and if the latter, was often referred to as "degeneration." Because climates were believed to influence all living things, acclimatization was thought to effect people as well as plants and nonhuman animals. In the first of Burnet's two references to acclimatization, he likened the decline of human bodies after the Flood to the "degeneration" of a plant transported from its native climate to a less salubrious one. "When some excellent Fruit is transplanted into a worse Climate and Soil, it degenerates continually," Burnet wrote, referring to a phenomenon with which all European naturalists, colonists, and bioprospectors were sadly familiar. In similar fashion, human "longaevity sunk half in half immediately after the Flood, and after that it sunk by gentler degrees." Humans after the Flood were, in a sense, "transplanted" from an "excellent" climate to a "worse" one, and so they have "degenerated" in much the same way as a plant withers and dies when replanted in foreign soil.[61]

Burnet's second reference to acclimatization following the Flood brings the context of colonialism and colonial science into even sharper focus. Burnet compared humanity's physical transformation from the Edenocene to the Fallocene to the racial transformations of contemporary humans as they moved from one of the world's postdiluvian climates to another. He argued that human longevity had declined gradually over the course of several centuries rather than all at once. In order to explain why, Burnet invoked the increasingly common belief that humans in the present could gradually change their race after relocating to a foreign climate. "We see the Blacks do not quit their complexion immediately by removing into another Climate," Burnet wrote, referring to the skin color, temperament, and physical constitution ("complexion") of sub-Saharan Africans.[62] Instead of changing immediately, "their posterity changeth by little by little, and after some generations they become altogether like the people of the Country where they are."[63] As the transatlantic slave trade began to boom around the turn of the eighteenth century, and slave traders forcibly removed sub-Saharan Africans from their homelands in ever greater numbers, the question of how they might change as a result of their forced relocation to American plantations was far from academic. Under the influence of early modern climate theory and a nascent ideology of white supremacy, white Europeans contemplated with a mixture of curiosity, hope, and horror the possibility that Black Africans living outside of Africa might lighten their skin or even become white, while white Europeans living far from Europe might "degenerate" to become more like the darker races in the colonies, ports, factories, or diplomatic outposts they now occupied.[64] John Woodward requested that Englishmen voyaging on colonial or commercial expeditions make note of "whether white people removing into hot Countries become by degrees browner, etc., and Blacks removing into cold Countries, paler." Woodward also requested that travelers make note of "the colour of their Infants when first born," reflecting the European suspicion that everyone was born the same color (probably white) and took on their skin color and other raced characteristics under the influence of their local climate.[65]

In the seventeenth and eighteenth centuries, European naturalists found it increasingly plausible that people, like plants, experienced profound physiological changes as they moved from one climate to another. Burnet invoked this process of biological change across climactic boundaries in the early modern world in order to explain how humanity degenerated en masse as they "moved" from the static and salubrious climate of the Edenocene to the changeable and unhealthy climate of the Fallocene. Burnet never used the phrase *climate change*, nor did any of his contemporaries, as far as I have been able to determine. Nor did he coin a term for the planetary complex of air, wind, weather, season, and temperature that, he argued, was transformed by the Flood and transformed humanity in turn. But it is clear that a change in climate—"climate" as it was understood in early modern Europe—is more or less what Burnet had in mind. His invocation of living things moving across oceans and being forced to adapt to new climates also shows how deeply his story of the ancient ruin of nature and humanity was shaped by contemporary issues of race, slavery, and colonialism in the British Atlantic world.

His appeal to contemporary theories of acclimatization is one of the clearest indications of how central the idea of climate was to his efforts to tell a story about the universal decline of humanity and nature on a global scale. Another indication is the several ways in which his theory of a vanished Edenocene and a ruined earth resonated with more recent changes in weather, climate, and environment in seventeenth-century Britain and Europe. Burnet's theory of degraded air shortening human lives might have been particularly resonant with his fellow Londoners, for example. Long-standing medical theories of air as a vector of disease grew out of centuries of experience with plague outbreaks, commonly attributed to the long-distance movement of "corrupt" air. Urban areas like London were sites of heightened apprehensions about the adverse health impacts of corrupted air. William Cavert has recently demonstrated the pervasive concern of early modern Londoners regarding the worsening air quality of their city, among the first in the world to make the transition to coal as its main source of energy. Burnet certainly would have noticed London's terrible air quality after settling there in the 1670s, and he might have drawn on this lived experience of air pollution in order to imagine how the deterioration of the world's air could have reduced human lifespans from nine hundred years to ninety.[66]

A significant number of Burnet's British critics agreed with him that a change in the earth's air was the likely cause of humanity's postdiluvian decline. The celebrated naturalist John Ray speculated in his own history of the earth that "a great and extraordinary Change at the time of the Flood . . . in the Temperature of the Air" may have caused the nosedive in human longevity.[67] In a critique of *Sacred Theory* published in 1697, the Scottish physician Robert St. Clair argued that massive earthquakes caused the earth's crust to break apart during the Universal Deluge, releasing toxic airs trapped underground. These subterranean airs unleashed on the human population, he argued, "might occasion that shortening of Man's Life, which happen'd quickly after the Deluge."[68] Whiston's *New Theory of the Earth* repeated Burnet's idea that the antediluvian air was superior, "without such sudden and violent changes in the Climates or Seasons from one extreme to another, as the present Air, to our sorrow, is subject to," and then specified that the placid air precluded diseases and plagues.[69] Popular theories about air and disease in Europe and the lived experience of worsening air quality in London may have formed part of the appeal of Burnet's idea that a change in the air was the main cause of human bodily decline following Noah's Flood.

Readers of *Sacred Theory* throughout northern Europe may have found its discussion of modern-day weather patterns particularly resonant with their lived experience of the Little Ice Age, which reached peak intensity in the decades between 1680 and 1730. Cold, wet summers led to large-scale crop failures and severe food shortages. Winters brought unusually low temperatures and large amounts of snow. The freezing of the River Thames was frequently cited by London's inhabitants as evidence that winters were colder than they used to be. The winter of 1683–84 that preceded the 1684 publication of the English-language edition of *Sacred Theory* was so cold that sea ice appeared in the English Channel and filled much of the North Sea. Burnet's description of the weather in his seventeenth-century present—"sometimes we are steept in Water, or in a misty foggy Air for several days together, sometimes we are almost frozen with cold"—is not a bad description of weather conditions in Britain at that time. Burnet's theory of a "perpetual Spring," by contrast, must have appeared fantastically appealing to his readers as well as intuitively plausible as an explanation for the good health and long lives of their antediluvian forebears.[70]

More research needs to be done on the possible connections between the idea of the world's decay, on the one hand, and the climactic disruptions associated with the Little Ice Age, on the other. At this stage, it is certain at least that Burnet was not the only scholar of the seventeenth century to negatively contrast prevailing climactic conditions in Europe with those they imagined to have prevailed globally in the past. Raleigh's *History of the World* draws a similarly stark contrast between the cold weather of the world its author knew with the warmer weather he believed was characteristic of biblical times. In Raleigh's formulation, the world actually increased in warmth and verdancy in the century immediately following the Flood before deteriorating to its present nastiness. Raleigh noted that winters in seventeenth-century England—or perhaps across the world, though like Burnet he was often imprecise about specifying whether the various elements of his *History of the World* were to be understood locally or globally—were long and severe. "In this our climate," he wrote, "the dead and destroying winter depresseth all vegetative and growing nature, for one halfe of the yeare in effect." He contrasted these long, "destroying" winters with the warm, long growing season of the immediately postdiluvian earth, when "all sorts of plants, reedes, and trees," flourished in "the climate of a long and warme Sommer."[71] Where Burnet focused on the air of the Edenocene, Raleigh focused on water, which he described as drenching the air and soil after the Flood in junglelike fashion and fueling the postdiluvian century of abundant natural growth. Although Burnet departed from Raleigh in arguing that climactic decline occurred immediately following the Flood, both philosophers drew on the concept of a changing climate as they sought explanations for the contrast between their cold, wet present and a warmer, better past.

Works by natural philosophers such as Burnet and Raleigh form a body of indirect and suggestive evidence that people in seventeenth-century England possessed some kind of historical awareness, however dim, that the climate was colder and wetter than it used to be. Further research could yield exciting new insights about the relationship between climate history and the history of ideas. Intellectual constructs such as the Golden Age and the world's decay, and intellectual debates like the Battle of the Ancients and Moderns, which likewise compared the splendors of the ancient world to the reality of present day, might very well appear in a new light when contextualized within the "seventeenth-century climate crisis," as Geoffrey Parker has called it. At the same time, climate history was absolutely not determinative of climactic discourse specifically or intellectual history more generally. Burnet's complaint about being "frozen with cold" was paired with a complaint about "fainting with heat at another time of the Year," which reflected the prevailing medical preoccupation with sharp changes in season, weather, and climate rather than the ill effects of cold weather per se, and hardly reflects the actual low temperatures of the Little Ice Age or of the typical English summer. Burnet's concern with excessive heat following one of the coldest winters in recorded English history serves as an important reminder that the lived experience of climate may have inspired, but by no means determined, the range of climate theories and histories that seventeenth-century people came up with in order to make sense of their world.[72] Nevertheless, the lived experience of climactic and environmental change formed an important backdrop to the histories of the world contrived by seventeenth-century English philosophers, while contemporary theories and discourses of climate exerted an undeniable influence as well.

## Understanding "Atheism": Burnet and His Critics

Burnet's sacralized climate history provided a philosophically satisfactory explanation for ancient gigantism, Virgil's Golden Age, Methuselah's longevity, and Britain's terrible weather in one fell swoop. As ingenious as Burnet's theory of global climate change may have appeared to some readers in Britain, others felt that it raised a whole host of theological problems that negated its philosophical virtues. The very same qualities that lent his speculative history of the Flood concrete plausibility—its appeal to colonial and medical theories of climate and its resonance with the lived experience of air pollution and the Little Ice Age—were also some of the most problematic, from a religious point of view. His effort to unite human and natural history via climate history—rather than, simply, via divine providence—provoked concerns about philosophy's encroach-

ment on theology's domain. Moreover, Burnet viewed humanity's present spiritual state as determined, to a large degree, by the climactic properties of the
Fallocene, which carried some fundamentally disturbing theological implications with respect to the perennial issues raised by the Flood: sin and salvation.
More generally, criticism of *Sacred Theory* reflected continued unease about the
synthesis of natural and human history that the Universal Deluge was supposed
to achieve.

Despite the onslaught of negative, often hysterical criticism *Sacred Theory*
would receive in the coming decades, the initial reception of the book was
largely positive.[73] It even attracted the attention of King Charles II, to whom
the English edition of volume I was dedicated.[74] Shortly thereafter, in May 1685,
Burnet was appointed to the prestigious post of Master of the Charterhouse, a
public school in London.[75] Burnet weathered the storm of the Revolution of
1688–89 quite successfully. He was one of the first clerics to preach before the
new monarchs after the revolution and was appointed King William's chaplain-
in-ordinary in 1691. *Sacred Theory* was not only a political success but a popular
one, going through at least ten English-language editions between 1681 and
1816, as well as two further editions in Latin. It was translated into Dutch in
1696 and German in 1703, suggesting that the work found its most eager and
receptive audiences in Protestant countries. It also found enthusiastic readers in
France and Italy, where it could be appreciated by a smaller demographic in the
original Latin.

*Sacred Theory* also provoked a torrent of criticism. More than thirty direct
responses to Burnet's book appeared in print before the turn of the century,
with many more attacks circulating in private correspondence and semiprivate
manuscript, igniting a controversy that put a definitive end to Burnet's political
career and a significant damper on the enthusiasm for his distinctive approach
to writing the earth's history.[76] At the very least, it meant that those who adopted
various aspects of his approach felt compelled to distance themselves from him.
This barrage of criticism was remarkably consistent in diagnosing what was
wrong with *Sacred Theory*: it was deistic and even atheistic; it ignored the words
of Moses and downplayed the power and providence of God. Edmond Halley,
who agreed with Burnet that the earth was a ruin, told the Royal Society that
"Dr. Burnet's Hypothesis . . . [is] as jarring as much with the Physical Principles
of Nature, as with the Holy Scriptures, which he has undertaken to reconcile."[77]
In other words, Burnet's attempt at Mosaic natural philosophy was a failure: he
had not done justice to either the text of Genesis or to the principles of natural
philosophy in his reconstruction of the earth's past and present.

One of the most common criticisms, and the one most relevant to the present discussion, was that Burnet had inappropriately "mechanized" sacred history. By enumerating the natural causes of the Flood and describing its effects

on both planet and people, Burnet downplayed divine agency and thereby diminished the Flood's theological significance. One of the very first attacks came from Herbert Croft, who denounced Burnet in 1685 as "a kind of Deist, acknowledging God as the supream Origin of all: But after his first Creation, he takes all out of his hands," so that "God is very near justled out of all."[78] Croft had an unusual biography: he was an ex-Jesuit who later became a rabidly anti-Catholic Anglican priest and subsequently the bishop of Hereford, near Wales. This unusual route to the upper echelons of the Anglican ecclesiastical hierarchy may help to explain the vigor and vitriol with which Croft tried to police the boundaries of acceptable religious and philosophical discourse within the Church of England, in which Burnet was also ordained. But men of the cloth who were seemingly more secure in their confessional and professional identities launched similar critiques. Thomas Robinson, the Cambridge-educated rector of Cumberland, warned readers about the dangers of "mechanizing" the Flood in his 1696 work of Mosaic earth history. "We ought to be cautious," he wrote, "of making such Grand Revolutions to rowl upon Machines."[79] Partisans of the new science launched broadly similar critiques. The physician John Beaumont asserted that explaining the Deluge according to natural principles "destroys the Miracle by lessening it, and makes it cease to be a Wonder."[80] The Scottish mathematician and astronomer John Keill accused Burnet of denying that the Flood was "the immediate work of the Divine Power."[81] Even Burnet's coworkers turned on him; one of his colleagues at the Charterhouse went so far as to publicly accuse him of "overturning all Religion" and endangering his readers' souls, putting them "in a fair way to Hell."[82]

To a modern reader, the frequent charges of deism and atheism directed at Burnet are puzzling. *Sacred Theory* is liberally strewn with frequent protestations that treating the Flood as a natural phenomenon with natural effects on human bodies was by no means intended to deprive the Flood of its religious significance. "Explaining the Deluge in a natural way, or by natural causes" was well within the purview of pious natural philosophy, Burnet wrote. The Flood was a miracle, ordained by God, who set its natural causes in motion and oversaw all of its natural effects. Not even the most mundane event happens without God willing it, "much less doth the great World fall in pieces without his good pleasure and superintendency."[83] Such a position was perfectly orthodox. Halley, who attacked *Sacred Theory* as irreligious, echoed Burnet's argument that enumerating natural causes of the Flood (or any phenomena) rested on an implicit acknowledgment of the First Cause of all things, "the Almighty generally making use of Natural Means to bring about his Will."[84]

The criticism that Burnet's account of ancient history was not completely harmonious with the one given by Moses was absolutely correct, but this was also true of nearly every work of Mosaic natural philosophy. Burnet's hypothesis of a flat, smooth antediluvian earth, for example, appeared to contradict certain passages in the book of Genesis referring to mountains. Burnet, who knew his scripture as well as anybody, was fully aware of this discrepancy between Moses's history of the earth and his own. He responded to his critics in an appendix to the 1697 edition of *Sacred Theory*, arguing that Moses, while an excellent and divinely inspired author of ancient history, did not "Philosophize or Astronomize."[85] Justifying his methodology to Isaac Newton in a series of letters exchanged just prior to the publication of the first volume in 1680–81, Burnet invoked Augustine's time-honored principle of accommodation in his defense. Moses had given "a short ideal draught of a Terraqueous Earth rising from a Chaos, not according to the order of Nature & natural causes, but in the order w[hi]ch was most conceivable to the people."[86] Newton, who warned Burnet in advance that he might be criticized for this apparent disregard of the text of scripture, nevertheless acknowledged the soundness of Burnet's methodology, agreeing with him that Moses spoke prophetically not philosophically, and "described realities in a language artificially adapted to the sense of the vulgar."[87] In sum, Burnet's approach to the Universal Deluge: a miracle that also had natural causes and effects—was widely shared, as was his approach to interpreting Mosaic history. Burnet's God was active, and his natural history of the earth was steeped in spiritual significance.

Burnet certainly did not fit the seventeenth-century definition of a deist, a person who believed that God created the world and then left it to run on its own. His discussion of "natural Providence" explicitly asserted the opposite. Burnet may have fit the early modern definition of an atheist, but only if we understand that term in historical context. As Sachiko Kusukawa argues, the epithet *atheist* in the age of the Reformations did not signify someone who doubted the existence of God but rather "someone who denied an essential element of faith."[88] By the late seventeenth century, the term was frequently levied as an accusation but was not yet a label that anyone owned. Given that Burnet was certainly not an atheist in the modern sense, what tenets of the faith—and which faith—was his *Sacred Theory* perceived to undermine?

For some critics, the charge of atheism (and deism, mechanism, and so forth) was a register of anxiety or discontent around bringing the Flood into the ambit of natural philosophy in the first place, or more generally, of collapsing the distinction between natural and human history. Such was almost certainly the case with Croft, who fully embraced the Flood's universality while disputing the notion that it changed anything in nature at all. While conceding that "the Animate World was destroyed," in the Flood, Croft strongly objected to Burnet's idea that "the Natural World, as he calls it (that is, the Body and Frame of the Earth) was also destroyed by this Flood."[89] (Indeed, the bishop's use of the phrase "as he calls it" seemed to question the very idea of any such entity as "the Natural World.") Croft and Burnet, as well as everyone else in their seventeenth-century world, agreed that the Flood was a major turning point in human history. The crux of their disagreement was whether it was also a major turning point in nature's history. More fundamentally, it was a debate about the possibility of fusing natural and human history together, of telling the story of humanity's changing fortunes and the history of nature's global transformations as part of the same, causally interconnected story. In Croft's eyes, any attempt to assign natural causes and consequences to an event of such providential significance as the Flood was an illegitimate transgression of disciplinary boundaries, philosophy treading onto theology's turf. The charge of atheism in this context might have signaled anxiety about what would happen if other biblical events were suddenly the domain of natural philosophers to study and to interpret.

Interestingly, this attempt to wrest the Flood out of the grip of natural philosophers and restore it to the hands of clerics and theologians was undertaken by diverse critics with agendas and positions very different from Croft's. An anonymously published critique of Burnet that was itself widely condemned as deistic launched a similar tirade against Burnet's naturalized Flood. *Two Essays Sent in a Letter from Oxford* (1695) revives the problematic theory put forward decades earlier by scholars such as Isaak Voss and Edward Stillingfleet that Noah's Flood did not cover the entire planet, though it still managed to kill everyone except Noah and his family. The author of *Two Essays*, who identified himself by the initials "L. P." in a possible allusion to Isaac La Peyrère, objected to the needless and unjust destruction of the natural world that a Universal Deluge would have caused. The notion of a global Flood, L. P. wrote, "is not agreeable to the usual methods of Providence, nor to the Wisdom of the Divine Nature; for what design could there be in destroying all the innocent dumb Creatures, and the Beauty of the Creation, in the Uninhabited Parts . . . for the sake only of a few Wanton and Luxurious Asiaticks, who might have been drown'd by a Topical Flood, or by a particular Deluge, without involving all the Bowels of the whole Mass, and the remote Creatures upon the face of the Earth, in their Ruin."[90]

L. P.'s complaint that a global flood would have killed off "the innocent dumb creatures" and destroyed the "Beauty of the Creation" points toward one of the biggest theological problems with a Universal Deluge, namely, that it would have punished parts of God's creation—nonhuman animals and inanimate Nature—that hadn't sinned, and, moreover, were categorically incapable of sin. L. P.'s Orientalist and probably anti-Semitic invocation of "Wanton and Luxurious Asiaticks" only served to highlight, by contrast, the moral purity of nonhuman animals and hence the injustice of a Flood that targeted them as well. By contrast, a universally genocidal but still less-than-planetary Flood would have spared nonhuman nature from punishment it did not deserve. His objection to "involving all the Bowels of the whole Mass" made clear that the physical earth, in addition to the animals, should likewise have been spared from the Flood's destructive effects. L. P., like his possible inspiration and name-sake La Peyrère, was widely decried as a deist for this and other allegedly hetero-dox assertions in *Two Essays*. But it is worth noting that his objection to a Flood that impacted nature as well as humanity was broadly the same as the one put forward by Bishop Croft.

This same critique of a naturalized Deluge from two very different English authors in the late seventeenth century helps us to understand why a pious author might be called an atheist. Both Croft's and L. P.'s critiques were, among other things, attempts to undo Burnet's synthesis of natural and human his-tory. Denying the Flood's globality and denying that it had any effect on nature both boiled down to saying that the Flood was an event in human history only. Saying the Flood had nothing to do with natural history, then, was a means of wresting the Flood out of the hands of natural philosophers and putting it back into the hands of priests and theologians. John Keill's critique of Burnet con-tains a proposed division of labor for writing the earth's history that would have reversed the century-long process of bringing the Flood into natural philosophy. "Theorists or Philosophers," he recommended, should concern themselves with "the common and ordinary phenomena of nature," leaving the study of miracles, catastrophes, and other instances of God's interposition in nature to the men who had been brought up and ordained to the study of God's works.[91] The charges of atheism, deism, and disrespect for the words of scripture that were brought against Burnet may very well have been a reaction to his perceived diminish-ment of the power of priests and theologians to interpret scripture and to nar-rate humanity's ancient history. Objections to Burnet's natural, global Flood were likewise objections to his synthesis of human and natural history. Splitting them back apart, conversely, was a means of maintaining or even building disci-plinary and professional boundaries between religion and philosophy.

A second possible factor behind the charges of atheism—which is to say, some-thing in *Sacred Theory* that might appear to undermine the Anglican faith—was the striking absence of Christ's first appearance on earth from Burnet's history of the world.[92] Christ's birth, death, and resurrection, though clearly of the ut-most importance in the providential history of humankind, do not activate global changes in nature, as do other events of world-historical significance. Burnet's entire approach to constructing a "sacred" history of the earth rested on con-tinual changes to global nature as a result of humanity's changing spiritual con-dition. Yet there is no "Christocene" in Burnet's history: no geo-climatological epoch signifying the new spiritual condition of humanity following the appear-ance on earth of God made Man, spirit turned flesh, offering redemption from sin. It is extremely unlikely that Burnet thought Christ's life and death were theologically unimportant. Rather, his omission of Christ's first coming suggests

Christ is not entirely absent from Burnet's epochs of world history, as shown in *Sacred Theory*'s famous frontispiece. A Christ-like figure straddles the globes representing the first and last stages of the planet's history, and Christ's second coming is associated with the Conflagration and subsequent millennium, represented by globes 5 and 6 (*clockwise from top*). However, there is no globally transformative event associated with Christ's first coming in this schema, nor is there a geological epoch associated with the Christian era following Christ's death. Charles Deering McCormick Library of Special Collections, Northwestern University Libraries.

that he believed that the salvation offered by Christ was not universal. Presumably, God's embodiment on earth did not activate profound enough changes in the spiritual condition of humankind as a whole to provoke a transformation in global nature, as the universality of human sin had been enough to provoke the Flood and usher in the Fallocene. Nothing prevented Burnet from inventing a global catastrophe that transformed the Fallocene into the Christocene. In fact, there was a long-standing Christian tradition regarding an earthquake at the moment of Christ's death or resurrection, which Burnet surely could have expanded into a global earthquake that ushered in a new geo-climatological and spiritual epoch. The fact that he chose to have his Fallocene persist during and after Christ's brief life on earth indicated a belief that not much about humanity's spiritual condition changed as a result of Christ's first coming.

Burnet's failure to fabricate a Christocene suggests that he held a deeply pessimistic view of the scale and scope of Christian salvation. His discussion of Noah's ark as a type for the church furnishes further evidence of Burnet's pessimistic soteriology. The ark, he wrote, "was a Type of the Church in this World," implying that only members of the one true church were eligible for salvation, just as only the ark's inhabitants were saved from the Flood.[93] Moreover, he claimed that the people on the ark survived only through the divine intercession of angels, who steered it through the Flood and kept it from sinking. There was nothing Noah and his family could do to save their lives from inside the ark, Burnet stressed, as it was tossed to and fro on massive waves. Likewise, he implied, membership in the church is a necessary but not sufficient path to salvation; God's grace is also necessary. There is absolutely nothing that individuals can to do to secure their own salvation. Surely Burnet would have admitted that certain individuals had been, and continued to be, saved through Christ, but the persistence of the Fallocene after Christ's death indicated (according to Burnet's logic) that humanity on the whole, as a single moral subject and as a natural kind, remained in a fallen state.

According to Burnet's account of world history, sin was universal, but salvation was not. This pessimism could have sprung from a generally Augustinian view of human sinfulness that exerted considerable influence on Protestant science in the seventeenth century.[94] But Anglican critics may have felt that the extreme selectivity of salvation implied by his history of the Flood veered too close to the Calvinist doctrines of predestination, total depravity, and the salvation of the Elect. In spite of the profound influence of Calvinism on the early Church of England, the place of Calvinists and Puritans within the church was hotly contested throughout England's tumultuous seventeenth century.[95] Alternatively, the absence of Christ from his history may have rung alarm bells in a milieu in which accusations of Arianism and Socinianism were in the air.[96] Generally speaking, official tolerance for a philosophical reinterpretation of biblical history that was both materialist and deeply pessimistic—even if grounded in an unimpeachable source like Augustine—was probably at a low ebb in the decade in which Burnet chose to publish *Sacred Theory*. The 1680s witnessed the death of the crypto-Catholic King Charles II; the brief reign of the openly Catholic King James II; the revocation of the Edict of Nantes, which sent thousands of French Protestants fleeing into England; and the Revolution of 1688–89, which brought the Dutch stadtholder William of Orange and his wife, James's daughter Mary, both Protestants, to the throne. This rapid succession of destabilizing events led to a tightening of Anglican orthodoxy, such as it was, and a correlative anxiety regarding the stability and integrity of Britain as a Protestant state.[97] Archibald Lovell, Burnet's coworker at the Charterhouse, compared *Sacred Theory* to "a Bomb, or Granado-Shell" that "only a Good Natur'd Peaceful, and Unaspiring Monarch" would have tolerated. As long as state power was letting him go free, Lovell enjoined the "Governours of our Church" to "have a watchful Eye" on Burnet.[98] At least some critics felt that the theological implications of Burnet's history of the world threatened the stability of the Church of England and perhaps, by extension, the kingdom.

Beyond the specific religious and political context of Britain in the 1680s, Christian readers across the Continent may have found additional reasons to be dismayed by Burnet's deep pessimism about the universality of salvation through Christ. It severely undermined one of the main things that people found appealing about the theory of the Universal Deluge: its usefulness in legitimatizing Christian evangelism beyond Europe. Burnet's version of the global Flood guaranteed universal sinfulness (as did the Flood and land bridge theories discussed in the previous chapter), but it actively foreclosed the possibility of universal salvation and thus undermined the Universal Deluge's utility as a justification for global evangelism. Burnet's deep pessimism about the global reach of Christian salvation was increasingly out of sync with broader trends in European Protestantism at the turn of the eighteenth century, when Protestant missionaries began competing in earnest with their Catholic counterparts for souls in the East and West Indies. The soteriology lurking around the edges of Burnet's history of the world thus robbed the Flood of one of its primary ideological and religious functions in early modern Europe.

A third and final reason for the perception of Burnet's *Sacred Theory* as atheistic was its pessimism about the scale and scope of salvation on offer even in Europe. Whereas Burnet's theory of the Flood implied that humans possessed a certain degree of geological agency—the totally involuntary yet incredibly powerful ability to ruin the global environment, and thereby their own bodies, through sin—his description of the Fallocene evinced a kind of environmental determinism that stripped humans of any agency to redeem themselves or the world. The condition of humanity in the Fallocene, Burnet wrote, is absolutely determined by the material structure and climate of planet Earth. "Nature is the foundation," he wrote, and "the affairs of Mankind are a superstructure that will be always proportion'd to it."[99] Not only did the state of nature in the Fallocene determine human social and economic structures (to be discussed in greater detail below), it also kept humans in a permanent state of depravity. In particular, the fabric of human bodies after the Flood constrained humanity both mentally and morally. "I call this Body a Prison," he wrote, "both because it is a confinement and restraint upon our best Faculties and Capacities, and is also the seat of diseases and loathsomness."[100] As Peter Harrison has shown, original sin was increasingly understood in seventeenth-century England to compromise human cognitive ability as well as bodily integrity.[101] Burnet seems to have shared this view while denying that these bodily constraints could be overcome in the pursuit of knowledge, as some of his more optimistic contemporaries in the Royal Society believed. He also apparently believed that the "prison" of the human body foreclosed any chance of collective moral progress; the body, like a prison, "commonly tends more to debauch mens Natures, than to improve them."[102] In this respect, he anticipated British climate discourse of the eighteenth century, which invoked climate as a constraint on human bodies and minds.[103] Following the Flood, Burnet appeared to be saying, humanity no longer possessed the power to transform nature or themselves, to restore what they had ruined.[104]

## Soil Degradation and the Saving Grace of Agricultural Labor

The troubling theological implications of Burnet's history of the world come into even sharper relief when compared with the one constructed by his rival John Woodward in *Essay toward a Natural History of the Earth* (1695).[105] Woodward's *Essay* borrowed heavily from *Sacred Theory* while changing key aspects of it in an attempt to redress the theological problems that had provoked charges of deism and atheism. Woodward championed Burnet's theory of a global and naturally transformative Flood, whose empirical basis he hoped to place on a firmer basis by establishing a global network of correspondence and fossil-exchange, as will be discussed in the next chapter. Woodward similarly saw the Universal Deluge as the pivotal juncture in a universal history of nature and humankind. But Woodward deliberately altered Burnet's theory regarding the climactic changes attendant on the Flood, and in so doing, he constructed a different account of sin and salvation, one that reopened the possibility that humanity in the Fallocene could work to achieve environmental and spiritual renewal. The Edenocene's natural abundance, in Woodward's telling, led men to sin by enabling idleness and vice. The earth's ruin by the Flood was, therefore, a net bonus for humanity. Even as the Fallocene weakened their bodies, it forced them to labor and lifted them out of their vicious indolence, thus saving their souls. The ruined earth was a product of sin but was also, at the same time, the means to salvation. Woodward's revised history of the Flood reflected an increasing optimism about the possibility of overcoming original sin in Protestant countries in the early Enlightenment.[106] His idea about diluvial changes in soil chemistry forcing a new regime of labor could even be read as a materialist history of prevenient grace, the idea that God's grace assisted people in doing good works. Woodward attempted to write a history of the Flood that would not only buttress a more optimistic and expansive theology of salvation but would also bolster the emerging economic ideology of agricultural improvement.

Although Burnet's history of the earth is perhaps the more famous of the two, Woodward's *Essay* was in its time equally controversial and influential. In his *Essay*, Woodward announced his determination to construct a history of the earth that was more faithful to the Mosaic history than Burnet's. Keenly aware that "Well-wishers to Moses" had expressed "Concern and Uneasiness to see him thus set aside only to make way for a new Hypothesis," Woodward hastened to assure his readers that his new account of the earth's history—the third major new telling to come off the English presses in a decade, after John Ray's 1692 *Miscellaneous Discourses*—would scrupulously adhere to the "Fidelity and Exactness of the Mosaick Narrative of the Creation, and of the Deluge." Trying to inoculate himself in advance from charges of deism, Woodward claimed that his account of the planet's history, from the Creation through the Flood, would demonstrate at every turn the "Superintendence and Agency of Providence in the Natural World." Although his title promises readers a "natural history" of the Flood, Woodward insisted that "a Deluge neither could then, nor can now, happen naturally" in an effort to avoid the criticisms Burnet had garnered for "naturalizing" the deluge.[107]

Woodward's *Essay* was widely read in Britain and Europe and was generally far better received abroad than at home (as will be discussed in the next chapter). At least some domestic readers believed he succeeded in his stated goal of writing a more Mosaic and less deistic account of the world's transformation at the time of the Flood. The physician John Fisher gushed over Woodward's "incomparable" essay, telling him in a letter of 1718: "All Mankind who are concerned for the truth of the Mosaick account of the Creation are highly oblig'd to You, for Your Defence of it against the Atheists & Deists of the Age."[108] Other readers found the exact opposite to be true. The Welsh naturalist Edward Lhwyd, a former friend turned bitter rival, told the eminent English naturalist Martin Lister: "When we come to consider him with some attention, we shall find him to differ from Moses litle lesse than the Author of the Sacred Theory."[109] Indeed, some observers felt that Woodward's history of the earth was even more irreligious than Burnet's. Lhwyd's friend and ally Tancred Robinson reported from London in 1696, the year following the *Essay*'s publication: "The Clergy say he hath done more mischief than Dr Burnet."[110]

Despite Woodward's stated goal of rectifying Burnet's "considerable Mistakes," the numerous commonalities between their theories makes it is easy to see why so many critics levied the same charges of deism and disregard of the text of scripture against Woodward's *Essay*.[111] Woodward's history of the earth mirrored Burnet's in its broad strokes, especially in making the Flood, whose universality he was likewise keen to prove, the key turning point in nature's history and in the history of humankind. In a similar vein, Woodward's account of the earth's history was undergirded by the same conviction regarding the synchrony of human and natural history as was expressed in Burnet's theory of "natural Providence." "The First Earth was suited to the first state of Mankind," Woodward wrote. His Edenocene was topographically similar to the present earth, not flat and smooth like Burnet's, but it was climatologically distinct, warm and verdant. Like Burnet, Woodward aligned the Fall with the Flood far more tightly than the text of scripture indicates, identifying the Flood as the Fall's delayed but inevitable aftereffect: "But when Humane Nature had, by the Fall, suffer'd so great a Change, 'twas but necessary that the Earth should undergo a Change too, the better to accommodate it to the Condition that Mankind was then in, and such a Change the Deluge brought to pass."[112] Woodward also adopted Burnet's view that the Flood was entirely responsible for the state of nature—including the form and fabric of human bodies—as it existed in the present day. Nothing significant had happened geologically or climatologically or physiologically since the Flood, indicating that nature and mankind were still in their fallen, ruined state, and that Christ's brief appearance on earth had not been sufficient to cause a universal rehabilitation of either one. Finally, Woodward agreed with Burnet that the Fallocene they both currently lived in would persist up until the second destruction and subsequent renewal of the earth and humanity after the Apocalypse.

Woodward's first step in elaborating a new theology of the Flood begins with a discussion of its effects on the global environment. Contra Burnet, Woodward argued that it was the degradation of the earth's soil, not its air, that was primarily responsible for weakening human bodies and shortening human lifespans after the Flood. Whereas Burnet's focus on air was tied to his theory that the onset of illness and debility was the leading cause of this specieswide change in mortality, Woodward's focus on soil leads him to identify food scarcity and the introduction of agricultural labor as more important causal factors behind postdiluvian mortality rates. Before the Flood, the earth's "Soil was more luxuriant, and teemed forth its Productions in far greater plenty and abundance than the present Earth does." This rich soil led to an abundance of food for human consumption and also to an abundance of free time. "The Earth requiring little or no Tillage" before the Flood, "there was little occasion for Labour," and antediluvian humans accordingly spent most of their days in the pursuit of "far more divine and noble" activities than agriculture. After the Fall, however, mankind abandoned these noble pursuits and instead "spent their time in Gluttony, in Eating and Drinking, in Lust and Wantonness." Whereas the Edenic earth's marvelous fertility initially promoted virtuous behavior, after the Fall it became a source of vice and "Corruption."[113]

At the point when human sin had become "universal" and "epidemical," God sent the Flood, which worked to punish humanity primarily by reconfiguring the chemistry and composition of the earth's soil. The Flood dissolved the entire mineral edifice of the planet, Woodward theorized, and in so doing, the super-fertile "Vegetable Matter" of which Eden's soil was principally composed became intermixed with "steril mineral Matter," making it far less fruitful overall. The sharp decline in soil fertility after the Flood forced humanity to spend the vast majority of their time engaged in agricultural labor, leaving precious little time for overindulging in food and sex. The Flood, then, was a blessing in disguise, at least for those fortunate enough to survive it. It was a punishment for sin, true enough, but it was sent by God in order to save humanity from themselves. The fertile soil of the antediluvian earth had caused them to sin, and so reducing the soil to near-sterility, "as might just sufficiently satisfie the Wants of humane Nature," caused them to be more virtuous, primarily by affording them less opportunity to indulge in vice. The reduction of human lifespans worked in tandem with the advent of agricultural labor as a means of "shortning the power of sinning." Not only was their time now "taken up in Digging and Plowing, in making provision for bread, and for the Necessities of Life," but humanity simply had less time overall, decades instead of centuries, to engage in sinful behavior, resulting in an absolute quantitative reduction in sin measured over the course of a single human lifetime. Woodward's theory of global soil degradation as a result of the Flood thus sanctified agricultural labor. The Flood killed most humans but saved humanity by forcing them to work.[114]

Woodward indirectly responded to many of Burnet's critics, from pious conservatives to liberal free-thinkers, who criticized the idea that the Flood transformed global nature. He acknowledged and then rejected the view that the Flood was intended to punish only humanity—or rather, he insisted that the Flood's destruction of nonhuman nature was an essential component of God's punishment of mankind. He conceded that God could have sent any number of global catastrophes that would have exterminated the human race (save Noah's family) just as effectively as the Flood. God certainly could have sent "Famine," "Pestilence," "Fire from Heaven," or "a thousand other Disasters" instead. Woodward even envisioned a scenario involving total global war, in which God implanted a murderous bloodlust into "the Heart of every Man . . . [and] made them Executioners of his Wrath upon one another" until no one was left standing (again, except Noah and his family, presumably surviving this total war in a hidden cave or providentially designed bunker). None of these premodern science-fiction scenarios satisfied Woodward, however, because they lacked one critical element that the Flood alone possessed: the transformation of nature itself. War, famine, plague, and so on were all equally effective as instruments of the near-extinction of the human species, but only the Flood was capable of transforming global nature as well.[115]

Yet in trying to prophylactically shield himself from the types of criticism that Burnet had received, Woodward ended up making several theological innovations of his own, which were not universally welcomed by his fellow Anglicans. He took the rather extraordinary step of insisting that the earth's ruin was in fact God's primary object in sending the Flood. "'Tis very plain," he wrote in contravention of nearly all exegetical precedent, "that the Deluge was not sent only as an Executioner to Mankind: but that its prime Errand was to reform and new-mold the Earth." In making this bold theological intervention, Woodward indirectly acknowledged the allegation by seventeenth-century critics of the Universal Deluge that it was a "needless miracle" because it required unnecessary labor and effort on God's part. "Mankind," Woodward conceded, "might have been taken off at a far cheaper rate: without . . . unhinging the whole frame of the Globe." Yet this superfluity of death and destruction and divine labor, Woodward argued, is precisely what reveals God's true purpose in sending the Flood: to transform global nature in such as a way as to usher in a new spiritual, social, and economic order among humankind.[116]

## The Theology of Improvement

The degradation of global nature as a result of the Flood was not merely an instrument of weakening human bodies and destroying human health as punishment for their sins, as Burnet had claimed and as the general notion of the world's decay seemed to indicate. It was also the means by which God instituted a new global regime of political economy. The Flood's destruction of Eden's green climate stopped men from sinning so much by forcing them to engage in time-consuming, physically demanding agricultural labor. Even as Woodward expanded on Burnet's theory about the Flood destroying the Edenic planet and climate, he broke with his predecessor in arguing that the Fallocene was better for people's morals even if it was worse for their health. With the simple move of swapping air for soil in his account of diluvial climate change, Woodward offered a naturalistic theory of human morality and of political economy. It was above all in his predictions regarding the joint future of the earth and humanity that Woodward departed most significantly from his controversial and influential predecessor. Woodward's soil-centered history of the earth's ruin and humanity's decline was, at the same time, a theory about what humanity could and should do in order to redeem itself. Labor, in his account, was the means to individual and collective salvation. Working in and on that sterile yet mineral-rich postdiluvial soil was the means by which people might redeem themselves and their ruined planet.

Woodward's theory of soil sterility functioned as a theological justification for the sanctity and necessity of agricultural labor in a time when agriculture remained the backbone of England's economy. It also participated in a growing literature on agricultural "improvement" in Britain, which drew on Mosaic natural philosophy, especially chemistry, as well as political economy in order to justify landowners' experiments with enclosure and new kinds of fertilizer.[117] Starting in the 1660s, a group of gentlemen-philosophers calling themselves the Georgical Committee (after Virgil's pastoral epic *Georgics*) began meeting to discuss ways to improve Britain's agricultural productivity through application of "the new science." The committee, whose members included such luminaries as Robert Boyle, John Evelyn, and Henry Oldenburg, married the Baconian ideal of practical natural knowledge to an emergent economic patriotism. They planned to use the latest advances in chemistry to enrich the earth itself, thereby making crops grow more abundantly, "for the common benefit of their Countrey."[118] Boyle hoped to identify the chemical principle—he called it a "vegetable Salt"—that made the soil fertile, distill and analyze it, and then reapply it back into the soil, making it even more productive than ever before. Evelyn compiled instructions for increasing soil fertility in *Philosophical Discourse of Earth* (1676). The earth, he wrote, possesses "a wonderful prolific virtue" that is susceptible to "loss and decay" but also to human "helps" in order to restore its "vigor."[119] The Georgical Committee expressed optimism that Englishmen, armed with chemical knowledge of soil fertility, could make even the most barren patch of ground yield fruit, thereby boosting agricultural yields and enriching their country.

Improving projects were often unpopular with tenant farmers, who resented aristocratic interference in their farming methods, and likewise with local peasants, who used unenclosed land for gathering firewood, grazing their animals, and a variety of other economic activities necessary for people living in a subsistence economy. Popular resistance to improvement sparked a need to legitimize it on grounds other than the strictly economic, especially since such projects tended to benefit landowners the most, if not exclusively.[120] Accordingly, the literature on chemical agronomy articulated a theology of improvement that sought to highlight the spiritual dimensions of agricultural labor. Woodward's contemporary, Henry Rowlands, in his philosophical agronomy treatise *Idea agriculturae* (1764; written 1704), argued that if the difficulty of making crops grow was the legacy of God's punishment of mankind, the discovery and use of chemical fertilizers was a sign of God's grace and a means to redemption. By "the Sweat of the Brow" and "the Toil of our Brain," humankind could begin "to soften that cursed inflicted hardness of the Earth, and to open and penetrate her contracted Bowels, with which we might hasten and facilitate her Labours, in bringing forth Plants, Fruits, and such other Products."[121] The abundance of the Edenocene might therefore be partially recovered through the development of philosophical knowledge of soil chemistry ("the Toil of our Brain") as well as through plain old agricultural labor ("the Sweat of the Brow"). Indeed, the circle around Evelyn and Boyle believed that finding the chemical principle of soil

The poet John Dryden's popular 1697 English translation of Virgil's *Georgics* (frontis-piece shown here) inspired theories and projects in political economy even as it gener-ated fantasies about the leisured Golden Age that preceded the invention of agricultural labor. Charles Deering McCormick Library of Special Collections, Northwestern Uni-versity Libraries.

fertility and using it to improve the English earth would be, in some real sense, to recreate Eden.[122] Agricultural labor then, and the philosophically grounded improvement of the soil, was a means to collective salvation, a way to work one's way back into a state of grace. Boosting agricultural yields would also, as Boyle's readers well knew, considerably enrich his already wealthy family, one of the biggest landholders in the British Isles. Philosophical agronomy, in the context of late-seventeenth-century England, promised salvation to laborers and hand-some profits to landowners. Working the soil was both a holy and a profitable business. Woodward's spiritual history of agriculture thus lent historical, theo-logical, and philosophical legitimacy to the emerging ideology of improvement.

In an unpublished treatise on mining written sometime before 1724, Woodward sought to extend his support for agricultural initiatives to Britain's fast-growing mining industry. Mining, too, could be a means to salvation as well as a source of profit. Previously, in the *Essay*, Woodward had argued that the Flood dissolved and then rearranged all mineral matter into the planet's present stratigraphic layers. (The collapse in soil fertility as a result of the introduction of "sterile mineral matter" was, therefore, one by-product of this massive structural transformation of planet Earth.) In the treatise on mining, Woodward explained that the Flood had also scattered precious minerals and metals beneath the earth's surface in patterns that were extremely difficult for humans, living on the surface, to discern. Deposits of precious ores, he wrote, are "so distributed that they can be discover'd but by Degrees: & not without great Pains & Industry."[123] The seeming randomness of these deposits, so frustrating for those involved in mining, was, Woodward argued, additional proof of God's providence. It was a good thing that there was no field test for locating subterranean minerals and metals, similar to the ones that were regularly employed for identifying rich soil suitable for agriculture before crops were even planted.[124] If there were such a test, humans would be able to find and extract the earth's entire stock of precious minerals and metals too quickly, which would make them all rich and cause them to sink back into antediluvian depravity. The slow and painstaking work of sinking mines and extracting ore, Woodward concluded, was "Wise & Providential" because it was perfectly suited to mankind "in their present State."[125] The challenges associated with mining, like agriculture, demonstrated God's providential wisdom in creating a natural environment that would benefit human souls through economic labor.

Woodward was careful to stress, in the treatise on mining, that the providential difficulty of mining worked to the benefit of the economy as well. The illegibility of subterranean deposits on the earth's surface guaranteed that "there will ever remain a Quantity behind, for the use & supply of Future Ages. . . . And Metalls will always keep up their Price & Worth." If gold and silver, "the Standard of the Exchange," were to suddenly become superabundant, their value would plummet, and "the very Life of Commerce and Business would be presently at an End." Woodward interpreted the stability of the global market, based on a small but steady supply of gold and silver, as a providential act. Conversely, he envisioned the crash of the global market as a kind of apocalypse. If God were to suddenly make these subterranean riches legible on the earth's surface—to make manifest that which was hidden, a kind of revelation—it would lead not only to a resurgence of lazy sinfulness but also to the collapse of the global economy.[126]

Woodward also sought to highlight the more practical and specific ways in which his natural history of the Flood could support contemporary economic growth. In the *Essay*, he argued that the Flood, among its many other fortunate and providential side-effects, had outfitted every country with exactly the climate, topography, and natural resources it needed in order to survive. The terrain and waterways created by the Flood, especially, were "suitable to the Necessities and Expences of each Climate and Region of the Globe."[127] This diluvial cameralism indicated that Britain, therefore, already possessed a rich and sufficient stock of natural resources. While the treatise on mining adopted a global view of the industry, it was also attentive to the particular challenges of developing Britain's rapidly expanding mining industry. The influx of silver and gold from New World mines into European markets went mostly to Spain, leaving Britain chronically short of specie and hunting for native deposits of coinable metals. Meanwhile, coal mining was becoming increasingly central to the English economy, as registered in London's increasingly smoky air. In the second half of his treatise on mining, Woodward pivoted from his previous declaration that there was no field test for identifying subterranean deposits. While giving thanks to God for supporting global commerce by keeping minerals and metals hidden, Woodward also argued that knowledge of the earth's history and structure, such as he himself possessed, could make mining easier and more profitable. In particular, natural philosophers could help landowners and mining engineers decide where to site mines, based on their knowledge of the superficial geological formations that were most likely to indicate precious metals hidden below ground.[128] Woodward's theory of the Flood thus supported not only the ideology of improvement but the growth of Britain's mining industry and the expansion of global commerce at the turn of the eighteenth century.

Woodward's more capacious and progressivist theology of salvation was thus tied to a philosophical argument in favor of emerging forms of political economy. By contrast, Burnet's pessimistic soteriology did not ennoble economic activity in any way. In keeping with his theory of the synchrony between natural and human history, Burnet sought to relate the social and economic order of his day to the natural state of the earth in the Fallocene, in what appears to us now as an early and forceful version of environmental determinism. Burnet's contention that "Nature is the foundation" of the superstructural "affairs of Mankind" expressed his sense that geological epochs played a more or less absolute role in determining humanity's spiritual state as well as its social organization and economic arrangements. The revolutionary shift from the Edenocene to Fallocene exemplified this principle of environmental determinism in especially dramatic fashion, since "every new state of Nature doth introduce a new Civil Order, and a new face and Oeconomy of Humane affairs."[129] In the temperate climes of the Edenocene, people lived in tents and wore little clothing, whereas in the Fallocene, much human labor had to be expended on constructing sturdy buildings and producing suitable clothing in order to protect people from the rough weather. The existence of oceans (lacking in the Edenocene) meant that "a good part of Mankind is busied with Sea-affairs and Navigation." Nature's declining bounty after the Flood led directly to the growth of "Arts, Trades, and Manufactures" in order to satisfy people's needs. Greed for superfluous material possessions had also flourished in the Fallocene, further driving the development of commerce and manufacturing. In other words, nearly all economic activity in Burnet's seventeenth-century world could be directly attributed to the state of the global environment characteristic of the Fallocene. His degraded earth was the material basis for the socioeconomic superstructure of the modern world. Of course, the earth was only degraded because of human sinfulness, so the power of sin to transform nature was compounded by its power to institute, through its natural transformations, new socioeconomic regimes.[130]

Burnet's Protestant ethic, then, was hardly in sync with the spirit of capitalism. The emerging global economy, far from being a sign of divine providence, was yet another by-product of human depravity. For Burnet, who came from a family of higher social rank than Woodward, the necessity of postdiluvian labor was, like illness and death and prison-bodies and ugly mountains and all of the Flood's other deformed progeny, nothing more than dreary and well-deserved punishment. The necessity of toiling on the earth in order to survive was not ennobling or redemptive; it was merely a sign of how severely humanity had ruined the earth as well as their own lives through sin, of how far they'd fallen from the ease of the Edenocene and the luxury of God's grace. For Woodward, who rose from an artisanal background to gentlemanly status largely by making a career for himself as a fashionable London physician, labor was a well-deserved punishment but also a sign of divine blessing and a means of improvement. Their different socioeconomic backgrounds help to explain the difference in their theologies and philosophies of the earth. It is also worth noting that Woodward's efforts to align himself with the landowning class by weaving the ideology of improvement into his theory of the earth was not entirely successful. The frequent charge that he had plagiarized Italian geology and paleontology was possibly an indirect criticism of his arriviste, tenuous class status.[131] Sir Tancred Robinson, writing privately to Lhwyd, contrasted Burnet and Woodward's philosophies of the earth by making an unsubtle reference to their difference in rank: "D[octor] W[oodward] is a plagiary and a calumniator. D[octor] B[urnet] is an Original and a Gentleman."[132] Burnet's earth history was perhaps legible to some readers as traditionally aristocratic in its denigration of agricultural labor. His economic theology aligned with landowning interests in more traditional ways.

## The Ruined Earth and the Argument from Design
Burnet and Woodward's divergent views on the spiritual significance of the early modern economy opens up a fresh vantage point on the politics of the "argu-

ment from design," an idea popularized in the late seventeenth century as a means of legitimating natural philosophy in reference to its (entirely benign and orthodox) theological ends. The argument from design was influentially articulated by Burnet and Woodward's contemporary John Ray, whose *Wisdom of God Manifested in the Works of the Creation* (1691) was one of the first best sellers of natural theology. Burnet articulated his theory of a ruined earth in deliberate opposition to the argument from design, and Woodward's contention that the postdiluvian earth was providentially designed as well as spiritually and materially beneficial to humanity was formulated in deliberate opposition to Burnet. The growing popularity of the argument from design reflected a transitional focus from an angry, Old Testament God to a more benevolent, New Testament God in Anglican theology after the Restoration and particularly after the Revolution of 1688–89, as well as a transition from a Calvinist to an Arminian view of sin and salvation.[133] But the argument over whether nature was ruined or designed, as should now be clear, was not just about God's disposition toward his human creation. Even more fundamentally, it was about whether the planet they all lived on was a product of human or divine agency. Was the earth orderly and well-designed by a wise and powerful Creator? Or was it a "dead heap of Rubbish," unwittingly yet irreversibly trashed by abject humanity?[134] In short, was the present earth created by God, or by people? Burnet's contention that the earth was fundamentally a human artifact, not a divine one, may ultimately have been the most controversial of all his philosophical and theological innovations.

Burnet knew very well that his theory of the Flood and Fallocene was defiantly out of step with the new fashion for deducing God's existence, omnipotence, and benevolence from the order, beauty, and variety of the Creation. Immediately after his evocative description of the ocean floor as a hellish pit, Burnet declared: "[W]e should not be so bold as to make them [i.e., the earth's features] the immediate product of Divine Omnipotence; being destitute of all appearance of Art or Counsel."[135] The point of Burnet's repeated descriptions of nature as ruined and disorganized was to emphasize that God had not fashioned the earth, the climate, or the human body in their present forms. Nature in the present day was the wreckage of the Flood and thus, indirectly, an artifact of human sin, which had obliterated God's original, artful design. Burnet was not rejecting natural theology per se, but he was offering a very different take on the religious lessons one would learn by studying nature. His vision of the Fallocene made nature a book that spoke not of God, of his design and creativity, but of humanity, of their ignorant, brutal, and irredeemable proclivity toward death and destruction.

Several of Burnet's critics took him to task on precisely these grounds. Mackaile complained that Burnet had written "most undervaluingly of the present

Earth," while Croft accused Burnet of "censuring the workmanship of God."[136] Woodward took great offense at what he perceived as a disparagement of God's benevolence and creative power, accusing Burnet of denying the "Art and Skill in the Make of the Present Globe." Even though Woodward used the language of destruction in order to describe the Flood's effects on the global environment ("broken," "dislocated"), he insisted that the world the Flood created was not a human artifact but a divine one. God had taken "great Care" in reassembling the earth after the Deluge, he wrote, which was apparent in the "very many real Graces and Beauties" that adorned the present earth. Even the "Limbs and Parts" of the earth's topography that Burnet had singled out as especially ugly were in fact "ordered and digested with infinite Exactness and Artifice." There was nothing at all "Artless, or useless and Superfluous, in the Globe." The post-diluvian state of the earth testified to God's power as a "Reasoning or Designing Agent" and also to his benevolence toward humanity, in designing a global environment that would suit and indeed benefit them.[137]

Woodward's argument contra Burnet that the earth shows evidence of "Art" and "Artifice" was a slight but crucial misreading. Burnet also recognized that the earth shows evidence of "Art" and "Artifice"—just not God's. As the epigraph to this chapter makes clear, Burnet read the earth as showing evidence of *human* handiwork. St. Clair's criticism hit nearer the mark when he claimed that Burnet "sees this our Earth so ugly . . . that he cannot think it to have come thus out of the hands of the All-wise God."[138] St. Clair took Burnet to task for denying God's role as the Creator of the (modern) earth, thus indirectly acknowledging that Burnet had turned the earth into the work of human hands.

Moreover, Burnet theorized that human activity continued to make the originally "Natural" earth ever more "Artificial" as time went on. He deplored the "Artificial Earth," the earth that is "inhabited and cultivated" by humans and divided up according to man-made units "of Countries and of Cities," and he called for the creation of "natural Maps" that would "leave out all that, and represent the Earth as it would be if there was not an Inhabitant upon it, nor ever had been."[139] Burnet here engages in a deep green fantasy about pristine wilderness, the world without us, the earth untouched by the Flood and by human agency. Like Bill McKibben lamenting the onset of a world thoroughly transformed by human activity in *The End of Nature* (1989), Burnet believed that humans no longer lived in "the Tract of Nature" but rather in a degraded and continually degrading "Artificial" landscape of their own making.

Works of art and works of nature were frequently juxtaposed in early modern Europe, but the playful elision of the boundary between them was an equally influential aspect of early modern culture, as when natural and man-made objects were jumbled together promiscuously in a cabinet of curiosity.[140] The master trope of nature as a/the Creation—such a common analogy in the early modern period that we now barely recognize it for the metaphor it is—was one such blurring of the distinction between the two. "The Creation" indicated that things which people were inclined to see as "natural"—the air, the trees, their hands and lungs and hair and eyes—were, in fact, artifactual, designed and crafted by God's own hand. Both the proponents of the argument from design and the partisans of the notion of a ruined earth agreed, then, on the notion that nature was in some literal sense *made*—a built, curated, and (following the eco-critic Ursula Heise's adoption of a popular science fiction term) "terraformed" environment.[141] They simply disagreed on who or what had done the making. It is possible, then, to read the controversy around Burnet's *Sacred Theory* as a test case for the acceptability of the idea that the earth was a human creation. In the final analysis, it may have been this rather shocking implication of *Sacred Theory* that earned its pious author the lasting epithet of "atheist."

# CHAPTER FOUR

# The Flood and the Apocalypse
## Building the Republic of Letters

### "Making Observations in All Parts of the World"

John Woodward's 1696 guidelines for collecting and observing natural phenomena begin abruptly with the following command: "Keep a Journal of the Ship's Course."[1] Unsurprisingly for a resident of an island nation, the English naturalist imagined that the first stage in a journey devoted to natural-history collecting would begin with embarkation on a ship.[2] Like other query lists popular at the time, Woodward's *Brief Instructions for Making Observations in All Parts of the World* contains lists of questions to be answered, observations to be made, and requests to be fulfilled by travelers with a taste for "Knowledg both Natural and Civil."[3] "Take an account of the more observable and peculiar Diseases of the Country," he enjoined his readers; "Observe whether some Seas be not salter than others." Woodward asked his travelers to collect waterfowl and agates, to keep records of the climate and weather ("Heat and Cold, Fogs, Mists, Snow, Hail, Rain"), and to observe the "Customs and Usages" of any "Natives" they might chance to encounter.[4] These were things of general interest to any early modern scholar.

However, Woodward's *Brief Instructions* was distinguished from the typical query list both in its global ambitions and in the nature of its specific requests. Many of the queries were oddly specific and obviously self-interested, most notably the repeated request for fossils. Following the observation that "Sea-shells, Teeth, and Bones of Fishes, etc. are found very plentifully in England," Woodward issued the following dictate: "'Tis very extremely desirable that careful search be made after these things in all Parts of the World."[5] By "Sea-shells, Teeth, and Bones of Fishes" he meant marine fossils, which, when discovered on land, especially far away from bodies of water or in high mountains, were prized by partisans of the Universal Deluge as evidence of the global reach of the biblical Flood.

Woodward was one of many hundreds of European natural philosophers who felt that proving the Flood's universality was a worthy intellectual pursuit and an urgent task for Mosaic natural philosophy, as the preceding three chapters have demonstrated. However, he was one of the first, and perhaps the most influential, during the fossil wars of the late seventeenth and early eighteenth centuries, to yoke the theory of the Universal Deluge to the organic theory of fossil origins, as opposed to the notion that fossils grew *in situ*, that is, in the rock or soil in which they were found.[6] "The vast Multitudes of Shells, and other Marine Bodies, found at this day incorporated with and lodged in all sorts of Stone, in Marble, in Chalk," he wrote in the preface to *Essay toward a Natural History of the Earth* (1695), could be "found in all Parts of the known World, as well in Europe, Africa, and America, as in Asia," thus proving the "Universality of the Deluge."[7] Woodward's request that English travelers make a concerted effort to collect fossilized mollusk shells, sharks' teeth, and fish skeletons wherever they went was a key strategy for amassing a global data set that would allow him to definitively demonstrate the global scale of Noah's Flood.

Woodward's *Brief Instructions* was also distinguished from the typical query list by the scale on which Woodward hoped and imagined his readers would collect these valuable relics of the Deluge. The repetition of the phrase "in all Parts of the known World," echoing the title *Brief Instructions for Making Observations in All Parts of the World*, underscores Woodward's thirst for a truly global data set. One of the first and best-known exemplars of the query-list genre was Robert Boyle's *General Heads for a Natural History of a Countrey, Great or Small* (1666), a set of instructions for collecting within the bounds of a single country, a scale that most naturalists regarded as already quite ambitious. Traveling across the Continent was difficult enough—topographically, financially, and socially—to inhibit most naturalists from making the attempt. It remained nearly impossible for an individual to travel the world on a scientific voyage. The Jesuits already had a more or less global missionary network, which doubled as a knowledge network, up and running by the seventeenth century, but this system was closed to non-Jesuits.[8] Woodward's plan to survey "all Parts of the World" reflected the desire on the part of many of his British Protestant contemporaries to study and to understand nature on a global scale. It was an elusive goal, but one that scholars living through an age of rapid globalization were increasingly determined to reach.[9]

Woodward never set foot beyond England's shores. He had tried unsuccessfully in his youth to find a tutoring job that would allow him to go on a Grand Tour, but he quickly realized that he did not need to travel himself in order to describe the natural world as it existed beyond England. He thus proposed a solution based on the Jesuit model to the problem of individual limits on mobility: build a global network of observation and circulation. He subtitled his query list *An Attempt to Settle an UNIVERSAL CORRESPONDENCE for the Advancement of Knowledg* [sic]. Woodward could write an empirically grounded natural history of the entire planet from his comfortable rooms in Gresham College as long as he could get colonists, diplomats, mariners, and other Englishmen traveling or living abroad to send him their data and specimens. *Brief Instructions* was designed to generate not just a global data set but a "global knowledge infrastructure," to borrow a phrase from the historian of science Paul Edwards.[10] Edwards finds that the establishment of modern climate science depended on achieving a fit between the scale of the object of knowledge and the scale of the network devoted to sourcing that knowledge, which, in the case of global warming, was necessarily planetary. So too, at the turn of the eighteenth century, did Woodward and his colleagues and collaborators imagine that the study of global disaster—a disaster that many of them believed involved a catastrophic transformation in the global climate, as discussed in the previous chapter—necessitated a social infrastructure that could extract and circulate knowledge from "all Parts of the World." Although Woodward's primary mission, amassing a personal collection of fossil specimens, was time-limited and self-interested, he imagined the network persisting beyond the accomplishment of this specific goal. The project of proving the Flood's global scale was to be the occasion for building a permanent and global network of natural inquirers collecting and sharing information with one another: a "universal correspondence."

The Republic of Letters is the key social context for understanding how and why the increasingly orthodox theory of a Universal Deluge continued to grow in popularity across Europe in the last decades of the seventeenth century and the first decades of the eighteenth.[11] This chapter explores several interrelated aspects of the Flood's social utility, demonstrating the various ways it was used to forge relationships and build cultural capital in this particular social network. From the perspective of a pluralistic community of inquirers seeking common ground, the Universal Deluge possessed many social virtues as a topic of inquiry. It appeared to offer men of different faiths, churches, and countries the opportunity to jointly pursue the shared enterprise of Mosaic natural philosophy. The Flood's potential to unite Protestant and Catholic scholars had been identified in the late sixteenth century and was furthered by the threat of polygenism in the 1640s and '50s, but the growing popularity of fossils from the 1660s onward opened up a new way for the Flood's irenic potential to be realized. Long regarded as natural curiosities, fossils were now ennobled as "relics" of the deluge, boosting their value in the gift economy that powered the Republic of Letters.[12] By stimulating the collection and exchange of these sacralized tokens of biblical history, the Flood had the potential to unite naturalists, physicians, and philosophers in a shared philosophical enterprise that was also laden with religious significance. The practice of fossil exchange endowed this irenic agenda with an empirical basis that further enhanced the Flood's ability to knit together a diverse and dispersed community of inquirers.

Perhaps the biggest social virtue of the Universal Deluge as a collective topic of philosophical inquiry was its demonstrated ability to expand the Republic of Letters, if not to global dimensions, then at least to transregional ones. Fossils discovered in Europe's provinces, peripheries, and colonies provided lower-status people at Europe's margins with a pretext to make connections with higher-status people in Europe's major cities and centers of learning by giving them something valuable to trade. Fossils were a form of social capital that could be sent through the mail in order to establish and maintain valuable relationships with foreign correspondents. By stimulating the long-distance exchange of fossils, the Universal Deluge proved remarkably effective at cementing transnational ties within Europe and at the same time expanding the Republic of Letters to encompass Europe's peripheries and overseas colonies. The Flood's irenicism and empiricism worked hand-in-hand to expand the size and scale of European scholarly networks.

The first half of this chapter focuses on the epistolary network that Woodward built with his Swiss translator, the Zurich naturalist and physician Johann Jakob Scheuchzer. Drawing primarily on manuscript letters from archives across Europe that were exchanged within and around Woodward's network of correspondence, I show how the Universal Deluge as a topic of collective philosophical inquiry aligned with the epistemic values and social practices of the Republic of Letters during a period of expansion of the scale on which knowledge was being created and exchanged. In addition to its irenicism and empiricism, the Flood's social utility flowed from its reinforcement of the Republic's self-defined image as a uniquely cosmopolitan, tolerant, egalitarian, and worldwide community. Because the Universal Deluge was supposed to have touched every part of the earth, theoretically everyone on earth had something to contribute. The planetary scale of this imagined disaster legitimated and indeed actively called for the participation of inquirers of all kinds from across the globe. The pursuit of the Universal Deluge allowed the Republic's members to imagine themselves as participants in a truly universal community, in all senses of the word.

The social value of the Universal Deluge to the Republic of Letters is brought into especially sharp relief by comparing its fortunes to that of its future doppelgänger, the Universal Conflagration. The second half of the chapter shows how the Flood and the Apocalypse, routinely paired as biblical catastrophes in the earth's past and future, decoupled in the decades around the turn of the eighteenth century as the Flood grew in popularity and the Apocalypse's popularity waned among natural philosophers. Their surprising and contingent divergence can best be understood, I argue, by recognizing that the Universal Conflagration lacked the irenic potential and the empirical basis of the fossil-based Flood. Because it had not yet happened, there were no physical traces, no material objects associated with the Universal Conflagration that could be exchanged across religious and political boundaries. Moreover, the Day of Judgment and the "flood of fire" whose future reality was nearly universally accepted, would necessarily have differential impacts on Christians of different churches and faiths, making it an extremely challenging subject to discuss across the religious divisions of post-Reformation Europe. The Universal Conflagration—that is, the Apocalypse reimagined as a global, natural disaster amenable to philosophical study—possessed none of the social virtues offered by the Universal Deluge, exacerbating tensions instead of bypassing them. The apparently secular drive to remove the Apocalypse from studies of the earth and its history needs to be understood, therefore, in relation to the social utility of the Flood during a period of expansion in the Republic of Letters.

### The Flood, Fossil Exchange, and the Republic of Letters

The push to build a "Universal Correspondence" was not realized in Woodward's lifetime, of course, but it was remarkably effective at building his own international reputation as well as a diverse and large-scale network of correspondents and fossil suppliers. Woodward's first step was to cultivate an international readership for his *Essay toward a Natural History of the Earth*. In order to do that, he would need to have it translated out of English, a language then understood by few people beyond Britain and its colonies. Shortly after the turn of the century, Woodward struck up a correspondence with Scheuchzer and began lobbying him to produce a translation of his *Essay*. When Scheuchzer proposed a German translation—his native tongue—Woodward rejected the offer, saying, "I had rather it were also translated in Latin, & in French: those two Languages being more generally understood."[13] Scheuchzer acquiesced and opted for Latin. The result, *Specimen geographiae physicae* (1704), was a success. From his location in Zurich, Scheuchzer became Woodward's main translator and propagandist on the Continent for a Germanophone audience as well as for a broadly European, Latin-educated one.[14] Woodward eagerly anticipated how this Latin translation would help his fame and his epistolary network to grow. "Since you are pleased to condescend to the Translating it into a Language that will make the Work known abroad," he told Scheuchzer in 1703, "I shall expect more and more Communications of Observations from Learned Men in all Places."[15] He constantly encouraged Scheuchzer to make sure copies were "dispersed to all Parts," instructing him in a letter of 1712 to have copies sent to France, Germany, Holland, England, Italy, Spain, and Portugal.[16] Woodward's fossil collection, reputed to be the largest in England and boasting thousands of specimens from around the world, testified to Scheuchzer's success in making sure that his book and his call for fossils reached an international audience.[17] His suppliers included a Capuchin friar who brought Woodward a fossil from Tartary, a German baron who oversaw mining operations in Saxony, and a British aristocrat whose "Curiosity, and Desire of Knowledge" led him to "Egypt,

Arabia, and several other Parts of Asia, where few Europeans besides have ever been."[18]

Scheuchzer's assistance in enlarging Woodward's reputation, epistolary network, and fossil collection worked to his personal advantage as well. Scheuchzer initially played the field when trying to secure English patronage for himself, reaching out to Woodward and also to Woodward's rivals, the natural history collectors Hans Sloane and James Petiver. Woodward eventually won exclusive rights to Scheuchzer's supply line in part by offering to get Scheuchzer's books printed in London instead of Zurich, which guaranteed Scheuchzer the international fame he likewise sought. Originally inclined toward a vitalist view of fossil origins in the 1690s, Scheuchzer thereafter became convinced not only of the organic origin of fossils but also of their diluvial origin.[19] He began publishing fossil catalogs with titles like *Museum Diluvianum* (Museum of the Flood, 1716) and *Herbarium Diluvianum* (Herbarium of the Flood, 1723), which polemically identified all of the fossils cataloged therein as relics of the Flood. Woodward's enemies in England complained bitterly about the degree of Scheuchzer's devotion. In a private letter of 1704, Ray wondered why "Dr Scheuchzer or any other considerate & inquisitive person should be so fond of Dr Woodwards bold & ungrounded Hypothesis . . . as to turn his book into Latine & to professe themselves his Proselytes."[20] But Scheuchzer's international fame, fossil collection, and publications grew apace, and he seems not to have minded being seen as the "proselyte" of his older and more famous British colleague. Hitching his wagon to Woodward's worked to their mutual advantage. The Dutch antiquarian Gisbert Cuper told Woodward in a letter of 1708, "I read your *Specimen Geographiae* with the greatest of pleasure, which was given to me by my very dear friend, the most learned Jacob Scheuchzer, who merits the name the Swiss Pliny."[21] Comparing the Swiss naturalist to the most lauded ancient writer of natural history was a high compliment. Scheuchzer's canny act of gift exchange secured him the correspondence and the respect of a famous man of letters in a country in which Scheuchzer aspired to find a university position, a strategy that his patron Woodward encouraged him to pursue. (Woodward offered to send Cuper a box of fossils in order to grease the wheels.)[22] The mutually beneficial though unequal relationship between Woodward and Scheuchzer worked to the benefit of both men's international reputations as well as to their mutual practices of fossil collecting, reputation building, and network expansion.

Woodward's call for fossils to prove the Flood's global scale was remarkably effective at building a scholarly network beyond his home country and across territorial boundaries. It also proved effective in bringing Protestants and Catholics together in a shared scholarly enterprise laden with religious meaning. Woodward deliberately sought readers and correspondents in Catholic France and Italy. He deputized Hugh Bethell to negotiate a French translation of the

*Essay*, which appeared in Paris in 1735 with a dedication to the royal librarian l'Abbé Bignon, a frequent correspondent.[23] In 1702, Woodward exulted to hear the news that an Italian translation was under way in Rome.[24] The news turned out to be unfounded, but Woodward's desire to see an Italian-language edition of his book printed in the heart of the Roman Catholic Church indicates that he sought not only an international audience but a multiconfessional one. (An Italian translation finally appeared in Venice in 1739.) In 1714, Woodward's *Essay* was among the first items of discussion at the inaugural meeting of a new learned academy in Bologna, the Istituto delle Scienze (Institute of Sciences).[25] Led by the naturalist, papal ambassador, and Hapsburg military general Luigi Ferdinando Marsigli, the Bologna Institute of Sciences modeled itself on London's Royal Society, of which Woodward was a member.[26] The institute also had strong ties to Switzerland and especially to the Scheuchzer brothers, without whom it is doubtful that an English work of natural history would have appeared on its agenda.[27] Many of the institute's members formed a positive opinion of *Essay toward a Natural History of the Earth*, indicating the willingness of at least some Catholics to participate in Woodward's call for a global network of fossil collectors to prove the Flood's global scale.[28]

Giuseppe Monti was one of the members who most enthusiastically embraced Woodward's global history of fossils and the Flood. Professor of natural history and head of the botanical gardens at the University of Bologna, Monti amassed a sizable fossil collection that he named the Museum Diluvianum, or Museum of the Flood, echoing the idea promoted by Woodward and Scheuchzer that all fossils were products of the Flood.[29] Five years after the institute's inaugural meeting, Monti published a short treatise, *De monumento diluviano nuper in agro Bononiensi detecto dissertatio* (Testimony of the Deluge recently discovered in the territory of Bologna, 1719), which was larded with praise for "the Great Woodward" as well as "the sacred historian Moses."[30] The "testimony" of the book's title was a fossilized skull that had been found in the hills outside of his city and that Monti identified as belonging to a walrus. Since walruses were not native to Italy, or indeed, to land, the presence of a walrus fossil in the hilly interior of the north-central Italian peninsula was indeed remarkable. Monti concluded that the walrus fossil could only have been the result of the Universal Deluge, which, having killed the animal, washed its remains into the Bolognese hills and left them there as the waters receded back to the ocean. The fossil of an exotic marine mammal was the perfect piece of evidence to prove that the Flood had been truly global.

Monti's enthusiasm for Woodward's synthesis of fossils and the Flood no doubt reflected his hope that it could ground philosophical collaboration between men of different countries and faiths. Like Woodward and Scheuchzer, Monti cultivated a large-scale network of correspondents as a strategy for collecting fossil specimens, telling a fellow Bolognese naturalist in 1719 that he expected to receive shortly "175 fossils from different countries" from correspondents in Germany and England.[31] Monti's unusual ability to read English gave him access to a fuller range of English works on earth history than was available internationally, including Ray's *Three Physico-Theological Discourses* (1693), which was never translated into Latin.[32] He corresponded frequently with Scheuchzer and Bourguet. The dream of Mosaic natural philosophy uniting Protestants and Catholics represented by the Universal Deluge was kept alive in part by the material benefits of having foreign correspondents and in part by the desire to leverage that correspondence to achieve a panoptic view of global nature. After learning about the existence of coal deposits in England, which he compared to similar deposits "found in almost all places in the territory of Bologna," Monti predicted that subterranean coal seams would eventually be found "in all mountains not only in Italy but in other countries as well."[33] It is likely that Monti shared Woodward's belief that coal was formed and deposited along with all other minerals and metals at the time of the Universal Deluge.[34] The presence of coal deposits in both Italy and England therefore guaranteed the worldwide distribution of coal reserves and their common origin in a key event in sacred history. Coal deposits, like fossils, were relics of the Universal Deluge buried underground, secretly connecting all parts of the world and forming the basis for transnational and cross-confessional collaboration, in addition to their economic benefit to the locales in which they were found. A shared set of natural resources and scientific specimens could be used to establish a shared history that justified philosophical collaboration and exchange.

It is also possible that Monti did not see Woodward's style of earth history as a foreign or distinctively Protestant import, as did his colleague and correspondent Antonio Vallisneri (the subject of the next chapter). Woodward's *Essay* borrowed heavily from Italian literature on fossils and the Flood from the 1660s and '70s, which led to charges of plagiarism in England but which may have endeared him to colleagues in Italy.[35] Monti favorably compared Woodward's *Essay* to Jacopo Grandi's *De veritate diluvii universalis* (On the truth of the universal deluge, 1676), which anticipated by twenty years Woodward's central argument that fossils were the remains of organisms killed by the Flood whose lingering presence across the earth's surface proved the Flood's universality and physical reality.[36] A Modenese naturalist and physician, Grandi was part of a trend that included two far more famous naturalists working in Italy, the Sicilian painter Agostino Scilla and the Danish experimental philosopher Nicolaus Steno (Niels Stensen).[37] The notion that fossils, especially marine fossils found in inland areas, were relics of Noah's Flood dates to the earliest centuries of Christian antiquity.[38] However, philosophers in Europe in the sixteenth and sev-

enteenth centuries tended toward different explanations, relying on astrological influences, vital principles in the earth (whether chemical or biological), or
simply the playfulness of nature in order to explain how and why certain rocks
exhibited unusual figurative patterns.[39] Scilla's paleontological treatise *La vana
speculazione disingannata dal senso* (*Vain Speculation Undeceived by Sense*, 1670)
broke new ground by offering the most potent defense yet written of the origin
of figured stones in the plants and animals that they so strongly resembled.
Steno anticipated Scilla's argument about the organic origins of fossils while asserting that at least some fossils probably had their origins in the Deluge in *De
solido intra solidum naturaliter contento dissertationis prodromus* (*The Prodromus
to a Dissertation on a Solid Naturally Contained within a Solid*, 1669). Certainly,
Steno thought, the landscape of Tuscany, which he had been surveying on behalf
of his patron, the Grand Duke Ferdinando II de Medici, contained geological
and paleontological traces of "a kind of universal deluge that is not rejected by
the laws of natural movements."[40] Steno, an honorary Italian and Catholic convert (he later became a Catholic bishop), was widely read by English naturalists
after the intelligencer Henry Oldenburg translated *Prodromus* into English in
1671. Woodward knew Steno's work well and purchased Scilla's fossil collection
upon the latter's death.[41] In a direct acknowledgment of Woodward's debt to
an earlier generation of Italian naturalists, a 1697 defense of Woodward's *Essay*
by the mathematician John Harris credited Grandi, Steno, and Scilla with advancing the organic theory of fossil origins and thus paving the way for Woodward.[42] Woodward's obvious debt to recent Italian literature on fossils and the
Flood—a debt not shared by Burnet, Whiston, or Ray[43]—no doubt played a
role in the warm reception of his book in Italy. Pious Catholic philosophers
like Monti felt that Woodward's version of the Universal Deluge, based on the
long-distance exchange of fossils, could build bridges across the divided terrain
of Christian Europe.

Woodward's tireless self-promotion and Scheuchzer's translation and epistolary
work on Woodward's behalf built up both men's international reputations while
increasing the size and scale of their fossil collections and networks of correspondence. These obviously self-interested motives did not necessarily detract from
their ability to gain readers, adherents, and fossil suppliers across Europe. Sending
fossils to foreign correspondents for the ostensible purpose of proving the global
scale of Noah's Flood was a philosophical and religious agenda that many scholars of different countries and confessions could get behind. Even the unoriginal
aspects of Woodward's theory of fossils and the Flood, borrowed from Italians,
helped to foster the perception that the study of the Universal Deluge, when
pursued through the collection and long-distance circulation of fossil specimens, could bring together a pluralistic community of philosophical inquirers.

### *Scaling from Local to Global Studies of Nature*

A second major factor behind the success of Woodward's call for a global network of fossil collectors was the way that it resonated with a major new trend in
the study of nature: pursuing local natural history with an eye toward producing
a composite portrait of nature on a large and possibly even global scale. Woodward followed the lead of his Italian predecessors in more ways than one, adopting a process of scaling articulated by Steno and others whereby many people
would perform small-scale, empirical studies of the Flood's effects on local nature with the ultimate goal of producing a patchwork portrait of global nature
and of the Flood's planetary effects. The Universal Deluge seemed to offer a
means of transforming local studies into global knowledge, offering both a conceptual framework and a methodology for how this kind of highly desirable
scaling might be accomplished.[44] The Universal Deluge might therefore also
be a machine for building the kind of "universal correspondence" that Woodward dreamed of.

The network that Woodward, Scheuchzer, and their collaborators managed
to build never came close to universal coverage. However, the idea of forming a
"universal correspondence" was in some ways more useful to them than the reality, allowing them to build a transnational network that brought in ever more
participants under the pretense of collectively studying a global phenomenon,
the Universal Deluge, on a global scale.

The seventeenth and eighteenth centuries witnessed profound changes in the
scale and scope of Europe's engagement with the rest of the globe: the expansion of European empires and of commercial, religious, and diplomatic networks, not only in the Atlantic but in the Indian and Pacific Oceans as well.
The knowledge network known as the Republic of Letters piggy-backed on the
expansion of these imperial, religious, and commercial networks.[45] It may very
well have been the rapid growth of these interlocking systems of circulation and
exchange from the late seventeenth century onward, particularly in northern
Europe, that made global natural knowledge appear to scholars in this period
as a tantalizingly achievable goal. The question was *how* to write an empirically
grounded account of nature on a planetary scale. Few had made the attempt.
Prior to the turn of the nineteenth century, truly global studies based on empirical observation and data collection were rare.[46] The challenge of working on
large scales and across long distances was exacerbated by the difficulty of travel,
which lagged behind the shipment of letters, newspapers, journals, and other
forms of information.[47] In the mid-seventeenth century, René Descartes, Johann
Joachim Becher, and Athanasius Kircher all offered portraits of the earth *in toto*,
but they were largely based on laws of physics or chemical principles that were
assumed aprioristically to be global.[48] These were no substitute, in the minds of

many naturalists, for the slow work of collection, compilation, and description that the study of the plant, animal, and mineral kingdoms required.[49] Doing careful, descriptive natural history on a planetary scale hardly seemed possible.

Local natural histories were therefore extremely popular in early modern Europe. Still, calls to compile local studies into national surveys increased from the late seventeenth century onward, often with the explicit understanding that these would be collaborative enterprises.[50] Query lists like Boyle's *General Heads for a Natural History of a Countrey* joined works like Robert Plot's *Natural History of Oxfordshire* (1677), whose subtitle, *an Essay toward the Natural History of England*, announced this county-level study as a contribution toward a nationwide project. The industrious Plot was only able to produce two such local histories in his lifetime (he also published a *Natural History of Staffordshire* in 1686), a clear indication that compiling a natural history of England would necessarily be a collaborative and long-term enterprise.

Many naturalists imagined correspondence networks as the means by which local studies could be turned into large-scale natural histories. John Ray encouraged his friend Edward Lhwyd, a fossil collector and keeper of the Ashmolean Museum in Oxford, to expand his planned fossil catalog from a local to a national scale. "I would not have you confine your self to so narrow a compasse as the neighbourhood of Oxford," Ray told Lhwyd in 1691, "but take in all of your knowledge that are found in England."[51] Ray acknowledged the difficulty of a single person undertaking a national survey of fossils, so he recommended that, in addition to "excursions" Lhwyd could make when he was not "confined to Oxford," he could also count on his correspondents to help him out: "by your friends you may procure a good number of such species as easily occurre in all the celebrated places."[52] Lhwyd's Welsh background inspired him to reach even further than Ray recommended, producing a celebrated fossil catalog of British, not just English, fossils, *Lithophylacii Britannici ichnographia* (Illustrated catalog of the formed stones of Britain, 1699).

Fossils might have seemed more suited to large-scale study than other collectible items of natural history. Compared with most plants, animals, and minerals, fossils were rare and difficult to find. Being small in number made them easier to describe. Moreover, their radically unequal distribution across any given patch of earth meant that the fossil collector need not to aspire to equal coverage of all parts of the territory he was surveying. A single specimen could stand in metonymically for a fairly large swath of earth. Ray's encouragement to Lhwyd to produce a national fossil catalog may have flowed from his assessment about the relative paucity of Lhywd's object of collection compared with his own. Ray despaired, for example, of compiling "a generall History of Insects . . . found in this Island," complaining that the sheer number of species "would be work enough for a man's life."[53] This sense of the relative ease of producing large-scale fossil surveys might also have motivated Woodward's ambition to move beyond the national scale to a global one. In a letter of 1691, before Woodward and Lhwyd's falling-out, Woodward complained to the Welsh naturalist about the lack of scalar ambition on the part of their predecessors and colleagues: "It hath been the great defect of those that have wrote on this subject that their observations have been confined to one Country, or some small part thereof perhaps." Woodward, however, boasted that "I have been already over the greatest part of England" on fossil collecting trips and "have at this time Queryes about these things, in many parts of Asia, Africa & America as well as Europe, to some whereof I have already received Answers."[54] Woodward, in other words, was following the exact same methodology that Ray recommended Lhwyd follow in compiling a national catalog—travel as far as you can yourself, and depend on your correspondents to send you whatever information they can collect in their own locales—only he claimed to be doing it on a global scale.

Many of the most ambitious plans for transforming local studies into global natural histories came from those studying fossils and the earth's history. In *Prodromus*, Steno argued that the evidence he had collected in his travels across Tuscany allowed him to reconstruct the geohistory of the entire region and perhaps even of the entire world, which he summarized thus: "Thus we recognize six distinct aspects of Tuscany, two when it was fluid, two when flat and dry, two when it was uneven." The "fluid" stages he identified with the Creation and the Universal Deluge, and the "uneven" stages as the interstitial periods immediately before and after the Flood. In other words, Steno believed he had unearthed evidence of global geological processes from his study of local nature. Such individual studies could therefore be a valid means of piecing together global history. But Steno also felt that other local studies were needed as confirmation. *Prodromus* concludes with the following methodological reflection on the process of scaling up from local empirical studies to claims about global history: "[W]hat I demonstrate about Tuscany by induction from many places examined by me, *so I confirm for the whole earth from the descriptions of many places set down by different writers.*" In effect, Steno proposed that his own study of the Tuscan landscape, when combined with local studies performed elsewhere, could establish the history of truly global events and epochs in the earth's history.[55]

The German polymath Gottfried Wilhelm Leibniz, who admired *Prodromus* and met Steno in Rome in 1689, formalized Steno's methodology into a more general principle in his influential account of the earth's history, *Protogaea* (First Earth, 1749), which circulated in printed synopses and extracts in the 1690s and 1700s.[56] "When everyone contributes curiosity locally," Leibniz enjoined, "it will be easier to recognize universal origins."[57] By "universal origins" (*origines communes*), he meant something like: a portrait of nature in its spatial and temporal entirety, which is to say also, the common home and point of origin of all of humanity. The first step in writing a global history of the earth, as Leibniz imagined it, would be to solicit local studies from far and wide. Situated knowers would perform studies of their own local natural environments, making topographical observations and collecting fossils, minerals, and other specimens, just as Leibniz had done in the Harz region in Germany, where he supervised mining operations on behalf of his patrons in the Houses of Brunswick and Hanover.[58]

As scholars like Steno and Leibniz imagined it, the study of global nature necessarily called for the participation of hundreds if not thousands of situated knowers, all contributing curiosity locally. Implicit in their methodological pronouncements about the scale of research necessary for such an undertaking was the need for some sort of network for gathering and circulating and compiling all of these local studies—one that could synthesize local knowledge into global knowledge. The Republic of Letters to which they all enthusiastically belonged (Steno until his conversion and philosophical retirement, at least) seemed to be the obvious solution, the very mechanism they needed to mobilize people the world over, to prompt them to conduct studies, to collect, and to circulate the fruits of their labors so that the history of global nature could be compiled.

The one glaring problem with the grand scheme of building a global network in order to write the history of global nature and to establish the universality of the Deluge was that the Republic of Letters fell far short of global coverage. Even as it grew by leaps and bounds in the seventeenth and eighteenth centuries, it remained stubbornly Eurocentric. It mostly connected Europeans in different parts of Europe to each other and to Europeans abroad, whether in colonies, missionary or diplomatic outposts, or commercial entrepôts. The Republic of Letters was indeed transnational, even transregional, but it was hardly global. It therefore could not produce a truly global network of inquirers or a truly global data set—which the Flood, as a global natural phenomenon, required. When Scheuchzer attempted to compile all of the local and national natural histories that had been heretofore published in *Bibliotheca scriptorum historiae naturali omnium terrae regionum inservientium* (Catalog of natural histories of all regions of the earth, 1716), apparently in the hope that they would jointly cover the entire planet, he was sadly disappointed. Coverage of Africa, Asia, and the Americas was noticeably scanty, and the natural histories from those parts of the earth were for the most part not produced by local observers but by European travelers. Woodward's motley crew of fossil suppliers, meanwhile, were all European, even if they achieved greater diversity than was typical for the time by hailing from different European kingdoms and territories, including ones hostile to England, and from different Christian faiths. Despite its global ambitions, *Brief Instructions* functioned mostly to stimulate the flow of specimens and information from within the British Atlantic world.[59] In these and other ways, the global ambitions of European naturalists in the early Enlightenment were constrained by the reality of the less-than-global networks to which they had access.

Woodward's youthful plans of traveling abroad on a fossil collecting tour, as relayed to Lhwyd in a letter of 1691, foundered first on the shores of lack of funds and subsequently, paradoxically enough, on the shores of his growing international fame.[60] By the time his *Essay* was published in 1695, he seems to have abandoned his plans of making a Grand Tour, perhaps realizing that he could stay in London, and the fossils would come to him. He could adopt the scaling methodology of Steno and Leibniz, achieving a panoptic view of nature through collaborating with or co-opting the labor of others. The overweening scale of Woodward's ambitions matched perfectly with Steno's and Leibniz's lofty ideas about setting up a global network. The way he put their methodological principles into practice provides us with an illuminating window into the inner workings of the Republic of Letters, as a collaborative, competitive, prestige-driven, hierarchical, and not-actually-global network of knowledge exchange. The Flood's alleged universality built bridges across the religious and political divisions of Europe and Europe's empires, suggesting that this, more so than an truly global network, may have been the underlying motivation all along.

### *The Republic of Letters and the Rhetoric of Global Participation*

While the Republic of Letters did not achieve global coverage, the *idea* of a global network to produce global knowledge was deeply compelling, and in fact did much to stimulate the growth of the network. The rhetoric of universal participation articulated by Leibniz and others dovetailed nicely with the rhetoric of the Republic of Letters as a radically inclusive and egalitarian community, as the French scholar Noël Bonaventure d'Argonne's utterly typical definition demonstrates: "It extends across the earth and includes people of all nations, all ranks, all ages, [and] all sexes . . . all sorts of languages, living and dead, are spoken there."[61] Distinctions of political, religious, linguistic, socioeconomic, racial, and gender-based affiliation, which mattered so much to daily life in early modern Europe, were supposed to vanish upon entering the magical space of this "Republic." This was hyperbole, verging on falsehood, as all members of the Republic of Letters knew very well. The women, non-Europeans, Jews and Muslims, and members of the merchant and artisan classes who were occasionally able to join the Republic did so without fundamentally altering its social composition, which remained overwhelmingly elite, male, Christian, and European.[62] But the rhetoric of inclusivity was crucial to the self-definition of this imagined community. As such, the idea of a network that "extends throughout the world" offers us a further reason why the global Deluge, with its demand for a global network of participants, would have been so eagerly embraced by the Republic of Letters as a subject of collective inquiry. It reinforced their highest aspirations for the scale, scope, and character of this community to which they belonged.

The Flood's universality also assured a kind epistemic leveling that aligned perfectly with the Republic's self-professed egalitarianism. The feature of the Flood that invited the participation of people from across the globe—the fact that it had touched every part of the earth—was the very thing which guaranteed that no one local, situated knower was any more epistemically privileged than anyone else. There was no best place on earth to study the Flood. You could do it from anywhere; any point on earth was equally good as any other point. This notion of a global phenomenon that had done roughly the same kind of work everywhere, shaping the earth, changing the landscape, leaving its detritus, on every single part of the planet, offered a radically egalitarian vision of equal access to this object of knowledge. The Universal Deluge rhetorically invited the participation of people across the world by implying that anyone, anywhere on earth could contribute toward the goal of finding physical evidence of the Flood. A global Flood fit marvelously well with the Republic's rhetoric of inclusivity and equality. And while this levelling was not accomplished in practice, what it did do was to enable men of different countries, confessions, and classes to form relationships in the Republic of Letters that would have been otherwise difficult or impossible in other spheres.

The Flood, as an object of inquiry, was remarkably effective at swelling the Republic of Letters—if not to global dimensions, then at least to include a growing number of provincial Europeans and Europeans abroad. It was this second part of the Flood's alignment with the self-definition of the Republic of Letters—the notion that anyone, anywhere, could potentially contribute—that helps us to understand why Woodward's call for fossils was so effective at expanding the boundaries of the Republic of Letters even as it failed to produce a truly global network. While the call for fossils was transparently self-interested, it also appealed to the self-interest of men of letters eager to make contacts with influential men in other countries.

In the gift economy that supported early modern scholarship, the rarer the gift—be it a book, an instrument, an antiquarian artifact, or a natural specimen—the higher its value. One of the quickest ways for provincial, colonial, or merely unknown and ambitious people to establish relationships with powerful men in the Republic of Letters was to send them specimens: natural objects that might be abundant in the sender's locale but rare, and hence, valuable, in the recipient's locale. Fossils were rare relative to other types of exchangeable naturalia, and hence, unusually valuable as scholarly gifts. If one happened to possess an abundance of local fossils, they could easily be traded for foreign fossils or other rarities. When Petiver offered Lhwyd dried butterflies in exchange for "duplicate fossils"—that is, fossils of which Lhwyd possessed more than one kind—he did so with the knowledge that naturalists would be loath to part with a type of fossil of which they possessed only one example but would be happy to trade "duplicates" away in exchange for a new type of fossil from a country or continent not currently represented in their collections, or for some other type of equally valuable natural specimen.[63] The value of fossils on the scholarly gift market was only enhanced by the growing belief that they were remnants of the Flood, which in effect turned them into sacred relics as well as natural curiosities. In a letter of 1724, Scheuchzer thanked a correspondent in Basel for "augmenting my Cabinet" with a fossil: "Your beautiful present, which I received with pleasure, serves equally well as proof of the Deluge and of your kindness towards me."[64] Sending fossils could establish scholarly relationships of reciprocity and further the diluvial research agenda at the same time. These intersecting forms of value made fossils excellent presents in the gift economy that powered the Republic of Letters.

The rhetoric of global participation served the interests of established naturalists in European centers of power and learning, and it also served the interests of men in relatively more marginal parts of Europe, and in Europe's colonies, who wished to join the elite network of the Republic of Letters. Established naturalists in metropolitan Europe needed foreign correspondents to supply them with observations of foreign nature and above all with exotic specimens that would enhance their collections and with it, their scholarly profile. Meanwhile, ambitious men in Europe's provinces and overseas colonies were eager to act as their suppliers in exchange for admission to the Republic.

Woodward was one of many established naturalists in Europe in need of exotic fossils. He competed fiercely with two of his Royal Society rivals, Sloane and Petiver, for access to suppliers of fossils in the British colonies in North America.[65] When the Boston minister Cotton Mather sent word to the Royal Society of fossil teeth and bones unearthed in New York, Woodward and Petiver quickly dispatched letters to Mather soliciting specimens and further correspondence. Each one requested that Mather send any future fossil specimens directly to him and not to the other. Woodward seems to have won the exclusive rights to the fossils Mather was able to gather from in and around New England. Petiver retaliated by capturing the supply stream of another colonial collector, Hugh Jones, a minister living in Maryland. Colonial elites like Jones and Mather were eager to act as suppliers to their metropolitan counterparts. Mather hoped for, and eventually won, membership in the Royal Society largely thanks to his willingness to supply his London counterparts with fossils and other American curiosities—making him one of the first British subjects born in the colonies to gain this honor. Mather framed his relationship with Woodward as furthering their joint goals of pursuing Christian natural philosophy, telling Woodward in his first letter that "the true Friends of Religion, and Philosophy" were "obliged and even commanded" to "serve you with as many communications as they can, that may be subservious unto your Noble Intention." No doubt Mather, deeply pious as well as philosophically inclined, believed this sincerely. But his reasons for wanting to form a relationship with Woodward and the Royal Society were also self-interested. Fossils exchanged in the joint pursuit of Mosaic natural philosophy served an important social function as well, affording colonists in the Americas a point of entry into the European Republic of Letters.[66]

The exchange of fossils thus offered a valuable pretext for the formation of new social relationships between people living far distant from one another, which resulted in the geographic expansion of the network. This same dynamic repeated itself within the bounds of Europe, as aspiring men in peripheral regions sought to leverage specimens in their own locales in order to establish contact with scholars living in centers of power and knowledge. Even as Petiver battled Woodward for access to colonial suppliers in British North America, Petiver gained an unexpected foothold in the Swedish fossil market when Magnus von Bromell, a professor of medicine in Stockholm, sent him a box of Swedish fossils. "My curiosity is at present principally occupied by the natural things, vegetable and especially mineral, of the Kingdom of Sweden," he told Petiver in a 1709 letter, "and above all I am working on a collection of Figured stones of our country." Von Bromell described the gift to Petiver as "little jewels of our country" (*petites bagatelles de notre pais*), no doubt aware that their value as a gift rested chiefly on their place of origin.[67] In the early eighteenth century, Sweden was a relatively minor player in the Republic of Letters.[68] After the abdication of Queen Christina, a great patron of the arts and sciences in the mid-seventeenth century, Sweden lost ground vis-à-vis the other nations of Europe as a center of learning. Von Bromell would certainly have been aware that it was precisely the underrepresentation of Swedish scholars within this transnational network which guaranteed that Swedish fossils would be rare—and highly valued— outside of Sweden.

Without people to circulate them, local objects did not travel. In a 1710 letter to Scheuchzer, Woodward expressed a desire to establish correspondence with someone, anyone, in France because he had hardly any fossils from there, laying bare the assumption that the best way to get a fossil from a place was to know someone in that place.[69] The fact that von Bromell offered in his 1709 letter to send Petiver more fossils from the peripheries of Sweden—the island of Gotland, he suggested, or Lapland in the far north—shows that he well understood that the more remote the location from which he could source specimens, the higher their value would be in a foreign country and in a center of knowledge like London. Von Bromell's offer of fossils from the Swedish periphery also increased the likelihood that a powerful naturalist like Petiver would repay the gift by sending von Bromell news, specimens, books, journals, and—perhaps the most valuable currency in the Republic of Letters—letters of introduction to other eminent men of letters in Petiver's network.

It was not just the rarity but the continuing connection to the place in which it was found that increased a fossil's value on the gift market of European learning. The social importance of a fossil's place of origin in the early Enlightenment is well illustrated in a passage from one of Scheuchzer's many fossil catalogs, *Piscium querelae et vindiciae* (Quarrels and claims of the fishes, 1708). Describing a fish fossil from the collection of his Italian colleague and correspondent Antonio Vallisneri, Scheuchzer identified its country of origin by declaring that "Italy produced it." While lamenting that "the memory of the particular place" where it was found "has been lost," he also brushed this loss aside, claiming that the lack of this information "does not diminish at all the faith of its testimony, a relic which clearly shows the remains of the animal's corporeal substance." The fossil's form and structure, in other words, demonstrated its origin in a once-living fish, no matter where it had been found. So why mention place of origin at all? If the goal was simply to prove the organic theory of fossil origins, the country where this fish fossil was found should not matter. Nor was Italy far way enough from Zurich to be useful in proving the Deluge's global scale. I would argue that the best way to make sense of this perplexing passage is to contextualize it within contemporary practices of fossil exchange. Telling readers that the fossil had been found in Italy and was currently held in a private collection in Italy was an indirect but unambiguous way for Scheuchzer to brag about the extent of his transnational connections. Scheuchzer only knew of this Italian fossil because he personally knew a naturalist in Italy, who also happened to be famous. Being from Italy did not change the philosophical meaning of the specimen, but it did change the reader's perception of Scheuchzer's social status within the Republic of Letters. Fossils' place of origin came to stand in for the geographic scale of a naturalist's network, yet another source of their social value in the Republic of Letters.[70]

Moreover, the social dimension of fossil exchange was to a significant degree independent of the philosophical significance of the specimens being exchanged. Commitment to the Universal Deluge was by no means the only force driving the global collection of fossils, nor does it seem to have been necessary, given that many of the most eager fossil collectors and gifters were openly opposed to the theories that fossils originated in biomatter and/or in Noah's Flood. Edward Lhwyd, for example, was adamantly opposed to the notion that fossils were relics of the Flood and was nevertheless at the center of fossil collection and exchange in the British Isles. One did not need to commit oneself publicly to any side of the fossil debate or to any particular theory regarding the Flood in order to exchange specimens and build relationships in the Republic of Letters. They made excellent presents of immense social value no matter what the sender and

*Psciculi Diluviani figura in fissili Lapide candido ex Museo Vallisneriano.*

The fossilized fish described in the text of Scheuchzer's *Piscium querelae et vindiciae* (1708) as a product of Italy is here identified in the illustration's caption as a product both of the Flood and of Vallisneri's personal collection. Linda Hall Library of Science, Engineering & Technology.

recipient believed about their origins and meaning as natural specimens.[71] That was the beauty of exchanging fossils in the early Enlightenment: even though their circulation was spurred by the popularity of the Universal Deluge and undertaken under the aegis of a pious research program, it was entirely possible to participate in the fossil gift economy while remaining completely agnostic about the debates that raged about the true nature of fossils and the Flood. Fossil exchange was an excellent way of side-stepping philosophical controversy while bridging religious and political divides.

Fossils were like trading cards. Naturalists sought to collect as many different kinds as they could, not just by species but by country of origin. As the example of Scheuchzer's Italian fossil demonstrates, bragging of having fossils from many countries was a means of boasting by proxy of having correspondents in many countries. Correspondents, too, were collected like trading cards; the *éloges* published in learned journals frequently mentioned the size and scale of the deceased's network of correspondence. The social dynamics of fossil exchange forged new relationships, solidified old ones, and drove the geographical expansion of the network. The Universal Deluge was remarkably good at expanding the Republic of Letters.

## "A Deluge of Fire": The Flood and the Apocalypse

The growing popularity of the Universal Deluge in the late seventeenth and early eighteenth centuries following decades of skepticism and controversy surrounding its inclusion in natural philosophy flowed from its increasing alignment with the social practices and values of the Republic of Letters. Many scholars in this period felt that the Flood could bring together men of different countries, faiths, and philosophical persuasions in the shared pursuit of Mosaic natural philosophy via the long-distance exchange of fossil specimens. The clearest indication of the Flood's social utility to the Republic of Letters comes by contrasting its fortunes with those of its doppelgänger, the Apocalypse. The Flood and the Apocalypse were frequently paired in the temporal imaginary of early modern Europe.[72] Evidence for the similarity of these two global catastrophes, one in the planet's past and the other in its future, abounded in scripture, the classics, and natural philosophy. But there were key differences between the two as objects of study that militated in favor of the Flood and against the Apocalypse. As a future biblical disaster, the Apocalypse had not yet provided relics that naturalists might exchange with one another in order to accumulate social capital in the form of collections and correspondents. There could be no gift economy of the Apocalypse comparable to the one that promoted the study of the Deluge. Moreover, the Apocalypse exacerbated religious difference of opinion both between and within confessions by inviting scholars to consider who among them would be damned and who would be saved. The diverging fortunes of Flood and Apocalypse reveal that the Flood's empirical foundation and irenic potential was crucial to its popularity as a subject for transnational science in the early Enlightenment.

In order to appreciate the significance of this divergence, it is first necessary to understand the extent to which the Flood and the Apocalypse were paired in early modern European scholarship. One example is provided by Henry More, one of the seventeenth-century Cambridge Platonists, who likened the Apocalypse to a "Deluge of Fire" in *An Explanation of the Grand Mystery of Godliness* (1660). This was not simply a poetic figure of speech. More believed that the Apocalypse would be quite literally akin to the Universal Deluge in several key respects. Theologically speaking, the Apocalypse was similar to the Deluge in that it would be provoked by human sinfulness and would mostly exterminate the human species as punishment for those sins, thereby creating a spiritually pure world. From a philosophical point of view, the Flood and the Apocalypse were both planetary natural disasters, with one key difference: their elemental composition. "Universal Deluges," More argued, "doe argue the probability of a Deluge of Fire."[73] The Apocalypse, in other words, would be a planetary flood in which the element of water was switched out for the element of fire. The historical Flood both resembled and presaged the future Apocalypse. Scheuchzer's multivolume natural history of the Bible, *Kupfer Bibel* (1731–35), concludes with a discussion of the Apocalypse, which he described as the third and final of the major transformations in planetary history, the first two being the Creation and the Flood. The parallels between Flood and Apocalypse are especially notable;

"The final destruction of the Earth by fire," in Scheuchzer's *Kupfer Bibel* (1735). Charles Deering McCormick Library of Special Collections, Northwestern University Libraries.

Scheuchzer referred to them as two "great changes, the first by water, and the second by fire."[74]

There was ample scriptural support for the analogy between the Flood and the Apocalypse, especially if one conceived of the Apocalypse as a global fire. The Scottish scholar Matthew Mackaile declared in his 1691 theory of the earth: "I look upon the Deluge, as the baptizing of the Earth in the Clouds. . . . Moreover, I look upon the last Conflagration of the Earth, as the baptizing it with Fire," citing for support Matthew 3:11, "He shall baptize you with the

Holy Ghost, and with Fire."[75] Mackaile may have also had in mind a similar passage in the book of Luke in which John the Baptist declares: "I baptize you with water; but he who is mightier than I is coming, the thong of whose sandals I am not worthy to untie; he will baptize you with fire."[76] Many early modern commentators likened the Flood to a baptism, because the world emerged from the Flood newly washed of its sins. The Apocalypse would do the same, killing sinners and preparing a new world for a new race of moral innocents. The book of Matthew makes the analogy between Flood and Apocalypse even clearer: "For as the days of Noah were, so will be the coming of the Son of Man."[77] This passage was also noted by early modern commentators, for example in a commentary on the New Testament, published anonymously in London in 1719–25, which declares: "As [the world] perished then suddenly by those very waters . . . so it shall likewise perish at the last day by fire," to which was added the further cautionary reflection: "the Son of God shall surprise men like the deluge."[78] The conception of the Apocalypse as the second coming of Noah's Deluge, with fire instead of water, resonated with scripture and with the commentary tradition.

More's idea that the Apocalypse would be just like the Deluge, but with fire substituted for the water, was anticipated nearly a century earlier by the Ragusan writer Nicolò Vito di Gozze in *Discorsi . . . sopra la Metheore di Aristotele* (Discourses on Aristotle's *Meteorology*, 1584). The main interlocutor in di Gozze's dialogue predicts that a second Flood will come "at the end of the world" and that it "will be of fire, as the one that already happened was of water."[79] As a Catholic, di Gozze may not have had in mind the specific biblical passages that would later inspire More and Mackaile, though he may very well have recalled the widespread fears of an apocalyptic global Flood from earlier in the sixteenth century, which almost entirely collapsed the distinction between the Flood and Apocalypse by imagining the latter as a repeat performance of the former. The Reformation-era predictions of an imminent global Flood were widely disseminated from pulpits and in pamphlets, and were thus readily available to those who were, for whatever reason, not careful readers of scripture. Yet, as discussed in chapter 1, these imagined apocalyptic floods were widely condemned by philosophers and clerics, not least because many of them took seriously the scriptural passage that stated rather plainly that Noah's Flood was to be the last of its kind.[80] In spite of these objections, or perhaps because of them, the "Flood of Fire," strongly similar to the Universal Deluge but different in its elemental composition, grew in popularity from the sixteenth century onward.

Early modern philosophers even found confirmation for the twinning of Flood and Apocalypse in the natural philosophy of pre-Christian antiquity. Jacopo Grandi defended the view that both events were global and geologically transformative by invoking the Aristotelian notion of the power of water and fire, writing in 1676, "Changes can be made of the entire earth only through water and fire."[81] The following year, the English writer Matthew Hale discussed the "Opinions of some of the Ancients," including Plato, Aristotle, and Seneca, regarding catastrophic floods and fires in *The Primitive Origination of Mankind* (1677).[82] While Hale objected to the pagan idea of "Natural and Periodical Floods or Conflagrations," he felt that they served as indirect proof of the reality of a singular and supernatural Universal Deluge followed by a similar, singular Universal Conflagration.[83] The classical notion of successive floods and fires in the earth's history was reworked to fit Christian theology and history by Protestants and Catholics alike, lending additional support to the idea that a global flood and a global fire paralleled and verified each other.

Perhaps the strongest and clearest exposition of the symmetry of Flood and Apocalypse comes from Thomas Burnet, who picked up where his Cambridge mentor Henry More left off in philosophizing about "flouds of fire" in the second volume of *Sacred Theory of the Earth* (1689–90). Burnet's theory of the Universal Deluge is well known among modern historians, but his theory of the Universal Conflagration, which he considered to be of equal importance to this theory of the Deluge, has been relatively neglected. The enterprise of writing the earth's history, as Burnet imagined it, was just as concerned with the earth's future as with its past—a conviction mirrored in the publication history of *Sacred Theory*, whose first volume concerns the planet's past and present and whose second volume concerns its fiery future. "There is a great analogy to be observed betwixt the two Deluges, of Water and of Fire," he observed. "Just as at the Deluge, the Abyss broke out from the Womb of the Earth. . . . So we must expect new Eruptions, and also new sulphureous Lakes and Fountains of Oyl, to boyl out of the ground; And these all united with that Fewel that naturally grows upon the Surface of the Earth, will be sufficient to give the first onset, and to lay wast all the habitable World, and the Furniture of it." In both cases, the devastation has similar causes and follows a similar pattern. The Flood had begun with an earthquake, which caused subterranean waters to flood the surface of the earth. The Apocalypse would begin when volcanoes even more massive than the ones responsible for the recently discovered ruins of Pompeii and Herculaneum spewed fire from beneath the earth's surface until the entire globe was consumed in flames. The Flood and the Apocalypse had structurally identical roles in earth history as well as very similar causes: the Flood had destroyed the original earth to make way for the present one, and the apocalypse would in turn destroy the present earth to make way for the future one.[84]

Finally, as with the Universal Deluge, the Universal Conflagration reflected contemporary concerns about weather, climate, and political economy. "There is so much Coal incorporated" into the "Brittish Soyl," Burnet wrote, that "when the Earth shall burn, we have reason to apprehend no small danger from that subterraneous Enemy."[85] Burnet imagined Britain's coal reserves, currently being mobilized as a useful heating source during the coldest years of the Little Ice Age, becoming in the future a malignant planetary force that could incinerate the planet and everything on it.[86] It is remarkable to see a seventeenth-century observer of the origins of the fossil fuel economy link that development in any way to the future destruction of the planet and of living things. However, there are important disparities between Burnet's understanding of the link between coal and planetary "warming" and our own twenty-first-century understanding of that relationship.[87] According to Burnet, it was the coal remaining underground, unburned, that posed the greatest danger to humanity. The lived experience of heat and cold, both natural and man-made, indirectly informed the imagination of the Universal Conflagration, as it did for the Universal Deluge.

### *Flood versus Apocalypse: De-coupling Twinned Disasters*

The classical and biblical evidence for the pairing of Flood and Apocalypse helps to explain why it was so frequently invoked by a range of scholars, as does their joint utility in imagining epochal shifts in weather and climate. Yet the Universal Conflagration did not become a popular topic of transnational conversation in the Republic of Letters as did the Universal Deluge. As the fledgling field of earth history matured into the eighteenth century, the Flood became more and more prominent, while the Apocalypse became increasingly less visible. Not much has been written about this dimension of early modern theories of the earth, perhaps under the assumption that the Apocalypse's disappearance from an emerging field of scientific inquiry was natural or inevitable. However, given the strong affinities between the Apocalypse and its double, the Deluge, it should be regarded as surprising that the latter thrived for several more decades in mainstream natural philosophy while the former faded from view. Several interlocking factors motivated this contingent development, but a decline in apocalyptic belief does not appear to have been among them.[88] Rather, I argue, the Flood's persistence and the Apocalypse's demise can best be understood as a network effect of the Republic of Letters.

Predictably, the Universal Conflagration was subject to one of the same criticisms as the Universal Deluge: naturalizing it took power away from God as the sole determinant of when and how sinful humanity should be punished. As with the Flood, this criticism acknowledged the Apocalypse as an established future reality, just not one that philosophers had license to pursue. The Anglican bishop Herbert Croft, in his 1685 critique deriding Burnet as a "Deist," predicted that the second volume—still four years away from being published, though its contents could be guessed easily enough from *Sacred Theory*'s frontispiece— would be just as deistic as the first: "I would fain be satisfied concerning this Fiery destruction to come, and know how the Heavens and the Earth, they both being of so very different constitutions should both take Fire at once. . . .

I suppose he will stick to his own method of having Natural Causes for all things, and will not allow God the liberty to use any extraordinary means, tho upon such an extraordinary occasion as the Deluge or Conflagration."[89] Croft raised a philosophical objection to a global Conflagration—the impossibility of the planet's diverse elements all catching fire at once from a single cause—not in order to cast doubt on its future occurrence—"I know Gods Word hath said [it] shall [happen], and therefore I believe it"—but to argue that the Conflagration, like the Deluge, was necessarily miraculous and therefore inexplicable according to the methods of natural philosophy.

The publishing history of John Ray's *Miscellaneous Discourses concerning the Dissolution and Changes of the World* (1692) offers an especially illuminating case study of how philosophers who believed strongly in the coming Apocalypse, and who furthermore thought that it could be considered as a natural phenomenon, nevertheless participated in removing explicitly apocalyptic predictions regarding the planet's future from natural philosophy at the turn of the eighteenth century. This collection of philosophical "discourses" began its life as a sermon on the Apocalypse preached at St. Mary's Church, Cambridge, over thirty years previously, before Ray resigned his ministry when he refused to subscribe to the 1662 Act of Uniformity. The sermon's subject was the text of 2 Peter 3:11: "Since all these things are to be dissolved in this way, what sort of people ought you to be in leading lives of holiness and godliness," a passage that invited reflection on the relationship between the spiritual state of humanity and the natural state of the earth. The sermon languished among Ray's papers until the controversy surrounding Burnet's *Sacred Theory* inspired him to revise and expand it for publication, at the urging of his friends.[90] To his original reflections on the future "dissolution" of the world—a term borrowed from chemistry that was frequently used to refer to the earth's future destruction by fire—Ray added two subsections on the Creation and the Flood. But as Ray made clear in a letter to Lhwyd in 1691, it was first and foremost a book about the earth's future, to which sections on the earth's history had been added, describing it as a "Discourse concerning the Dissolution of the World" containing "Two large Digressions, one concerning the general Deluge in the days of Noah, another concerning the Primitive Chaos & creation of the World."[91] It was only in the second edition, when *Miscellaneous Discourses* (1692) was renamed *Three Physico-Theological Discourses* (1693), that the Flood approached equal stature with the Apocalypse. An equal number of pages were now devoted to the Flood and to the Apocalypse, and Ray reordered the material chronologically so that the Creation and Flood were clearly marked as separate sections that preceded discussion of the Apocalypse, bringing it into line with the narrative structure of Burnet's *Sacred Theory*. By this time, Ray was referring to it (again to Lhwyd) as "my Discourses concerning the Chaos, Deluge & Dissolution of the World."[92] By the time the posthumous third edition appeared in 1713, the expanded section on the Deluge was considerably longer than the original section on the "Dissolution."[93] In short, Ray's *Discourses* went from being focused on the Apocalypse, to being equally about the Apocalypse and the Deluge, to being focused on the Deluge.

There is no evidence to suggest that this shift in emphasis was caused by a weakening of Ray's convictions regarding the future reality or spiritual necessity of the Apocalypse. There is, however, evidence to suggest that the shift was propelled in part by a growing skepticism about whether the Apocalypse was a subject that could be approached through natural philosophy. "Scripture and Tradition" assures us, he wrote, that the future "dissolution" of the world would be effected by "that Catholick Dissolvent, Fire."[94] While insisting on the one hand that the world's dissolution by fire would be "the Work of God, extraordinary and miraculous," he also insisted that the "Means and Instruments" of that fiery blaze "shall be natural."[95] Ray discussed several possible natural causes of a Universal Conflagration, including the eruption of the planet's fiery core, floods of liquid fire, or the ignition of the hot, dry air in the torrid zone. But he refused to endorse any of these hypotheses, claiming that there was insufficient evidence to decide among them. As far as Ray was concerned, the Apocalypse would be a natural phenomenon, in the sense that it would have both natural causes and effects. But it was also a phenomenon about which very little could be known in advance. Its date was "absolutely uncertain and indeterminable." Equally uncertain were the natural qualities and characteristics of the Apocalypse, beyond the fact that it would be fiery. Ray ridiculed those who devoted time and energy to trying to find out anything about it, "For since this Dissolution shall be effected by the extraordinary Interposition of Providence; it cannot be to any Man known, unless extraordinarily revealed."[96] He reserved special scorn for the prophets whose apocalyptic deadlines had come and gone: "But yet notwithstanding this, many have ventured to foretel the Time of the End of the World, of whom some are already confuted, the Term prefixt being past, and the World still standing."[97] Ray had plenty of contemporary examples to choose from, such as the Dissenting preacher Thomas Beverley, who for years predicted the Apocalypse would begin in 1689 and then moved it to 1697 when 1689 failed to witness the world's end.[98] Ray's condescension toward those who claimed to be the lucky recipients of revealed knowledge about the Apocalypse implied that predicting the future was something a respectable philosopher should approach with extreme caution and humility, and perhaps not at all.

Ray was not alone in his belief that the obscurity of the earth's future meant that it could not be a proper subject of philosophical inquiry. Mackaile readily admitted that the Apocalypse was coming and that it would be a profoundly disruptive event in nature's history. He also recommended that philosophers stick to talking about the earth's history, "It being much easier to give an Account of what is past, than of what is to come."⁹⁹ Mackaile's pronouncement about the ease of learning about the past relative to the future may sound like a banal observation to twenty-first-century ears, but it would not have seemed quite so obvious to a seventeenth-century reader. Chronologers regularly included future events in their timelines of universal history.¹⁰⁰ The future was difficult to discern, undoubtedly, but so was the past. Ancient history, in particular, was widely thought to be shrouded in mystery and perhaps unrecoverable.¹⁰¹ Where previously the past and future had both been considered obscure, they were increasingly seen as epistemically distinct, with knowledge of the past capable of attaining a kind of certitude that knowledge of the future could not hope to match.

In the 1670s, a young Isaac Newton recorded the following observation in his private notebooks, which captures well the ambivalence about whether the past and the future were on a similar epistemic footing: "For if Historians divide their histories into Sections, Chapters and Books at such periods of time where the less, greater, and greatest revolutions begin or end; and to do otherwise would be improper: much more ought we to suppose that the holy Ghost observes this rule accurately in his prophetick dictates since they are no other than histories of things to come."¹⁰² In this brief passage, part of an unpublished treatise on Revelation, Newton sought to clarify the similarities and differences between history and prophecy. On the one hand, his definition of prophecy as a kind of history—specifically as a collection of "histories of things to come"—suggested that history was a continuum, stretching from the past into the future. Writing a history of the world that began with the earth's creation and ended with its future destruction made perfect sense in this way of thinking about things. On the other hand, Newton distinguished between the temporal scope and authorship of history and prophecy. History was written by humans and was concerned with the past. Prophecy was dictated to human writers by the Holy Spirit and was concerned with the future. The epistemic privilege enjoyed by the Holy Ghost made his "prophetick dictates" more trustworthy than anything a human historian might produce, but it also suggested that mortal historians possessed neither the knowledge nor the authority to write the history of the future.¹⁰³

By 1720, the unknowability of the future was coming to seem more categorical, and the distinction between past and future history began to overlap with a reemerging divide between philosophy and faith. Samuel Catherall's *Essay on the Conflagration in Blank Verse* (1719), a dialogue between a human and an angel, is specifically directed against the inclusion of the Apocalypse in earth history. The unnamed human speaker desires to know more about the coming Conflagration, and the angel appears to him in order to warn him away from his search for knowledge on the subject. Catherall was clearly sympathetic to Burnet's treatment of the Apocalypse, praising him for "having shown such a masterly Genius in his Account of the Conflagration" and describing this future event as

a time when the Earth "lies Delug'd, but in Fire."[104] The human speaker is eager
to know and understand the precise contours of the earth's future, wanting to
know exactly how the earth will catch on fire. The angel repeatedly reprimands
the man for his philosophical curiosity and for his correlative lack of faith, com-
manding him to "Control thy Appetite to understand / Things too sublime, and
Truths involv'd in Clouds."[105] Trying to determine the date of the Apocalypse
or the form it would take was an unacceptable transgression into the domain of
divine providence.

The early decades of the eighteenth century witnessed philosophers in Catholic Europe join the chorus of public denunciations of philosophical inquiry
into the earth's future. The French engineer Henri Gautier offered a speculative
theory of the earth's ancient destruction by water while casting doubt on the
knowability of the earth's future destruction by fire in *Nouvelles conjectures sur
le globe de la terre* (New conjectures on the globe of the earth, 1721). Repeat-
ing the increasingly common sentiment regarding the obscurity of the earth's
future, Gautier declared: "There are limits on what we are able to know."[106] The
Flood, on the other hand, was open to philosophical investigation in spite of its
ancient murkiness. "The Deluge," Gautier wrote, "was one of the epochs that
has most greatly disfigured the earth since its Creation by the Lord." Gautier
did not claim to know exactly what the Flood had done to the face of the earth,
but, significantly, he felt free to speculate that it had "formed continents where
there were perhaps only vast seas, and forged great expanses of sea where perhaps there were only continents."[107] The obscurity of the earth's ancient past
was not an impediment to philosophizing about it; the obscurity of the earth's
future, however, was. A fossil collector and professor of medicine at the ultra-
Catholic University of Würzburg, Johann Beringer, cautioned the readers of
his *Lithographiae Wirceburgensis* (The Figured Stones of Würzburg, 1726) about
"the ever-present danger that their ardent studies and their scrutiny of the future
may lead them to illicit practices of divination, to pry into the hidden counsels
of Providence." The future was not simply obscure; it was the province of God
and the realm of sacred mystery, to which humans were rightly barred. Beringer
therefore recommended that philosophers limit their investigations to the earth
as it was in the present day, leaving aside their "conjecturing of future events."
Beringer, Gautier, and Catherall's words of caution testify to a growing disas-
sociation of the past and future, Flood and Apocalypse, the former firmly em-
placed in the sphere of knowledge while the latter was relegated to the realm of
faith and mystery.[108]

## Scholarly Sociability and the Fate of the Apocalypse

The emerging sense that the earth's past was more certain than its future seems
to have played a key role in persuading scholars that discussions of the Universal

Conflagration, and of nature's and humanity's joint future more generally, were not suitable subjects for natural philosophy. The superior knowability of the past over the future was not, however, self-evident to early moderns. Rather, I would argue, the uncertainty of the earth's future had to be carefully constructed and persuasively propagated. A more fundamental driver behind this shift in epistemic values was the superior social function and utility of the Flood relative to the Apocalypse as a subject for transnational science. The Apocalypse lacked the empirical foundation and hence the social utility that made the Deluge such an attractive object of inquiry in the Republic of Letters. Fossils, while eloquent on the subject of the earth's past, were mute on the subject of its future. Nor was there anything to take their place. As a future biblical disaster, the Apocalypse had not yet left traces in nature for people to discover. The absence of natural evidence for the history of things to come, in Newton's evocative phrasing, left the natural philosopher with very little to say on the subject of the earth's future and nothing at all to trade within the gift economy of the Republic of Letters. Studying the Apocalypse could not enhance one's natural history collection or personal reputation. Nor did it appear to have the power to bring ever more people into the system of circulation and exchange on which expansionist natural history depended.

Equally problematic as its lack of associated gift-objects, the Universal Conflagration was religiously divisive, separating men of different religious persuasions, whereas the Flood was able to bring them into an uneasy alliance. This was just as true within as between confessionalized spaces, as evidenced by the heated debates around Burnet's Apocalypse between members of the Church of England and also between Anglicans and Dissenters. Although the first volume of Burnet's *Sacred Theory* garnered several admirers on the Continent, including the French philosophers Père Nicholas Malebranche and l'Abbé de la Pluche, the second volume, in which he claimed that the Universal Conflagration would start in Rome because it was "the Seat of Antichrist," did not find the same warm reception.[109] Burnet's jab at Roman Catholicism—typical in English religious discourse, which frequently identified the pope with the Antichrist—would not have appealed in the same way to Catholic readers or to the many Protestants who were trying to distance themselves from the religious and political radicalism with which apocalyptic predictions were increasingly tainted.[110]

As the previous chapters have been concerned to demonstrate, one of the main draws of the Universal Deluge as a topic of philosophical inquiry in the sixteenth and seventeenth centuries was its utility in exploring religious questions about sin and salvation. As this chapter has argued, the Flood became even more popular at the turn of the eighteenth century, when its growing association with fossils allowed naturalists to leave some of those religious polemics behind in the shared pursuit of an irenic Mosaic natural philosophy. The Apocalypse divided

Christian scholars of different churches and confessions by raising the uncomfortable question of who would be saved or resurrected into new life; its futurity meant that the division of all humanity into the saved and the damned, the living and the dead, was bound to correlate with present-day religious divisions. By contrast, there were no Protestants or Catholics or indeed any Christians in the days of Noah, meaning that philosophers from diverse Christian churches could safely discuss the salvation of Noah's family and the killing of everybody else without ruffling any feathers.

The Apocalypse's religious divisiveness and lack of material traces, contrasted with the Flood's irenic potential and abundance of tradeable tokens, reveals how the forces of long-distance intellectual sociability shaped discussions of the earth's history in the early Enlightenment. The transnational context in which discussions of the Universal Deluge and Universal Conflagration took place is critical to understanding why one of these biblical disasters was deemed too obscure and too religious for philosophical conversation, while the other—equally obscure and equally religious, not to mention equally imagined—was enthusiastically embraced by many. Foreclosing this future-oriented research agenda was a momentous step in the evolution of geology as a discipline in the eighteenth century, when it came to be defined as the study of the earth's past alone. One can imagine that the science of predicting the joint future of nature and humanity might have flourished under a different set of social circumstances, one that might have made it less contentious for the earth and atmospheric sciences to reorient themselves back toward the future in the modern era.

As a subject of inquiry, the Universal Deluge aligned in multiples ways with the norms, goals, and social dynamics of the Republic of Letters. It allowed Europeans to imagine they were studying the history of nature on a global scale and to cooperate with each other across deep divisions in European society in pursuit of that goal. Noah's Flood was not just an opportunity for pious scholars to demonstrate the harmony of their faith with the methods of the new science. Nor was it only an opportunity to clarify for themselves and to argue with one another questions about sin and salvation. It was also an opportunity to build relationships, to build social capital, to talk to themselves about themselves, to define communities of knowledge-making, and to further expand the networks that made natural knowledge possible in the seventeenth and eighteenth centuries. This imagined global natural disaster was, in many ways, an ideal subject for this imagined global community.

The disappearance of explicitly apocalyptic content from histories of the earth around the turn of the eighteenth century foreshadow the subsequent waning of the Universal Deluge as a popular topic for transnational philosophy in subsequent decades. The early Enlightenment witnessed a renewed commitment, harkening back to the Renaissance, to policing the boundaries between "natural" and "supernatural" phenomena and between philosophy and faith.[111] As the next chapter shows, the Universal Deluge, like the Universal Conflagration, began to slowly fade from natural philosophy, not because people lost faith in it but because they no longer deemed it useful as a means of transnational intellectual exchange. The promise of the Flood as a means of irenic collaboration no longer shined as bright. Once the network was up and running, the specific content that helped to build it was no longer necessary.

# CHAPTER FIVE

# Catholic Climate Change
## Heritable Sin and Strategies of Toleration

One of the last large-scale plague epidemics in western Europe broke out in 1720 via the French port city of Marseille. In January 1721, the Italian naturalist Antonio Vallisneri, a practicing physician and professor of medicine at the University of Padua, drew an explicit parallel between the biblical Flood and the present-day plague in a letter to one of his many correspondents, the Swiss Huguenot naturalist Louis Bourguet.[1] "I hear bad news about the Contagion, which spreads in all directions," Vallisneri wrote. "I tremble with fear, because the world is so polluted [*isporcato*], and in place of the Flood God is accustomed to send a Contagion every so often in order to purify it."[2] The pollution Vallisneri referred to was, of course, spiritual.[3] The world was polluted by sin, and only a plague, or a Flood, could wash it clean. Like most Europeans of the early modern period, Vallisneri regarded recurrent outbreaks of plague as a consequence of sin, one of several means by which God punished humanity via the medium of the natural world.[4] By virtue of their sin, humans had in some real sense provoked the plague that was presently afflicting them, he suggested, just as humanity in the days of Noah had provoked God to destroy and thus cleanse the world with a global deluge.

Vallisneri's comparison of the 1720 plague to the biblical Flood reflected the influence of the recent theories of Thomas Burnet and John Woodward, British scholars from a previous generation who influentially argued in the 1680s and '90s that sin caused the Flood, which in turn caused the precipitous and long-lasting ruin of the planet, its climate, and human health. Vallisneri and Bourguet were both familiar with Burnet's *Sacred Theory of the Earth* (1681–89), but they were especially close readers of Woodward's *Essay toward a Natural History of the Earth* (1695), which became popular across the Continent after being translated into Latin in 1704 by the Swiss naturalist Johann Jakob Scheuchzer. The resonance between Vallisneri's views on the Flood and those of his older British counterparts reflected the epistolary labor and translation work of Swiss scholars like Scheuchzer and Bourguet, who acted as key nodes in the transnational exchange of knowledge between Britain and Italy via Switzerland.

At the same time, Vallisneri was a critical reader of the British tradition of earth history, which he consumed through the mediation of his Swiss interlocutors. Vallisneri reworked the diluvial theories of his Protestant colleagues in Britain in his own equally influential work of earth history, *De' corpi marini che su' monti si trovano* (Of marine bodies found in the mountains, 1721), whose title refers to the philosophical problem of marine fossils discovered in land-locked mountains. Published in Venice as the plague began to subside, Vallisneri's book echoes Burnet and Woodward's idea that humanity was responsible for the Flood, via sin, and consequently for their own postdiluvian decline. The Flood transformed the global climate and thereby unleashed sickness and debility on the world, physically transforming the human species. But Vallisneri took issue with the idea that the Flood radically and permanently transformed nonhuman nature. He argued that the Flood did not affect the earth geologically at all and only changed it climatologically for long enough to ruin human bodies, which remained damaged even after the global environment recovered. Vallisneri's expertise as a medical professor, practitioner, and experimentalist informed his novel theory about the heritability of embodied sin and trauma, which rendered the human body, not planet Earth, the site of divine punishment and the continuing reminder of humanity's sinful nature. Whereas Burnet and Woodward imagined global nature as a human artifact, Vallisneri imagined the mortal human body as humanity's greatest unintended creation.

The previous chapter showed how the Flood seemed to become a promising foundation for transnational philosophical exchange around the turn of the eighteenth century, especially when based on the long-distance exchange of fossil specimens. This chapter shows how that cosmopolitan promise began to unravel, as confessional and national differences insinuated themselves into long-distance conversations about the Flood and the history of the earth. I focus on the rich epistolary archive, contradictory writings, and fascinating lifeworld of Antonio Vallisneri, one of the most celebrated naturalists of his generation in Italy, paying particular attention to the ways in which his strategies for negotiating philosophy and faith were shaped by his epistolary relationships with Swiss Protestants and Italian Catholics. I situate Vallisneri's rewriting of Protestant British histories of the Flood in the context of three major intellectual, political, and religious forces affecting northern Italy in the early eighteenth century: Italian cultural nationalism, Italo-Swiss transnational collaboration, and the Catholic Enlightenment. The first half of the chapter examines Vallisneri's alternative theories of bodily decline following the Flood and Fall in light of his religious and political commitments. Arguing that human sin ruined humanity alone, Vallisneri attempted to undo the synthesis of human and natural history that British philosophers writing about the Universal Deluge had tried to achieve.

In unpublished correspondence with colleagues from Catholic Italy and Protestant Switzerland, Vallisneri frequently argued for ejecting the Flood from natural philosophy altogether—less a revision of British theories than a complete rejection of their foundational premise. The second half of the chapter situates the reemergence of the medieval stance against a natural and global Flood in the advent of the Catholic Enlightenment in Italy and the continuing challenge of figuring out how to write an orthodox history of the Flood, a problem no less vexing in 1720 than it was in 1580. Renewed skepticism about the Flood also arose out of the complicated social dynamics of the Italo-Swiss network that brought together Catholic scholars from northern Italy and Protestant scholars from Switzerland in the early decades of the eighteenth century. This "Alpine Republic of Letters" worked on different terrain and according to different logics than did the network of fossil exchange detailed in the preceding chapter, which connected Britain to other Protestant places, including Switzerland, Sweden, Germany, and the British colonies in North America. The confessional as well as national diversity within the Alpine Republic of Letters bound these scholars together but also structured the conversation that could happen within that space. Much as Vallisneri valued his Swiss correspondents and shared their depth of religious feeling, he frequently tried to cajole, argue, silence, or shame them on subjects that veered uncomfortably toward religion. In correspondence with his coreligionists and compatriots, Vallisneri embraced the Flood as a point of Catholic faith while disputing its place in cosmopolitan natural philosophy. If his revised climate history of the Flood again separated human and natural history, Vallisneri's arguments against the Flood as a topic for philosophy made a broader case for disentangling philosophy from faith.

Catholic Enlightenment meant searching for new accommodations between faith and philosophy within the Catholic community and between the confessionalized states of Europe, as scientific exchange grew increasingly international in the eighteenth century. The diversity of opinion about the Flood within Catholic Italy attests to the challenges of pursuing pious natural philosophy in local and transnational contexts. This chapter reveals the essentially religious motivations behind the campaign in Enlightened Italy to displace the Flood from its prominent position in histories of the earth. While the religious controversy of the Reformations and the emergence of irenic Mosaic natural philosophy had helped to bring the Universal Deluge into natural philosophy in the late sixteenth century, the emergence of new forms of confessional and patriotic feeling in the early Enlightenment helped to push it back out again.

## Patriotism, Transnational Exchange, and the Catholic Enlightenment

Like his Paduan predecessor Camilla Erculiani, Vallisneri formulated his thoughts about the Universal Deluge in dialogue with foreigners and Protestants. *Of Ma-*

*rine Bodies* in its full title specifies that the book is a collection of "critical letters" regarding "the state of the world before, during, and after the Flood." Like Erculiani's *Letters on Natural Philosophy*, Vallisneri's *Of Marine Bodies* is dialogic, the record of exchanges of ideas between the author named on the title page and his far-flung correspondents, writing to one another across space and time about the planet and its history. Although the letters contained in *Of Marine Bodies* were all written by or addressed to other Italians, they make clear that British histories of the world from the 1680s and '90s were Vallisneri's main point of departure. Through his extensive correspondence with the Swiss naturalists Scheuchzer and Bourguet, Vallisneri learned about and debated key works on the Flood by Burnet, Woodward, and Whiston. He was not in touch with Woodward directly, but he received regular updates on the elder naturalist from their mutual correspondents Bourguet and Scheuchzer.[5] Scheuchzer's Latin translation of Woodward's *Essay* was the means by which Vallisneri and his colleagues on the peninsula accessed this key work, probably the most influential account of the Universal Deluge in Europe in the first three decades of the eighteenth century.

Vallisneri's letters about the Flood in *Of Marine Bodies* were written in conscious opposition to his British predecessors and to Woodward in particular, whom he viewed as profoundly mistaken about the Flood's effects on the world and unjustifiably popular among their contemporaries. Vallisneri complained to Bourguet in a letter of 1719 that "almost everyone living north of the Alps, and also many Italians," believed in Woodward's theory of the Flood.[6] Germany, he asserted in a subsequent letter, was being overrun by "sworn Woodwardians."[7] But he was especially disappointed in his countrymen who embraced Woodward, lamenting in a 1721 letter, "Here in Italy, Signor Monti, Signor Zanichelli, and many others hold as infallible truth this business of marine fossils all coming from the Flood."[8] The Bolognese naturalist Giuseppe Monti was one of several Italians in the early eighteenth century who saw Woodward's circuit of fossil exchange as a promising means of studying natural philosophy across political and religious boundaries. Monti, a correspondent of both Scheuchzer and Bourguet, confidently attributed a fossilized walrus skull found near his native city to the Universal Deluge in his *De monumento diluviano nuper in agro Bononiensi detecto dissertatio* (Testimony of the Deluge recently discovered in the territory of Bologna, 1719). The Venetian apothecary Gian Girolamo Zannichelli was somewhat more ambivalent, discussing Woodward's theory of the Universal Deluge as one of several possible explanations for the origin of figured stones in *De lithographia duorum montium Veronensium* (Description of the stones of two mountains of Verona, 1721).[9] Monti and Zanichelli indicate that there was a range of opinion on the British theory of the Flood in Catholic Italy, rather than the factionalized camps of pro- and anti-Woodwardians that Vallisneri imagined. Woodward's obvious borrowing from Italian natural histories of the late seventeenth century perhaps made it seem less like a foreign import to some early-eighteenth-century Italians.

Vallisneri's paranoid fantasy that his colleagues in Italy and Europe were all turning into "sworn Woodwardians" nevertheless highlights important aspects of the intellectual, religious, and cultural milieu of northern Italy in the early decades of the eighteenth century. First, it indicates the strength of a rising form of cultural nationalism in northern Italy, especially in and around the Veneto. Vallisneri was part of a movement gathered around the Venetian *Giornale de' Letterati d'Italia* (Journal of the learned of Italy, 1710–40) and including Scipione Maffei, Apostolo Zeno, and Lodovico Antonio Muratori, who sought to reinvigorate Italian scholarship and establish Italy's preeminent position within the Republic of Letters.[10] A founding editor at the *Journal*, Vallisneri in *Che ogni Italiano debba scrivere in lingua purgata italiana* (Every Italian should write in good, clear Italian, 1722) exhorted his countrymen to write in the vernacular and to thereby promote "the fame and glory of Italy, the former dominatrix of all nations who is now herself dominated, and abjectly servile."[11] While much of his patriotic ire was aimed at the French—denouncing the rising use of the French language as Europe's new lingua franca, he complained in *Every Italian* that "the mangy French . . . want to make everyone wear their clothes and speak and write in their language"—Vallisneri's fears about increasing numbers of "sworn Woodwardians" may have also reflected anxiety about undue British influence on Italian scholarship.[12] *Of Marine Bodies* might help return Italy to the glory days of the 1660s and '70s, when Scilla, Boccone, and Steno (the honorary Italian) were the acknowledged European authorities on fossils and the Flood. Not all Italians were enthused about this patriotic agenda; the antiquarian and astronomer Francesco Bianchini, declining Muratori's invitation to join his proposed pan-Italian academy, accused the scholar-patriots of creating a "Literary League of nation against nation" that would alienate their colleagues *oltramarini* ("across the sea") and *oltramontani* ("across the mountains").[13] But many of their compatriots welcomed the move. Giovanni Giacinto Vogli, a physician and professor at the University of Bologna, assured Vallisneri in 1721 that *Of Marine Bodies* would bring "as much honor to Italy as it will benefit the world of Letters."[14]

Second, Vallisneri's efforts to check the international popularity of Woodward and his British cohort by establishing his own international reputation highlight the paradoxical ways in which cultural nationalism fostered transnational exchange.[15] The logic of the Republic of Letters dictated that foreigners were needed to judge the value of a country's intellectual output.[16] Vallisneri's philosophical writings could not bring honor to Italy as Vogli promised unless they were read and praised by men of other nations, and that would be difficult as long as Vallisneri and his compatriots insisted on publishing in the vernacular. In order to raise Italy's profile in the Republic of Letters, they relied heavily on their correspondents in the polyglot and polycentric Swiss Confederation, where the absence of a common vernacular made its inhabitants unusually good linguists and able translators. Scheuchzer, his brother, and his son all helped to bring Italian scholarship to the attention of British scholars by translating and sending works by Italian authors to the Royal Society.[17] Bourguet and several of his colleagues in western Switzerland founded a journal, *Bibliothèque Italique* (Italian library, 1728–34), for the sole purpose of informing a broad European readership, in French, about the best new scholarship coming out of Italy.[18] The Alpine range that helped give Italians their sense of distinction from the rest of Europe—a common term for foreigners in early modern Italian, *oltramontani*, literally means "the people who live on the other side of the mountains"—also formed a physical barrier that made Swiss contacts critical in physically moving the objects of knowledge out of Italy, across the Alps, and into northern Europe. The Swiss scholar Jean Le Clerc instructed Vallisneri to have his publisher in Venice send copies of his books to a bookseller in Turin who would send them on to Le Clerc in Geneva; Le Clerc promised to relay them to a scholar of his acquaintance in Paris whom Vallisneri wished to impress.[19] But Italian reliance on Swiss intermediation cut both ways, bringing scholarship from northern Europe into Italy that did not always gel with the political, intellectual, or religious commitments of Italian savants.

Finally, Vallisneri's fears about Woodward's influence spreading south of the Alps reflected the challenge of crafting a distinctively Catholic science, as well as a distinctively Italian science, at the dawn of the Catholic Enlightenment.[20] Woodward was a Protestant as well as a foreigner, as were his Swiss promoters; a mistrustful observer might perceive not only creeping British hegemony in earth history but an emerging Protestant alliance that was successfully recruiting Catholics to their cause. When Monti told Vallisneri how, after much deliberation, "I finally gave myself over to the opinion of the Universal Deluge, as the celebrated Dr. Johan Jakob Scheuchzer and many others have also done," Vallisneri's response was so chilly that Monti never mentioned his allegiance to Woodward, Scheuchzer, or the Flood again.[21] In many ways, the project of Catholic Enlightenment in Italy was defined by the quest to find a new accommodation between science and religion, one that would take into account changing religious sensibilities in the new century as well as the scientific advances of the previous one. The Mosaic natural philosophy of previous centuries was still alive, but it had to change to survive. The spread of Newtonianism across the Continent in the first decades of the eighteenth century challenged Italian philosophers to decide whether and how to accept the Newtonian synthesis of science and religion along with Newtonian physics.[22] In similar fashion, Italian philosophers seeking to understand the earth's history were forced to contend with the synthesis of science and religion that Newton's British colleagues aimed to achieve with their histories of the Universal Deluge.

Vallisneri was deeply critical of the ways in which his British predecessors related science and religion in their histories of the Flood. In a letter to the mathematician Guido Grandi, he dismissed the most Newtonian of the British world histories, Whiston's *New Theory of the Earth* (1696), as "a dream, or a romance, like that of the soothsayer Burnet."[23] Writing to Bourguet, he accused Woodward of having "very nearly forgotten to look at the Book of Nature" when composing his *Essay toward a Natural History of the Earth*, so preoccupied was he with "the words of Sacred Scripture." This flawed methodology led Woodward into "ridiculous errors," because the Book of Nature, the unequivocal "voice of God," "never errs and is less difficult to understand than the Book of Sacred Scripture," which "can be interpreted in many different ways."[24] Galileo made a very similar argument, citing St. Augustine, about the ease of understanding the Book of Nature relative to the difficult-to-interpret Book of Scripture in *Letter to the Grand Duchess Christina* (1615). Channeling his famous forebear and predecessor at the University of Padua, Vallisneri accused his British colleagues—and implicitly, his Swiss ones—of unduly relying on scripture when writing about nature.[25]

But that did not mean that Vallisneri's own history of fossils and the Flood was secular or irreligious. The Veronese physician Sebastanio Rotari, to whom one of the letters in *Of Marine Bodies* was addressed, wrote to Vallisneri after the book's publication to congratulate him for having "perfectly inhabited the person of the Philosopher and of the Catholic." In Rotari's estimation, Vallisneri had spoken "with the freedom of a Professor of Philosophy" and at the same time "with modesty and Christian honesty."[26] In his study of the Catholic Enlightenment, Massimo Mazzotti situates the pious Milanese mathematician Maria Gaetana Agnesi's admiration for Vallisneri within a broader group of "Enlightened Catholics [who] believed that Vallisneri's work might provide a framework for harmonizing religious orthodoxy with the modern natural sciences."[27] Many of Vallisneri's compatriots and coreligionists regarded him as a role model of an Enlightened Catholic philosopher.

In seeking to advance a uniquely Catholic and Italian perspective on earth history, Vallisneri used letter-writing as the primary means of challenging the theories of his British predecessors regarding the Universal Deluge. In his published letters in *Of Marine Bodies*, Vallisneri articulated a substantially different history of the impact of the global environment on human bodies after the Flood. In unpublished correspondence with colleagues both Catholic and Protestant, Italian and foreign, Vallisneri sought an accommodation between his philosophy and his faith that extended so far as to argue for exiling the Flood from philosophy altogether.

### Catholic Climate Change: Vallisneri's Waterworld and the Heritability of Sin

Burnet and Woodward both tried to unify human and natural history on a planetary scale, as discussed in chapter 3, and both of them appealed to the Flood as a motor of environmental and human decline in order demonstrate that unity. Vallisneri's effort to craft a pious and patriotic account of the world's history began with revising key aspects of the climate histories of his British colleagues in order to challenge their synthesis of human and natural history. In a long letter in *Of Marine Bodies* addressed to his fellow *Journal* editor, abbot, and librarian Girolamo Lioni, Vallisneri rejected British claims that the Flood wrought permanent changes in the global environment. Instead, he argued, it did lasting damage to human bodies. The Flood temporarily altered the earth's air and water for just long enough to cause permanent changes in human health and reproduction. By emphasizing the human suffering in the wake of the Flood, Vallisneri shone a spotlight on the human responsibility for causing it. By emphasizing the full recovery of global nature after the Flood, he insisted that nonhuman nature should not be made to suffer for human sins. He thereby challenged the unity of human and natural history that both Burnet and Woodward sought to demonstrate in their histories of the Flood. The Italian Catholic philosopher absolutely rejected Burnet's notion of the earth as a human artifact, but he underscored more deeply than any of his Protestant colleagues the degree to which humans were responsible for the sick, weak, and vulnerable bodies they had been forced to live in since biblical times.

Burnet had previously identified the degradation of the planet's air as the key driver of illness, mortality, and debility on the postdiluvian earth. Woodward altered Burnet's theory to focus on the degradation of the earth's once-fertile soil as the major cause of postdiluvian human decline. Vallisneri chose to focus on changes in the planet's water and to explore their effects on human health after the Flood from within the framework of practical and academic medicine. In his long letter (really a short treatise) to Leoni, Vallisneri puts forward an elaborate hypothesis about the health hazards of the environmental disaster zone left behind by the Universal Deluge. The floodwaters receded slowly, leaving the ground "soggy," "filthy," and "corrupted." The earth's surface was dotted everywhere with stagnant pools, many of which were contaminated by the "corpses of men and beasts." In yet another implicit parallel between Flood and plague, Vallisneri the physician drew attention to a problem ignored by his colleagues but obviously entailed by the theory of a global deluge, namely, that there must have been hundreds of thousands of dead bodies left behind and almost no one to bury them. The effluvia rising from the fetid, corpse-filled lakes further degraded the quality of the air. "After the Flood," Vallisneri speculated, "the Atmosphere was filled for a few years at least with putrefying, agitated, and very heterogeneous particles," causing the planet to be blanketed by "ubiquitous, putrid, and unhealthy fogs." The muddy water, waterlogged soil, and humid air made the postdiluvian global climate significantly less healthy.[28]

This theory illustrates well how medical discourses on local climate in relation to health and disease among specific human populations informed thinking about the global climate and its effects on the human species as a whole. Vallisneri's vision of a swampy postdiluvian earth in *Of Marine Bodies* bears a marked similarity to a book on swamps published several years earlier by his friend and correspondent Giovanni Maria Lancisi, a physician in Rome. Lancisi's *De noxiis paludum effluviis* (On the noxious effluvia of marshes, 1717) intervened in an environmental debate currently raging in Cisterna, a town near Rome, about whether to clear the forest surrounding a nearby swamp.[29] Lancisi argued that the swamp's soggy soil and "putrid water" produced unhealthy air that in turn spread disease among the local population. Preserving the forest, he argued, would filter and purify the unhealthy air, thus reducing disease among the local people. Lancisi's understanding of the connection between water, air, earth, and disease was typical of environmental medicine, with roots in the Hippocratic tradition of "airs, waters, and places." Within this tradition, a dry climate with clear, pure air was regarded as the healthiest place to live. Conversely, a watery climate with impure air shortened the lives of people living there. "Healthy air," Lancisi asserted, has "moderate dryness and habitual fair weather, as well as gentle flow and movement, which as it were wards off the effects of putrid water."[30] This general medical wisdom informed not only Lancisi's work on a contemporary local climate but also Vallisneri's work on an ancient global one.

Vallisneri's focus on water—both the contamination of the world's water supply and the intermixing of water with the soil and air to produce humid and marshy conditions—enabled him to tell a strikingly different story from those of his British predecessors. The Flood's effects on global nature were temporary, he speculated. The water retreated, dry land emerged, and the air was gradually cleansed of impurities. However, the damage to human bodies was permanent. The swampy, humid climate, during its temporary reign on earth, was responsible for "weakening the fertile spirit of men and rendering the tiny frames [*machinette*] of future children that are enclosed in the eggs of women more feeble, more pliable, and more easily dissolved." Moreover, Vallisneri imagined these reproductive harms to be heritable, and therefore, permanent. As he put it, rather fatalistically, "[T]his hereditary malady will last until the end of time." By acting on the reproductive organs, environmental factors permanently degraded the reproductive capacities of human beings, such that all children born after the Flood would share in the same physical debility as the Flood's survivors. It was this "hereditary malady" inherent in human bodies—not the continuing impact of a degraded global environment on each new generation of human beings born after the Flood—that transformed humankind from a race of nearly invincible giants to a group of stunted weaklings, newly vulnerable to death and disease.[31]

Early modern climate theory would normally dictate that human health should improve along with an improved climate, as in the case of the swamp of Cisterna. Vallisneri preempts this objection by appealing to the classical medical theory of heritable diseases. "Why then, if the air, fruit, herbs, and grains returned to their original state, did longevity not return naturally" to its antediluvian level, Vallisneri asked rhetorically. "I respond, speaking as a professor, that according to the opinion of Hippocrates the hereditary diseases are indelible, even if the next generation eats food that is not only good but better, breathes air that is equally, if not more, perfect." Vallisneri positioned hereditary disease as a bodily limit on climactic influence, recalling the theories not only of Hippocrates but of Spanish Creole philosophers of a century earlier who hypothesized that innate and inherited racial differences imposed bodily limits on climatological influence, as discussed in chapter 2. The physical embodiment of sin guaranteed that it would have a very long afterlife, ensuring that human health and longevity would not recover along with the global environment.[32]

This divergence between people and climate also signaled a divergence between human and natural history that represented a sharp and deliberate break with British precedent. Whereas Burnet and Woodward imagined Edenic bodies and Edenic nature declining in tandem and in equal measure, Vallisneri argued that the Flood transformed human bodies more profoundly and more lastingly than it transformed nonhuman nature. That he would imagine the brunt of the Flood's effects as falling on human bodies, and, in particular, on human reproduction, makes a great deal of sense in light of his professional expertise as an embryologist. As a physician who lived through the 1720 plague and a father who was predeceased by fourteen of his eighteen children, Vallisneri may also have been trying to make sense of the illness and death he encountered in his personal and professional life. As an Italian Catholic philosopher writing with and against Protestant scholars from northern Europe, he may have also been trying to craft a history of the world that accorded better with his own understanding of free will, the wages of sin, and the limits of humanity's ability to change the world around them.

Vallisneri's grim portrait of the sinfulness into which the world had fallen prior to the Flood is matched by his quasi-gothic description of the horrors suffered by the Flood's survivors as a consequence of that sin. "Faith teaches us," he wrote, "that antediluvian men had an excess of arrogance, and were guilty of every kind of vice." They "gorged themselves to excess" and were "submersed in vice up to their eyes." But salvation from this world of sin and vice was paired with unprecedented suffering. In fact, the trauma experienced by Noah's family, the lucky survivors, the chosen ones, played a major role in the ruin of the human body. Having witnessed the deaths of nearly all of their friends and family as well as the utter destruction of the only world they had ever known, they were left to grieve in a dark and rudderless ship for an extended and unknown period of time. After reaching dry land, they emerged from the ark only to witness the "tragic, ghastly, horrible, and disgusting spectacle of so many corpses of men and animals." Vallisneri wrote movingly of "the horror of finding themselves alone after the Deluge, as if abandoned in a desolate desert." Likening the situation of Noah and his family to the wanderings of Moses and the Hebrews in the desert following their escape from slavery in Egypt, Vallisneri suggested that being spared by God was a mixed blessing for Noah's family and indeed for humanity. Certainly, the suffering of the Flood's survivors took a psychic toll that amplified the consequences of the climactic devastation. The spectacle of "horror," he claimed, "undoubtedly played a part in destroying the peaceful harmony of their blood, and disturbing the regular movement of their spirits and humors." The traumatic turbulence of their blood, spirits, and humors in turn "caused the principles of generation to be weakened and corrupted as well, so that these defects [*vizii*] were passed on to future generations." The shock to their interior emotional state produced a physical response in their bodies that caused additional harm to their reproductive organs.[33]

The notion that the Flood's survivors suffered from something like what we would now call intergenerational trauma, which in turn was a causal factor in producing a heritable succession of birth defects, makes sense in light of the common premodern idea that mental or emotional shocks from external stimuli could produce physical changes in the human body.[34] In the case of pregnant women, startling or disturbing experiences could be passed on to gestating fetuses, producing misshapen monsters whose physical features provided permanent testimony of the mother's shock, as in the case of the lobster-like baby whose pregnant mother had fainted at the sight of a lobster for sale in a seventeenth-century London market.[35] Vallisneri actively participated in the ovist-spermist debate that was raging within the life sciences and had clearly given much thought to the mysteries of human and animal reproduction. Although he likely would have dismissed the "monster" theory of reproductive anomalies as old-fashioned, the notion that external stimuli impacted conception and gestation tallied with Vallisneri's hypothesis that global climate change was a big enough "shock" to its human survivors, both physically and psychologically, that it could produce lasting damage to human reproduction. It also harmonized with the latest medical literature on the relationship between sin, sexuality, and the body. Fernando Vidal has shown how medical writers in the eighteenth century solved the moral problem of masturbation—namely, that there was no reliable means of detection and hence of punishment—by theorizing that the sin of onanism produced immediate and long-lasting effects in the body of the sinner, thus obviating the need for religious authorities to inflict any further punishment. "The sin included its own punishment," Vidal writes, "and a rapid one at that." Vallisneri's theory of ingrained sin in the bodies of the Flood's survivors exhibits the same "continuity of medicine and moral theology" that Vidal observes in other medical authors of the first half of the eighteenth century.[36]

The medical understandings of heritable disease, reproductive anomalies, and self-inflicted moral punishment that informed Vallisneri's account of postdiluvian human decline also participated in a theology of sin that took Dante's *Inferno* as an important point of departure. *Of Marine Bodies* is liberally strewn with quotes from the great Italian poet, and Vallisneri's description of human suffering in the wake of the Flood is steeped in Dante's vision of hell as a place where sins were made legible on the body of the sinner.[37] This theology of sin proposed the material rendering of the spiritual state of individuals in and on their physical bodies as a function of divine justice. Many of the words that Vallisneri chose to describe the misfortune visited on the Flood's survivors—*disgrazia, vizii, contaminazione*—carried the dual connotation of both moral and physical corruption, as did his reference to the "polluted" world that the 1720 plague and the biblical Flood helped to wash away. A reader confronted with Vallisneri's assertion that the "*vizii* were passed on to future generations" would understand by those words that children were inheriting both their parents' physical deformities as well as their moral turpitude.[38] Sin was simultaneously a state of one's soul and a heritable medical condition.

In sum, Vallisneri's account of postdiluvian human decline and environmental recovery undid the synthesis of natural and human history that Burnet and Woodward had worked so hard to achieve. In his *Essay*, Woodward claimed that God's purpose in sending the Flood was the "ransacking of Nature, and turning of all things topsie-turvy," to which Vallisneri responded: "Our Lord God wished to castigate Mankind, not to turn the entire earth inside out."[39] Human bodies, not the physical planet, served as the main site of punishment, the physical manifestation of humanity's moral failings, and the continuing reminder of past sins. Of course, Burnet had referred to the postdiluvian human body as "the seat of diseases and loathsomness" and a "Prison" of man's own making.[40] However, Burnet also argued that the after-effects of sin were principally manifested in the ruined planet and climate, which in turn degraded human health in each new generation. In Vallisneri's version of the story, sin was manifested immediately in the bodies of sinners and passed down from each generation to the next with no environmental input required. Although the climate played a crucial role in producing the initial damage to human bodies, it was not the primary target of divine judgment, as shown by the fact that it recovered while human vitality was permanently degraded. As a nonhuman actor, the earth could not properly be said to sin, and therefore could not properly be punished for sinning. It was humanity's sinful behavior that unleashed the Flood on the world, and therefore it was humans, not the earth, who rightly suffered from its devastating consequences.

Vallisneri's revised climate history in *Of Marine Bodies* represented a sharp difference of opinion with his colleagues in northern Europe on the question of sin and its realization in nature. It took aim in particular at the optimistic theology of improvement that undergirded Woodward's vision of the postdiluvian world. Displacing God's punishment for human sin onto the environment effectively opened the possibility that humanity could secure their own redemption from their fallen state if they worked hard enough to renovate the planet they had ruined. Woodward's *Essay* suggested that modern men might redeem themselves and the world through agricultural and extractive labor. Vallisneri's insistence that the Flood's punishment fell directly on humans foreclosed the possibility of a renewed earth producing a renewed humanity. If sin was an indelible stain in the fabric of the human body, then improving the earth would not do anything to improve people. Humans had not ruined the earth; they had only ruined themselves.

### Parasites in Paradise: The Flood, the Fall, and the Problem of Orthodoxy

We are now in a better position to appreciate the ways in which confessional divisions within Europe's Republic of Letters shaped premodern understandings of global environmental change. The range of stories about the Flood reflected divergent confessional notions of human agency vis-à-vis the environment, as well as several points of convergence. By arguing that human bodies and not the earth were the Flood's primary victims, Vallisneri shone a spotlight on something that was common to all of the accounts of climate change in the global histories of the early Enlightenment: the role of human sin in bringing it all to pass. At the same time, emphasizing the human culpability in causing climate change in this particular way worked to de-emphasize the potential of human action to reverse it later on. It also challenged the very idea that the Flood was, in Burnet's words, "a true piece of natural history."[41] In arguing that the Flood's long-term effects were limited to human bodies, Vallisneri tried to show that the Flood was an event of paramount importance in human history but of relative insignificance in nature's history.

In a work of natural philosophy published several years earlier, Vallisneri put forward a history of human decline that featured a similar theory of embodied and heritable sin but dispensed with the Flood altogether. In *Nuove osservazioni ed esperienze intorno all'ovaia scoperta ne' vermi tondi dell'uomo* (New observations and experiments on the eggs of parasitic worms found in the human body, 1713), Vallisneri argued that the ruin of human bodies occurred at the Fall, long before the Flood, and did not involve environmental causes at all. Like *Of Marine Bodies, New Observations* is a collection of long, scholarly letters on subjects in natural history and medicine that also touch on religious matters. One of the letters, dated September 1711, was from the Reverend Father D. Antonio Maria Borromeo in Rome, asking Vallisneri for his opinion on the vexed theological question of whether the bodies of Adam and Eve contained parasites.[42] The perfection of God's original Creation militated against their having parasites, Borromeo wrote, because "everything must have been clean and pure in that most innocent of places." The notion that humans would have been free of disease and even the threat of mortality had it not been for Adam and Eve's choice to sin was an especially important item of faith in post-Reformation Catholicism. Affirming the existence of human parasites in the Garden of Eden veered far too close to the heretical idea that humanity was created mortal by God, destined to die owing to their bodily constitution instead of freely choosing to sin and thus to die a mortal death, as Erculiani had argued in *Letters of Natural Philosophy* and later retracted in front of the Inquisition in the 1580s. It was therefore "difficult to understand," Borromeo wrote, how "the Author of Nature could have established worms in the body of Adam prior to Adam's sin." On the other hand, Borromeo reasoned, to say that parasites appeared for the first time on earth after the Fall would require "a new creation of [parasitic] worms," which would contradict Moses's account of all things being created by God before the creation of Adam and Eve.[43] The idea of "new creation," moreover, was rapidly falling out of fashion among philosophers armed with the latest microscopes and intent on destroying the theory of spontaneous generation.

In his response to Borromeo, Vallisneri attempted to solve this insoluble problem by proposing that parasites were created by God along with the rest of the animals and were originally a benign presence in Adam and Eve's bodies. The parasites could have lived in harmony with their human hosts and then become a malignant threat to human health and longevity after the Fall. Just as Adam "rebelled against our supreme and beneficent Father" through his sin, so in turn Adam's parasites "rebelled against him, [acting as] the just and righteous ministers of God's wrath."[44] God punished Adam and Eve for their sin by turning their intestinal parasites against them, thereby punishing all their progeny who would inherit those parasites with vulnerability to illness and death. The sin of Adam and Eve in the Garden of Eden, much like the sin of onanism in the Enlightenment and the sins of the damned in Dante's *Inferno*, were made physically manifest in the bodies of sinners.

In some ways, Vallisneri's theory that parasites were benign before the Fall and malignant afterward paralleled Woodward's theory in his *Essay* that the planet's fertile soil was beneficial for humans before the Fall, when the non-necessity of agricultural labor left them time for nobler pursuits, but became terrible for them after the Fall, when idle hands led to the devil's work. In other ways, it was a decisive rejection of Woodward and more generally of Swiss and British theories about the destructive impact of original sin on the world. In this letter to a Catholic priest and scholar, Vallisneri offered a way of understanding the inheritance of sin and bodily ruin without involving the Flood or the global environment. Human sin ruined human bodies. The scale of the agents and the victims were the same.

Several years later, Vallisneri proposed yet a third version of biblical history in which the Flood was totally irrelevant to the natural history of sin and human decline. In the very same letter to Lioni in *Of Marine Bodies* that put forward the history of the Flood outlined above, Vallisneri counterargued that the Flood perhaps had changed nothing at all in human bodies or in the global environment. As was in fact typical for a book that he proudly described to his friends as "skeptical"—one in which he frequently presents both sides of a thorny philosophical question without coming down in favor of either one—Vallisneri prefaced his history of Flood-driven environmental change by suggesting that no such thing had ever happened.[45] Perhaps the postdiluvian world was exactly the same as the antediluvian world. Nature was the same, the air and water and soil and seasons and food were the same, the same plagues and maladies afflicted humankind, and, most important, "the fabric of our bodies then was the same as it is now." Perhaps the Golden Age of robust men living for eight hundred years was nothing more than "dreams, fables, and romances . . . brought to us by Burnet and others." To write a history as Burnet and Woodward had done, in which humans flourished for many generations after Adam and Eve's catastrophic fall from God's grace, seriously diminished the significance of original sin by delaying its consequences. It was, in Vallisneri's words, an insult to "the upright justice of the Lord God, who wanted to make Adam and his posterity pay the price for their disobedience," to imply that "the Antediluvians" were somehow exempted from this punishment, in spite of being "closer to the sin wrought than us, and equally, if not more than us, sinners." It was theologically problematic, he argued, to imagine that God allowed humans to live for so long and to grow to such enormous size after the Fall.[46]

On some level, Vallisneri seemed to prefer the version of events in which the Fall was paramount and the Flood was relatively insignificant; it magnified the importance of original sin by showing how swiftly and decisively it transformed life on earth. On another level, he seemed deeply invested in the tale he had crafted of the traumatic experiences of Noah's family trying to survive in a soggy and "corrupted" world after the Flood. A race of supermen brought low by global disaster was an equally appealing way of underscoring the magnitude of humanity's fall from grace. Vallisneri's ambivalence about which one of these stories to favor may have flowed from the fact that both versions effectively highlighted the gravity of sin and the magnitude of its destructive effects on the sinners. His ambivalence may have also sprung from an uncertainty as to which version was the more orthodox and hence more acceptable to the Inquisition's censors. Vallisneri reported to Lioni that Giovanni Antonio Orsato, whom he respectfully described as a "great ornament to the Benedictine order, a great theologian, [and] a wonderfully learned scholar," instructed him that while it was licit, even necessary, to seek deeper explanations for the events of biblical history as recorded in scripture, it was illicit to dispute the fact of those events.[47] If antediluvian longevity was an item of faith recorded in the Mosaic history, then the version of sacred history in which the Edenic earth with its salubrious climate was destroyed by a global Flood must be the orthodox one. Framing his history of Flood-driven climate change as a response to a command from a colleague in holy orders was perhaps meant to signal to his readers that his true sympathies lay with the other version, the one in which the Flood was irrelevant. However, the command to hew to orthodoxy was not easily accomplished, as Vallisneri probably recognized. Claiming God's curse was effective immediately after the Fall was orthodox, but so too was claiming that the antediluvian patriarchs were incredibly long-lived, which required postponing God's curse until the Flood. So long as climate was employed as the principal explanation for ancient longevity and modern debility, scholars were caught in a bind of picking one or the other of these two items of orthodoxy without being able to accommodate both. Either the Fall took effect right away, in which case the Edenic climate would be gone, leaving philosophers with no means of accounting for the fantastic longevity of post-Adamic, pre-Noachian humanity. If, on the other hand, humanity and the global environment were granted a stay of execution until the Flood, human longevity could be satisfactorily explained by philosophy, but theology would then have to answer the question of why the Fall would not result in the immediate ruin of humankind.

The existence of this theological bind, which undergirded all debates about the relative importance of Flood and Fall in earth history, may help to explain the unresolved tension between Vallisneri's three versions of the biblical history of human decline in the letters collected in *New Observations* and *Of Marine Bodies*. It is impossible to tell from the text of these published letters which account he truly believed or preferred. He frequently adopted the same skeptical attitude in his manuscript correspondence, so his unpublished letters are no more reliable a guide to his true thoughts and feelings than the published ones. It might also be methodologically fruitful, even necessary, to take Vallisneri at his word when he claimed not to know which version to prefer. It is entirely possible that he truly could not decide between the three conjectural histories he proposed. Perhaps we might see Vallisneri's differing versions of the Flood story as reflective of his best efforts to craft an account of the world's history that satisfied the demands of both philosophy and faith as he understood them.

### Talking across Confessions in the Alpine Republic of Letters

Vallisneri's difficulties in determining what an orthodox natural philosophy of the Flood would look like helps us to better understand the self-proclaimed "skeptical" mode in which Of Marine Bodies was written, and to see his skeptical stance not simply as the disingenuous dodge of a libertine Enlightenment philosopher keen on avoiding the Inquisition's censors or the opprobrium of Padua's theology faculty. It is equally possible that skepticism was Vallisneri's response to the difficulty he experienced reconciling his philosophical and his religious commitments publicly, in front of both his Catholic colleagues in Italy, lay and ordained, as well as his Protestant colleagues north of the Alps. The ambivalence about the Flood that marks his main published work of earth history also extended to his private philosophical correspondence, in which he variously critiqued and defended the Flood's place in a natural history of the earth.

The apparent contradictions in Vallisneri's private correspondence, as well as the discrepancies between his published and unpublished writings, make more sense in the context of the Alpine Republic of Letters, the transnational and cross-confessional network of scholarly exchange that connected northern Italy and Switzerland in the early eighteenth century. The decades-long correspondence between Vallisneri and Bourguet offers an unusually rich and illuminating archive of unpublished letters that display affection and respect as well as bullying, misunderstanding, and occasional flashes of religious bigotry and patriotic pride, making them an ideal lens through which to explore how men of different countries and faith did and did not find ways to talk with one another about religiously charged topics. The strategies of toleration that characterized their epistolary exchanges produced what looks in retrospect like a secularizing trend within the field of earth history. Reading these letters to Bourguet alongside letters to compatriots and coreligionists reveals Vallisneri's desire to shield Catholicism and Catholic scholars from Protestant critique by reestablishing faith and philosophy as separate domains. The disappearance of the Flood from a field it had once dominated then appears in a new light: not as the sign of a decline in religious belief among Europe's naturalists, but as the unintended by-product of the conversational dynamics of a deeply pious and divided community.

The need to move knowledge over the Alps incidentally created a network of circulation and exchange within the Alps, bringing Swiss Protestants and Italian Catholics into regular contact with one another. If patriotic pride and the need for intermediaries drove Italian scholars to seek out contacts in Switzerland, the Swiss were eager to gain access to the ancient universities, new academies, and strong scholarly networks of northern Italy. In return for their help in making Italian scholarship known abroad, Vallisneri helped Bourguet and Scheuchzer gain a foothold in scholarly circles in the Veneto. He publicized Scheuchzer's books by distributing them to his correspondents in Italy.[48] He gave Bourguet similar exposure by circulating his letters and manuscripts among their mutual friends in the Veneto. "I sent your letter to Lord Abbot Conti," Vallisneri told Bourguet in 1727, "so that he may also enjoy it, and show it to his friends who are men of letters."[49] Vallisneri worked to further his Swiss friends' careers in more directly material ways as well, putting Bourguet in touch with his wealthy patron in Milan, Clelia Grillo Borromeo, and trying unsuccessfully to get Scheuchzer hired to a chair of philosophy at the Lyceum in Padua, telling him: "There will be little pay, to speak candidly, but much glory."[50] Coming from a country with relatively weak support for scholarship, Scheuchzer and Bourguet might have regarded access to the cultural capital and knowledge networks of the Veneto as a fair trade for the work of translating and disseminating its scholarship.

Even as they performed crucial mediative work between Catholic and Protestant Europe, and even as they worked to cultivate contacts in Italy, Swiss scholars maintained a strong sense of confessional identity and, in some cases, pursued highly partisan religious agendas. This was especially true of Switzerland's Huguenot community, whose firsthand experience with religious persecution —the revocation of the Edict of Nantes, which sent French Protestants scattering into exile across Europe, was still within living memory—put them in the delicate position of needing to converse with Catholic scholars even as they worked to propagate their own faith at the expense of their conversants'. Bourguet, who fled southern France with his family at the age of seven, came close to becoming a missionary but was prevented from doing so by poor health.[51] Instead, he gave generously to Protestant evangelical missions in the East Indies and North America and regularly published reports on the progress of these missions in the second journal he founded, the *Mercure Suisse* (Swiss Mercury, later the *Journal Helvétique*, 1732–69). *Swiss Mercury* was the partisan doppelgänger of the transnational and cross-confessional *Italian Library*: the publication in which Bourguet wrote to and for a confessional audience. In the preface to the first issue of *Italian Library*, the editors promised to avoid discussion of any topics "that unhappily divide the Christians of today."[52] Several years later, *Swiss Mercury* became the venue in which such divisive topics could indeed be raised.

Bourguet's evangelical zeal strongly informed his desire to publish his own account of world history and the Flood. The year before publishing his "*Mémoire sur la* théorie de la terre" (Notes on the theory of the earth, 1729), Bourguet wrote to his friend Georges Polier, a rector and professor of ancient Greek and Hebrew in Lausanne, to describe his religious goals in publishing this work: "The demonstration that I hope to give of the formation of the globe on which we live; of its dissolution by water, which Scripture calls the Deluge; and of its destruction by fire, which the Apostle [John] calls the Conflagration: this demonstration, I contend, is one of the strongest proofs that can be given of Providence and of Revelation."[53] Earth history could serve evangelical ends, persuading unbelievers to return to the right worship of God by demonstrating the consonance of natural history with biblical history and biblical prophecy. Bourguet felt just as passionately about furthering the spread of global Protestantism as he did about maintaining correspondence with Italian Catholics.

Vallisneri's fears about an Anglo-Swiss alliance in earth history were grounded in his correct assessment that his two main Swiss correspondents, Scheuchzer and Bourguet, joined their British colleagues in the conviction that the Universal Deluge was, in Woodward's words, "the most horrible and portentous Catastrophe that Nature ever yet saw."[54] The Flood's utility in building long-distance relationships in the Republic of Letters (as detailed in chapter 4), worked better in largely Protestant networks like Woodward's than it did in cross-confessional ones like Vallisneri's. When *Of Marine Bodies* was published in 1721, with its open criticism of Woodward, Vallisneri launched a diplomatic campaign designed to preserve his Swiss friendships. In August of that year, Vallisneri sent Scheuchzer a letter of warning that *Of Marine Bodies* was critical of the "theory of the Deluge" that "you, along with the celebrated Woodward and hundreds of others . . . uphold." He hastened to assure his Swiss friend that he had publicly named only Woodward, as "the principal head of this sect."[55] Four days later, Vallisneri wrote to Bourguet to solicit his help in pacifying Scheuchzer. "I don't know what there is to be said about not holding the same opinion as him on the subject of the Deluge. I have, however, treated his opinion with the utmost respect, not wanting to impugn any aspect of it on account of the friendship that passes between us."[56] The language of respect and friendship was part of a discourse of gentlemanly cosmopolitanism that Vallisneri clearly hoped would neutralize the very direct ways in which his latest book took aim at the Woodwardian consensus (and thus indirectly at Woodward's transnational network). These were magnanimous words, ones that seemed to elevate scholarly comity above the fray of confessional strife and philosophical difference of opinion.

Writing to his compatriots, however, Vallisneri spoke of his Swiss colleagues with far less respect and with open religious animosity. Just days after dispatching these magnanimous missives across the Alps, Vallisneri sent a very different kind of letter to a friend in nearby Castelfranco, Jacopo Riccati, in which he mocked Bourguet for being "supremely passionate about Sacred Scripture, as are all of the heretics"[57]—just as he had criticized Woodward to Bourguet the week before for being too preoccupied with "the words of Holy Scripture."[58] By "heretic," he meant "Protestant," and by deriding what he saw as an inappropriate and distinctively Protestant penchant for scripture, he thereby signaled his own deeply confessionalized understanding of natural philosophy as a space in which scripture had no purchase. This remark made to a compatriot and coreligionist reinscribed the differences of country and confession that his friendship with Scheuchzer and Bourguet had apparently overcome. In a gesture of communion with Riccati, Vallisneri then contrasted the heretical passion for scripture with the tenets of Catholicism, or, as he put it, "*Our* Holy Faith."[59]

It is no secret that religious hatred was a prominent feature of life in early modern Europe. Yet scholarship on the Republic of Letters has, until recently, too often assumed that its participants were somehow more tolerant and broadminded than the people living around them. These brave cosmopolites, the story goes, were the ones uniquely able to rise above the religious and political divides that made early modern Europe such a fractious and oftentimes frightening place to live. This interpretation is one that the Republicans of Letters themselves would have embraced. Vallisneri's gallant gesture of toleration and respect was one of many that gave rise to the early modern ideal of a Republic of Letters: an international community of scholars sustained by their ability to set aside differences of nation, language, confession, gender, and class when conversing with each other across the continent and increasingly around the globe.

On the other hand, Vallisneri's jab at heretics reminds us that this ideal Republic overlapped with other early modern networks, many of which defined themselves along linguistic, gendered, geographical, and confessional lines. Indeed, much of the recent scholarship on the Republic of Letters has revealed the existence of many smaller networks embedded within it that were organized in precisely such exclusionary ways. Epistolary networks of Jesuit scholars, for example, were defined along religious lines, while the handful of women scholars who were able to access the mostly homosocial world of the Republic of Letters also organized themselves into a smaller, all-female Republic.[60] So it should not surprise us that Bourguet gave as good as he got, treating Vallisneri with the same polite respect while telling a Huguenot friend in nearby Lausanne that the ultimate aim of his work as a scholar was "rendering justice to our Churches and at the same time shutting up the mouths of the Papists."[61] Despite the high-flying rhetoric of radical inclusiveness, many scholars recognized that a chasm separated the rhetoric of toleration from the reality. In fact, Bourguet and Vallisneri's seemingly Janus-faced behavior was perfectly acceptable and even expected in the Republic of Letters. As their contemporary Pierre Bayle, the famous skeptic, acknowledged: "One writes differently according to the person with whom one is exchanging letters."[62]

## Strategies of Toleration

By looking closely at the style and content of the letters that were exchanged in this Italo-Swiss network, we see a complicated dynamic at work, characterized by a range of strategies for avoiding, confronting, or neutralizing religious differences of opinion. Vallisneri probably thought Scheuchzer was a lost cause, but he directed considerable effort to maneuvering Bourguet toward his way of seeing things or, failing that, countering Bourguet's own efforts to persuade Vallisneri over to his point of view. Their philosophical correspondence was shaped in equal measure by their particularist agendas—Bourguet's evangelism and Vallisneri's nationalism—and by their mutual commitment to the Republic of Letters. The periodic flare-ups of hostility and dissent, and the uneasy truces that followed, point us toward a deeper understanding of the partisan religious motivations behind toleration and secularization in the early Enlightenment.

As scholars have recently argued, toleration is best understood as a practice rather than an ideal in the centuries prior to its widespread enactment into law in the nineteenth century. Bourguet and Vallisneri are interesting avatars of toleration precisely because they were not theorists or propagandists of toleration: they simply did it, according to what C. Scott Dixon has called the "pragmatics of diversity."[63] Strategies of toleration, as I have chosen to call the variety of rhetorical strategies that Bourguet and Vallisneri employed in their letters to one another, did not overcome difference so much as channel it into a negotiated and fairly narrow range of responses. In practice, toleration was characterized less by respect for difference than by attempts to suppress or resolve it. More often than not, toleration involved avoiding hot-button issues like miracles or scripture; politely ignoring someone else's breach of the philosophical compact if and when he did bring up such a topic; or, finally, shaming him for having done so. If they could not compel the other's agreement, they could at least try to compel his silence. This last, and ultimately highly successful, strategy resulted in a marked secularizing trend within earth science in the early decades of the eighteenth century, as two of its key proponents scrubbed their published works on earth history of things they feared might offend the other and his respective communities.

One of the strategies they employed was caginess. "You know my opinion, but I do not know yours," Vallisneri told Bourguet pointedly in 1710, the year they began exchanging letters. The matter about which he solicited Bourguet's opinion was "the discovery of marine bodies in the mountains," the same subject on which he would write *Of Marine Bodies* eleven years later. Often these fossils were found in the very same mountains that divided and united Italian and Swiss scholars, making them a popular topic of conversation in this corner of the Republic of Letters in particular. Bourguet already knew that Vallisneri had reservations about this theory, which he hinted at in the second letter he sent to Bourguet. So when Vallisneri declared himself "tractable, and yielding, so much so that I will gladly come over to your point of view, as soon as I find your theory to be more plausible than my own," Bourguet initially refused to take the bait. He was happy to exchange fossil specimens with Vallisneri but not his opinions about their origins.[64]

As Bourguet began to be more open and argumentative with Vallisneri, a second strategy emerged: the gentlemanly displays of respect for difference of opinion like the one that Vallisneri used to test Bourguet, claiming to have treated Scheuchzer's different opinions "with the utmost respect." "I see many points on which we agree, and many on which we do not agree," Vallisneri told Bourguet with admirable even-handedness.[65] But matters came to a head in 1715, when Scheuchzer published a book arguing that the mountains themselves, and not only the fossils found there, were products of the Universal Deluge and thus of divine agency. In *Terrae structura* (The structure of the earth, 1715), Scheuchzer echoed Burnet's theory of a smooth Edenic planet in his relation of how the earth's strata, which were "flattened and spherical" in the Edenic Age, were "broken up, dislocated, moved around, raised up here, and pushed down over there" by the violence of the Flood.[66] Vallisneri wrote to Bourguet to express his displeasure. "[Scheuchzer] would have it that after the Flood, all the sediment and strata remained horizontal . . . but then God in his divine omnipotence broke them all up and overturned them, from which they became mountains. I would prefer a natural cause, so as not to force God's hand or his omnipotence."[67] To Vallisneri, explanation-by-miracle was bad philosophy. Telling this to Bourguet, however, was a test. Vallisneri strongly suspected that Bourguet shared Scheuchzer's desire to highlight the active interventions of God in the making of the natural world—and perhaps even in the making of Switzerland. Would Bourguet side with his compatriot and coreligionist, or with his Catholic friend on the other side of the mountains?

Bourguet tried to split the difference. Scheuchzer and Vallisneri were both right, he replied. "That God in his omnipotence broke up the strata in order to raise the mountains does not seem to me a very philosophical assertion," he agreed. Nevertheless, he gently but firmly pushed back against Vallisneri's hard-line stance on miracles, writing that "the only miracle was the origin of the Flood itself, after which all phenomena followed in a natural manner."[68] Vallisneri was right that the chain of events set in motion by the Flood were natural, but Scheuchzer, Bourguet implied, was right that the Flood itself was indeed a miracle. This subtle challenge was met with a deliberate show of disregard. Simply ignoring things that one found objectionable was yet another strategy of toleration. Vallisneri's next letter warmly praised the part of Bourguet's assertion that accorded with his own ideas and passed over in silence the part that did not. "I am glad that we are in agreement that God did not perform a miracle in making the mountains," he declared, "but rather that everything, however rare or strange, has proceeded in a natural chain of events."[69] Bourguet's response to this rather coercive misrepresentation of his views is lost.[70] But their argument over what was natural continued.

Defining what was "natural" was another way of defining the scope of what they were allowed to say to each other. On another occasion, Vallisneri demanded that they discuss the earth's history "as natural philosophers"—which, he clarified, meant leaving scripture, God, and miracles out of the conversation.[71] Vallisneri thus attempted to shame Bourguet into silence on subjects Vallisneri considered religious (and thus inappropriate) and Bourguet considered absolutely central to their philosophical discourse. Chastising or shaming one another was a popular strategy of toleration that both men employed when they found themselves unable to agree or to magnanimously disagree. It was particularly popular during the time when *Of Marine Bodies* was first published. In 1720, Vallisneri commanded his zealous friend: "When you write to me, do not speak of religion anymore."[72] Bourguet shamed Vallisneri right back in a letter of 1722, goading him, not into silence, but into admitting that religion had a rightful place in their philosophical conversations. "I think that whoever recognizes the existence of a God cannot deny the possibility of miracles," he chided his friend. "I esteem too much your good heart and sublime genius to persuade myself that you subscribe to this proposition."[73] Appealing to Vallisneri's heart was a sentimental gesture, and to his genius, a humanist one. But these compliments, however sincere, were also coercive. Vallisneri did not deny the existence of God or of miracles, as Bourguet knew perfectly well. (Indeed, Vallisneri once bragged—to a fellow Italian, not to Bourguet—about sending his son all the way to Naples to witness the miracle of St. Gennaro, whose blood, preserved in a glass vial, miraculously liquefied on Catholic feast days.)[74] The implication was this: If Vallisneri believed in miracles, and the Flood was a miracle, then he would have to admit that it was not only permissible, but indeed necessary, for them to touch on the supernatural if they were going to continue talking about the earth's history. Instead of shaming him into silence, Bourguet would shame him into speech.

But Bourguet may have sensed that this was a losing strategy. He also worried that the book on the earth's history that he was writing in the wake of *Of Marine Bodies* might not be received well by his many Italian correspondents. In a series of letters to Polier, his Huguenot friend in Lausanne, Bourguet agonized over how much religious content to leave in and how much to take out. He felt torn, he told Polier, between his desire to convince unbelievers and his worry about giving offense to some unspecified readers.[75] Finally he resolved to marginalize the book's explicitly religious content: "I won't speak to my readers of Revelation until after I've already demonstrated the facts and the phenomena of Nature. . . . The Application of my discoveries to Scripture will only appear as a corollary."[76] In the end, he did even less than that. The Flood, one of his most beloved topics of epistolary conversation, received a bare two mentions in his "Notes on the Theory of the Earth" when it finally appeared in print in *Lettres philosophiques* (Philosophical letters, 1729). Bourguet waited until after Vallisneri's death and shortly before his own before he ventured to publish one of his philosophical letters defending the Flood, addressed to a fellow Protestant in Geneva, in *Traité des petrifications* (Treatise on fossils, 1742).[77]

It would appear that Vallisneri's strategies of toleration had finally succeeded in getting Bourguet to write the kind of purely "natural" philosophy he had insisted on. Vallisneri's repeated warnings about how Bourguet's book would be received in Catholic Italy made Bourguet anxious about losing his Italian correspondents, whom he depended on for patronage and scholarly community. Meanwhile, Vallisneri's program of national renewal made him especially worried about how northern Europe received Italian scholarship. Both felt that the eyes of scholars in the other's countries and religious communities were on them.

Vallisneri's letters to Muratori show that he was keenly sensitive to Protestant stereotypes about Catholic credulity, especially when it came to miracles. The official position of most Protestant churches was that the biblical age of miracles was over, whereas the Roman Catholic Church continued investigating and validating new miracles as part of the process of beatifying saints. Vallisneri himself, in his capacity as a medical doctor, was once called upon by the Congregation of Rites to examine the corpse of a seventeenth-century cardinal being considered for sainthood. After the autopsy, he wrote to Muratori to express his doubts about the body and his fear of what would happen if he were to recommend the man for sainthood regardless, as he knew was expected of him. His fear revolved around what the Protestants would say. Canonizing someone who was not truly worthy "gives an opening to the heretics," he objected, "who mock us, and make jokes at our expense about the lives and the miracles of our Saints."[78] Vallisneri's fears of Protestant mockery were not without a basis in reality.[79] Bourguet ridiculed Catholic credulity and cast doubt on the authenticity of Catholic miracles in a letter to one of his fellow editors at *Italian Library*, that vehicle of Italo-Swiss collaboration.[80] Vallisneri may very well have concluded that defending the cult of the saints, and Roman Catholicism more generally, would require shielding it from outside scrutiny.

### In the Shadow of Galileo: Separating Philosophy and Faith

The correspondence between Vallisneri and the Italian mathematician Jacopo Riccati illuminates the differing strategies by which laymen in the Catholic Enlightenment attempted to reconcile their philosophical inquiry with their personal religious faith and their standing in the Catholic community. Vallisneri was, on the whole, consistent in maintaining his position about the unknowability of the Universal Deluge and its unfitness for natural philosophy in his correspondence with Bourguet and Scheuchzer, his two main Protestant correspondents. A striking contrast comes in a series of letters exchanged with Riccati in the summer of 1719. The subject of the Flood arose in their very earliest letters to one another, when Riccati opened a letter by venturing the following argument against the universality of the Deluge. "Notwithstanding the learned proofs of many weighty authors, it seems to me that it is possible to say, at least in the mode of research and disputation, and meanwhile always affirming that I defer in everything to our Holy Roman Church, that the deluge was not universal across the entire earth, it being sufficient, in my opinion, to-flood only the part of our hemisphere that was already inhabited."

The string of qualifiers with which Riccati prefaced his denial of the Flood's universality indicates how thoroughly accepted this idea had become in Italian philosophy by the early eighteenth century, despite its being a relatively recent development. Riccati acknowledged that the weight of scholarly tradition and the authority of the Roman Catholic Church were arrayed against him, and so he cast his claim as a stance ("it seems to me"; "in my opinion") that he ventured only in the protected space of academic "research and disputation." Riccati was reclaiming, perhaps deliberately, the Renaissance tradition of denying the Flood's universality from within academic natural philosophy, as discussed in chapter 1.[81] Vallisneri's response was swift and severe. While claiming to "like your opinion that [the Flood] could have been particular to Asia," he immediately raised a "very strong objection," namely, that marine fossils had been discovered "in the New Indies and in all parts of the world, as Woodward also noted."[82] (Riccati had named both Woodward and Scheuchzer in the previous letter as partisans of the theory he was criticizing.) In fact, Vallisneri went on to say, as if this clinched the matter, he himself possessed a petrified crab brought to him from China by a Jesuit among the specimens in his personal collection. The Flood was universal, and fossils proved it.

Given the "strong objections" that Vallisneri continually raised to Woodward's theory of the deluge in his letters to Scheuchzer and Bourguet, it is rather astonishing to see him defend that same position, as well as Woodward himself, in a letter to a compatriot and a coreligionist—not to mention one with whom he would feel free to ridicule "heretics" in a letter several years later. Riccati's reply registered genuine surprise at Vallisneri's reaction as well as an awareness that he had gone too far with his new correspondent. The letter begins, "I am surprised at the strong objection you made to me" but immediately conceded the point of contention with a gentlemanly bow of courtesy: "I unabashedly retract what I said in the other letter." Moreover, given the evidence that Vallisneri had presented in favor of the Flood's universality, "I now come to be completely convinced." Riccati's total, generous, and almost certainly disingenuous about-face displayed the requisite level of deference necessary to keep this fledgling epistolary relationship alive.[83]

Riccati may have felt emboldened to question the Flood's universality after reading Vallisneri's previous letter, which contained the boldly Galilean declaration that "the words of Sacred Scripture . . . were written in order to teach the ways of heaven, not to explain the phenomena of the earth."[84] Riccati's anti-universalist discourse was likewise crafted in superbly Galilean fashion, reinterpreting various passages in scripture in order to make a philosophical argument over the objections of current best thinking from Catholic theologians on a subject concerning the natural world. This is exactly what Galileo had done in the *Letter to the Grand Duchess Christina* in order to make the case for the motion of the earth around the sun—a topic, that, not coincidentally, Riccati raised in the very same letter in which he questioned the Flood's universality.[85] Riccati clearly thought he had found a sympathetic interlocutor in whom he could confide, and Vallisneri did partially relent in a subsequent letter, admitting the existence of valid philosophical objections to the Universal Deluge and announcing that his forthcoming book on the subject would be "skeptical."[86] Vallisneri had learned a different lesson from reading Galileo's *Letter*, the same one he tried to teach Bourguet in critiquing Woodward: the difficulty of understanding the Book of Scripture made it a poor resource for philosophy, especially when the Book of Nature was so much easier to read. Two years later, Vallisneri articulated more strongly to Riccati his sense that the words of scripture were insufficient grounds for natural philosophy. "Moses did not try to explain the phenomena of Nature," Vallisneri stated, "but instead everything was directed toward teaching the ways of heaven, and teaching the populace to fear God."[87] That the goal of scripture was to inspire religious fear and more generally to instill religious values remained a constant in Vallisneri's writings, public and private, to both Protestants and Catholics. So while he certainly positioned himself differently vis-à-vis Protestant and Catholic interlocutors, it is possible to trace a common thread in letters sent both near and far advocating separate spheres for philosophy and faith.

Skepticism is a key framework for understanding the range of opinions Vallisneri espoused on the Flood in conversation with a diverse range of interlocutors and publics. It was an intellectual and religious position that served him well in the contexts of the Catholic Enlightenment and the multiconfessional Republic of Letters. "My book about the Deluge is, then, as you have seen, skeptical," he wrote in a a 1721 letter to Muratori, "because no opinion satisfies me" and because it was impossible to "explain with natural reasons such extraordinary phenomena."[88] His skepticism is best characterized as fideistic rather than atheistic. Early modern Catholic fideism justified doctrinal belief in light of the weakness of human understanding when confronted with religious mysteries. Skeptical fideism was also an anti-Aristotelian and neo-Augustinian theology of epistemic humility in the face of humanity's fallenness. In *Of Marine Bodies*, Vallisneri contrasted his belief in the Flood as a matter of faith with his incomprehension of the Flood as a natural phenomenon. "I confess, I grasp, and very well understand that terrible effect of the most righteous anger of almighty God, which is to say, the Deluge," Vallisneri declared. "But the mode in which it occurred, still I find difficult to understand, without supposing it one of those miracles not intelligible by us. . . . In short, I understand only that I do not understand it."[89] Throwing up his hands in the face of a sacred mystery, Vallisneri relegated the Flood to the category of the miraculous and the unknowable. While absolutely affirming that the Flood really happened, Vallisneri effectively removed it from the purview of natural philosophy by saying that it was a miracle and therefore not something that natural philosophers could learn anything about.

This skeptical posture might look cynical to a modern observer, an obvious plea for secular rationalism and nonbelief, but it was echoed by many of Vallisneri's pious Catholic colleagues in Italy, both lay and ordained. Muratori and Conti, Catholic men of the cloth and two of Vallisneri's close correspondents, agreed with him that scripture and natural philosophy were best kept separate.[90] Muratori, a reform-minded theologian admired by Pope Benedict XIV and the future Clement XIV, was a strong proponent of philosophy and learning who also believed that the weakness of humanity's fallen intellect in the face of the divine warranted a separate realm for faith that would be based not on reason but on revelation.[91]

Meanwhile, two different Catholic priests in the Veneto extended Vallisneri's diatribe against the philosophical Flood in their own books on fossils and earth history from the 1730s and '40s. Giovanni Giacomo Spada, a priest and avid fossil collector in a small village near Verona, forcefully argued against treating the Flood as a subject for natural philosophy in a book with a very Vallisnerian title, *Petrificati corpi marini, che nei monti adiacenti a Verona si trovano* (Petrified marine bodies found in the mountains near Verona, 1737). In it, Spada argued that fossils were not "relics of the Deluge, as many believe, but infallible signs of the action of the Sea over the course of centuries."[92] Moreover, the Flood was beyond human understanding. "Not being able to really comprehend the great workings of God in the fabric of the world," Spada stated with pious humility, "we remain in the dark, like a blind man."[93] Three years later, a village priest

# The Flood Subsides

## From Erculiani to Vallisneri

In the 1570s and '80s, the Paduan apothecary Camilla Erculiani daringly proposed to bring the biblical disaster of Noah's Flood into the space of natural
philosophy. In the 1710s and '20s, the Paduan physician and professor of medicine Antonio Vallisneri daringly proposed to push it back out again. Much had
happened in the intervening century, not least the trial of their fellow Paduan,
Galileo Galilei, in front of the Roman Inquisition in 1633.[1] And yet, some things
stayed the same. Padua remained a vibrant center of intellectual debate, medical
learning, and religious questioning across the decades separating Vallisneri from
Erculiani. A plague, which had preceded the writing of Erculiani's *Letters on
Natural Philosophy* (1584), struck the Veneto again in 1720–21, just as Vallisneri's
*Of Marine Bodies Found in the Mountains* appeared in print. Erculiani and Vallisneri were both, therefore, survivors of epidemics as well as medical profession-
als who directed considerable attention to the causes and cures of plague.[2] One
wonders how deeply their experience of a large-scale epidemiological disaster
informed their mutual belief in humanity's embodied punishment for sin, as expressed in the stories they crafted of the Flood and Fall. One wonders especially
whether Vallisneri may have read Erculiani's small collection of philosophical
letters, which was never put on the *Index of Prohibited Books* but likely never
found its way into many private collections in Italy owing to its small print run,
publication in far-away Kraków, and lingering association with heterodoxy after
her interrogation by the Paduan Inquisition. Two copies of Erculiani's *Letters*
survive in Paduan libraries in the twenty-first century, suggesting that several
more copies than that would have been held in private libraries in Padua during
the years when Galileo and later Vallisneri were professors at the city's renowned
university, near where the Erculiani family's apothecary shop once stood.

Had Vallisneri managed to read Erculiani's *Letters*, he might have been inspired by her materialist, medical theories of embodied sin and divine punishment. He might also have been inspired by its dialogic and epistolary format. *Of
Marine Bodies*, like Erculiani's *Letters*, was a collection of letters written by and

from the Friuli named Antonio Lazzaro Moro published an even more stri-
dent case for banishing the Flood from natural philosophy. Praising "the most
learned Vallisneri" in his book with an equally honorific title, *De' crostacei e degli
altri marini corpi che si truovano su' monti* (Of crustaceans and other marine
bodies found in the mountains, 1740), Moro argued that the Flood's miraculous
nature made it off-limits for natural philosophy.[94] The first 175 pages of Moro's
book are devoted to refuting Burnet and Woodward's theories that "the Deluge
happened . . . according to the natural disposition of secondary causes," when
in fact it happened "according to the disposition of the vengeful Divine Justice
. . . to punish the prodigal wickedness of wayward men." Moro believed in the
Flood as a matter of faith—"I embrace, I believe, and I confess the truth of
the Universal Deluge"—but he could find no evidence for it in nature and
therefore could not allow it any place in a natural history of the world.[95] In spite
of his own ambivalence and polyvocality on the subject of faith, philosophy, and
the Flood, Vallisneri was remarkably successfully in launching a pious movement to return the Flood into the hands of theologians and men of the cloth.

Vallisneri's collaborators and followers in the Veneto launched a marked
secularizing trend in earth history, one in which the Flood played no significant
role and in which the Mosaic history found little purchase. Moro, the Friulian
priest, proposed an influential theory of volcanic activity as the main driver of
geologic change instead of the Flood, helping to establish the vulcanist school
of geology that flourished in the second half of the eighteenth century. Italian
philosophers of the Catholic Enlightenment deliberately loosened the ties between natural and human history forged by an earlier generation of British Protestants. In dialogue with their Protestant counterparts in Switzerland, a group of
pious natural philosophers in northern Italy came to believe that disentangling
philosophy from faith might be a better means of protecting Catholicism from
Protestant critique and promoting Italian scholarship abroad.

Pursuing the earth's history as if it were a purely *natural* history, distinct
from the spiritual history of humankind, could be a better way of preserving
the transnational intellectual sociability that the Flood had promised, and to
a certain extent achieved. While some Italian philosophers like Monti still believed in the Flood's potential to build communities dedicated to Mosaic natural
philosophy across religious and political divides, many of his colleagues in Italy
and beyond came to think that it merely irritated those divisions. The Flood's
surge of popularity as a subject for transnational science had reached its limit.
Attending to the social dynamics of long-distance knowledge-making and the
persistence of deeply felt religious divisions within European Christianity com-
plicates historical narratives of secularization in Enlightenment science. Religious convictions, no less than political and intellectual ones, shaped the terrain
on which transnational conversations about the earth's history could unfold.

to the author named on the title page, all circling around the subject of Noah's Flood. He probably would not have appreciated Erculiani's use of specific passages from scripture as evidence to bolster her philosophical claims, novel for the time and anticipating the far bolder reinterpretation of scriptural text that would land Galileo, Vallisneri's hero, before the Inquisition fifty years later.[3] With the benefit of hindsight, and with an eye toward his Protestant colleagues to the north, Vallisneri was skeptical that such imbrications of sacred history and natural philosophy would redound to the benefit of natural knowledge, Italian learning, or the Catholic faith.

While on the one hand attracted to the biblical stories of Flood and Fall as a spiritual point of origin for the death and debility he witnessed in the world around him, Vallisneri also seems to have felt that Mosaic natural philosophy, the project of relating the spiritual history of mankind to the natural history of the earth, had run its course. In *Of Marine Bodies*, he called the Flood "one of those miracles not intelligible by us," a real historical event that was nevertheless beyond human understanding.[4] In a letter to his Swiss Protestant correspondent Louis Bourguet, he emphasized that miracles, inexplicable according to the laws of nature and reason, were thus also beyond the purview of natural philosophy: "When we have recourse to miracles, all natural history is at an end."[5] That the Flood's miraculous status excluded it from natural philosophy was an argument that had been deployed by generations of philosophers in the later Middle Ages and Renaissance, prior to the arrival of Mosaic natural philosophy in the wake of the Reformations. In *Of Marine Bodies*, Vallisneri repeated the medieval nominalist principle that later came to be known as Ockham's razor: "It is not necessary to multiply miracles."[6] Vallisneri was ventriloquizing, perhaps deliberately, the Renaissance meteorologists who had argued that the Flood in particular, and biblical history in general, stood well outside of their disciplinary sphere of expertise. Significantly, the praise he received for "perfectly inhabit[ing] the person of the Philosopher and of the Catholic" signals that this secularizing trend was hardly impious.[7] Catholic priests like Antonio Lazzaro Moro praised Vallisneri for recognizing that the Universal Deluge was an indispensable item of the Catholic faith and at the same time off-limits for philosophical investigation. "Notice of the Universal Deluge comes *from no other source* than Holy Scripture," Moro declared, and it therefore belonged to faith alone.[8]

### Enlightened Floods and the Return of Renaissance Meteorology

The Enlightenment's revival of the classical tradition of Renaissance meteorology was key to the development of geohistory in the eighteenth century. Aristotelian and Stoic traditions envisioned successive local floods and fires as the main forces that shaped the earth's surface over the course of history. Vallisneri used one of his letters in *Of Marine Bodies* to advance a theory of how "numerous local floods" were responsible for the landscape of northern Italy and for the marine fossils found in its mountains.[9] Local floods were physically possible and therefore nonmiraculous, because they did not require the supernatural interposition of God in the natural world, nor did they necessarily have any relation at all to human sin. Local floods were therefore utterly safe for natural philosophy. Vallisneri's stated preference for local floods as the key resource for explaining the state of the earth was tied to his rhetorical commitment to empirical fieldwork and also to his particular brand of intellectual patriotism. "We know for certain that *our seas* once flowed over *our mountains*," he told Bourguet in 1718.[10] The certainty that an Italian flood once covered his Italian mountains was grounded in his travels through the Apennines. "This system of mine can perhaps be verified only in Italy. . . . I speak only of that which I have observed, and will not speak of that which I have not observed."[11] The epistemic limits of his study of ancient floods coincided with the boundaries of his native land.

The idea of ancient floods drowning and thus creating particular European territories received increasing attention in the early Enlightenment. Vallisneri's belief that "my country [*patria*] Reggio," like the Italian mountains he knew so well, was formed through the ancient action of an overflowing sea strongly recalled the theories of the Modenese naturalist Bernardino Ramazzini, who argued in *De fontium Mutinensium admiranda* (The marvelous springs of Modena, 1691) that the Po River valley, which underlay his native city, had once been underwater.[12] The "invisible Ocean" that undergirded the region stretching between the Alps and the Apennines was the vestige of an ancient sea, Ramazzini argued in this widely read treatise dedicated to the Duke of Modena.[13] In *Petrificati corpi marini, che nei monti adiacenti a Verona si trovano* (Petrified marine bodies found in the mountains near Verona, 1737), the Veronese priest and fossil collector Giovanni Giacomo Spada transposed Ramazzini's ancient sea that filled the Po River valley slightly northward. Spada claimed that the sea near Venice used to reach all the way to the mountains of Padua and Verona, covering the entire territory that was now rightfully part of the Venetian Empire.[14]

The overtly patriotic overtones of the theories of ancient local floods, as hypothesized by Vallisneri, Ramazzini, and Spada, also characterized similar geohistories formulated by several French naturalists in the 1710s and '20s, who used the evidence of marine fossils in order to find that the kingdom of France had been born from the waters of an ancient flood. In 1718, the French naturalist Antoine de Jussieu published an article in the *Mémoires* of the Royal Academy of Sciences in Paris considering the curious plant fossils of Saint-Chaumont, a town near his native city of Lyon. The most striking thing about these fossils was that they showed the impressions of ferns and palms that were "unknown in Europe" and therefore "could only have come from the tropics." Jussieu surely knew that his readers expected him to conclude, based on this evidence, that the tropical ferns had been carried to Saint-Chaumont by the waters of the Universal Deluge, just as Giuseppe Monti's walrus skull had been carried by the Flood to the hills of Bologna. Jussieu mentioned this explanation only to reject it, proposing instead that the ferns had been carried northward by one of the many floods that periodically traveled between different parts of the earth. Imagining a global circulation of ocean water "from North to South" and back again, he proposed that a huge wave swelling northward from the tropics had submerged France for a period of time, leaving behind the tropical plants now preserved as fossils in the Lyonnois when it receded southward. "The daily experience of the vicissitudes that occur in certain countries, where the Sea alternately inundates and uncovers various pieces of land, shows us only too well how it could have been that these waters that we assume transported these Plants could have also covered these places in the Lyonnois . . . without being obliged to have recourse to the inundation of a universal Deluge, an earthquake, or any other big catastrophe." Jussieu was deliberately trying to craft an account of the physical creation of the landscape of Lyon that dispensed with the Universal Deluge and leaned instead on classical theories of endless "vicissitudes" of floods and fires as the key agents of change in the natural landscape. Although his fossil evidence came from the region around Lyon, Jussieu also cited paleontological and geological evidence from other parts of France—"diverse places at the heart of this Kingdom"—that proved this ancient flood had covered the entire geophysical territory of what was now gathered under the rule of the French monarchy. Aristotelian meteorology was repurposed in the Enlightenment in order to downscale from the Universal Deluge to the national framework of European states.[15]

Jussieu was explicit about the patriotic motives behind his hydrological history of the birth of France. He began the article by listing several non-French naturalists—Lhwyd and Woodward in England, Leibniz in Germany, Scheuchzer in Switzerland—who had "done honor" to their countries by discovering fossils that proved their countries' antiquity. He then launched into a discussion of his own fossil finds by declaring, "France has equal advantage in this regard as these countries."[16] Jussieu's French patriotism led him to interpret marine fossils not as evidence for a Universal Deluge but as evidence for the geohistorical creation of France as a natural territory.

Several of Jussieu's French compatriots proposed nearly identical histories of an ancient sea-flood creating their countries. The naturalist René Antoine Ferchault de Réaumur argued that an ancient flood had created his native province, La Touraine, in an article published in 1720. A layer of seashells several feet below the surface of the soil was found to extend roughly the length and width of the province, which Réaumur interpreted as the vestiges of an ancient sea bed. Like Jussieu, Réaumur deliberately differentiated the flood that covered La Touraine from the Universal Deluge: "The sea that covered these plains remained there much longer than the Deluge would have."[17] Like Jussieu's theory of the diluvial creation of France, Réaumur's theory also had clear patriotic overtones. Boasting that his natural history cabinet contained fossils from all over France, Réaumur concluded by suggesting that not only La Touraine but all of France had once been covered by the sea, which slowly receded over the course of centuries, laying bare a new land. Collecting fossils could be a way of forging transnational connections and strengthening the Republic of Letters; it could also be a means of strengthening ties within one's own political and religious spheres.

Rejecting the Universal Deluge in favor of provincial, imperial, or national deluges resuscitated classical theories of geological vicissitude that had gone out of fashion with the Reformations and combined them with emerging forms of geopatriotism. The turn toward local floods in the early Enlightenment was not, therefore, a simple return to Renaissance meteorology. These French and Italian diluvial geohistories showed the clear influence of the Mosaic natural philosophy that provided the dominant framework for writing the earth's history in the intervening long seventeenth century. They clearly adopted the sacralized notion of a flood as a world-making agent, a baptismal and purifying force that destroyed what existed before and left in its place something new and better. Instead of creating the planet, its climate, and the human species, these French and Italian floods left behind a nation in their wake. Classical meteorology received a new lease on life in the Catholic Enlightenment, now overlaid with incipient forms of European patriotism and haunted by the specter of biblical geohistory. French and Italian naturalists jettisoned the universalism of sacred history while retaining its world-making floods in order to mobilize geohistory for fabricating origin myths of their natal lands.

In a further sign that localist epistemologies and nationalist sentiment were driving the ostensible secularization of earth history, even pious Protestant philosophers who were committed to the Universal Deluge partially adopted the patriotic floods that increasingly populated Catholic accounts of earth history. Vallisneri's correspondent Scheuchzer, when he was not publishing fossil catalogs or works on the Universal Deluge in the style of Woodward, devoted considerable energy to documenting the unique glories of Swiss nature, many of which he attributed to the aftereffects of the Flood. In addition to a three-volume *Natur-Historie des Schweizerlandes* (Natural history of Switzerland, 1706–8), he published many shorter works devoted to specific subjects under the umbrella of Swiss natural history, including *Hydrographia Helvetica* (Swiss waters, 1717), *Meteorologia Helvetica* (Swiss weather, 1718), and *Cataclysmographiam helvetiae* (Swiss catastrophes, 1733). In 1699, he circulated a questionnaire, the *Charta invitatoria* (Invitational letter, 1699), to solicit information about the weather, topography, flora, fauna, and economic output of every city, town, and village in the Swiss Confederation. In this *Invitational Letter*—like Woodward's query list, a means of gathering information across long distances—Scheuchzer contested Burnet's characterization of Flood-made mountains as "heaps of Stone and Rubbish" in order to present a far more positive picture of the Alps, which he considered to be uniquely and wholly Swiss. "We confess that our country is jagged, and rough in its appearance," Scheuchzer told his compatriots, "but foreigners should be aware that it is neither rude nor uncultivated, nor is it positioned in an abject part of the world, where Nature, as if weary of her Labors, placed it thoughtlessly and far from any order, as if it were a nuisance to other lands, or useless, or hardly important." Foreigners who regarded the Alps as nothing more than a dangerous impediment to cross-Continental travel were blind to the beauty and utility of the Swiss landscape. "Nature has bestowed many diverse gifts on our Switzerland," Scheuchzer bragged in his *Invitational Letter*, "and many miracles." Instead of the Flood's ugly stepchildren, mountains —especially the Swiss mountains—could be seen as some of it most noble and sublimely beautiful progeny. In the hands of patriotic Protestant naturalists, even the Universal Deluge could be used to show that one's own country was a distinct, and divinely created, natural space.[18]

Perhaps because the Flood was flexible enough to accommodate these patriotic trends in earth history, as Scheuchzer demonstrates so well, serious philosophical study of the Universal Deluge did not end in the Enlightenment. European naturalists and natural philosophers continued writing flood-centric narratives of earth history well into the latter half of the eighteenth century and into the early nineteenth. The majority of these writers, however, were British and Swiss Protestants. The Swiss naturalists Élie Bertrand and Jean-André de Luc, who inherited Bourguet's cabinet of natural history, proudly continued in the footsteps of Scheuchzer and Bourguet. Meanwhile, numerous British philosophers, including Francis Walsh, Richard Kirwan, John Whitehurst, and William Buckland, wrote histories of the earth that continued to feature a global Deluge identified as the biblical Flood as a pivotal moment in the earth's history.[19] Some of them, like Edward King, even continued to study the future effects of the Universal Conflagration along with the past effects of the Universal Deluge.[20] The Flood's survival in the Enlightenment was uneven across the religious and political terrain of Europe, largely limited to the Protestant countries where the Flood had been a popular (if controversial) topic of natural philosophy in the long seventeenth century. The Universal Deluge, once seen as a means of building bridges between European Christians of different countries and confessions, grew ever more starkly confessionalized.

## Deep Time and Human History

The largely Catholic rejection of the Universal Deluge in favor of nation-sized floods hardly stopped philosophers from speculating about nature's history on a planetary scale. The comte de Buffon's history of the earth, as outlined in the first volume of his magisterial *Histoire naturelle* (*Natural History*, 1749), began with the planet's birth from the collision of a comet with the sun, succeeded by six geological epochs that figuratively mapped onto the six days of the Creation, but took immeasurably longer. Meanwhile, natural philosophers continued to produce planetary histories featuring a global flood that they explicitly refused to identify with the biblical Flood of Noah, emblematic of a process that Martin Rudwick has described as an "amicable dissociation" of the biblical Flood from a fully secularized Universal Deluge.[21] If the biblical Flood once enabled the imagination of global nature, it was no longer necessary for that line of planetary and historical inquiry to continue.

But the explicit and deliberate secularization of the Universal Deluge as distinct from Noah's Flood led to the loss of human history as a constitutive part of narratives of the earth's history. These global floods-that-were-not-Noah's lacked the human actors that had been a key feature of early modern histories of the earth centered on Noah's Flood. Once stripped of its religious significance, the Universal Deluge no longer served to unite natural and human histories. Once the planet's timescale deepened into the prehuman past, and the religious link between sin and flooding was weakened, there was no longer a clearly defined role for humanity in earth history, either as the agents of geological change or as the victims or beneficiaries of its aftermath. Without humankind in the story, climate lost its key explanatory role as the means of linking the histories of people and nature.

The arrival of deep time also undermined the idea of nature as a human artifact, most forcefully expressed in early modernity by Thomas Burnet. If the earth's history started eons before the first humans lived, and if Noah's Flood, once considered humanity's signal intervention in the history of the planet, was no longer afforded geological relevance—even if most Europeans continued to believe in it as both historical fact and as an item of faith—then humanity could not have played much of a role in shaping the earth as it was in the present day. The Universal Deluge had been a lynchpin keeping the course of natural and human history in sync. The slow de-emphasis of the Flood in mainstream earth science thus paved the way for Romantic conceptions of nature as wild, alien, and completely exclusive of the human.[22]

Buffon's innovative account of the "earth's deep history," in Rudwick's felicitous phrasing, was undoubtedly a major step forward in the development of an accurate understanding of the forces that have shaped our planet's surface, interior, and atmosphere over the course of the past several billion years.[23] It is only by appreciating the naturally caused cycles of global warming and cooling in the earth's deep history that current and future anthropogenic changes to the climate have become intelligible. The Enlightenment idea of deep time and the apparent secularization of geohistory from the mid-eighteenth century onward opened up a new and necessary way of understanding the relationship between humans and the global environment, but these intellectual and religious developments came at a cost that took nearly two centuries to overcome. The triumphantly modern self-consciousness of the eighteenth- and nineteenth-century geologists who ridiculed their early modern forebears for religious credulity and insufficient empiricism obscured the insight of an earlier period—weak in data but with the strong conviction that sin could change the world—about the inseparability of human and natural history.

A complete bibliography containing all of the works cited in the notes is available online at the Johns Hopkins University Press website, www.press.jhu.edu.

### Abbreviations

| | |
|---|---|
| ASRE | Archivio di Stato di Reggio Emilia |
| BACR | Biblioteca dell'Accademia dei Concordi, Rovigo |
| BE | Biblioteca Estense, Modena |
| BL | British Library, London |
| BOD | Bodleian Library, Oxford |
| BPUN | Bibliothèque Publique et Universitaire de Neuchâtel (Fonds Louis Bourguet) |
| BUB | Biblioteca Universitaria di Bologna |
| CUL | Cambridge University Library |
| RS | Royal Society Library and Archives, London |
| ZBZ | Zentralbibliothek Zürich |

### Introduction · A Natural History of Sin

1. Pope Francis, *Encyclical Letter* Laudato Si' *of the Holy Father Francis: On Care for Our Common Home* (Vatican: Libreria Editrice Vaticana, 2015), sections 66, 11, 21.

2. *Laudato Si'*, sections 8–9.

3. Rick Santorum, interviewed on "The Dom Giordano Show," June 1, 2015, quoted in Dom Giordano, "Rick Santorum on Pope Francis' Letter on Climate Change: 'Leave Science to the Scientist,'" CBS Philly, June 1, 2015, https://philadelphia.cbslocal.com.

4. Hartnett White appealed to her own religious background and credentials, describing herself as "a lifelong Catholic with graduate degrees in religious studies," in order to authorize her argument that Francis's religious background and credentials de-authorized him from becoming "embroiled in science." Kathleen Hartnett White, "Pope Francis's Poverty and Environment Ideas Will Worsen Both," *Federalist*, June 25, 2015, https://thefederalist.com.

5. Santorum has called anthropogenic global warming "bogus" and a "hoax," while Hartnett White disputed the scientific consensus regarding the human causes of global climate change in her Senate confirmation hearings to head the White House Council on Environmental Quality in November 2017 and in her 2015 *Federalist* op-ed, in which she referred to people who accept that consensus as "warmists." Emily Schultheis, "Santorum: I Never Believed Global Warming 'Hoax,'" *Politico*, February 7, 2012, https://www.politico.com. Hart-

nett White, "Pope Francis's Poverty and Environment Ideas"; Chris Mooney, "Trump's Top Environmental Pick Says She Has 'Many Questions' about Climate Change," *Washington Post*, November 8, 2017, https://www.washingtonpost.com. Hartnett White's denialist statements to the Senate during her confirmation hearing in 2018 may have played a role in the withdrawal of her nomination. Lisa Friedman, "Trump to Withdraw Nomination of Climate Skeptic as Top Environmental Adviser," *New York Times*, February 4, 2018, https://nyt.com.

6. Jim Yardley, "Pope Francis to Explore Climate's Effect on World's Poor," *New York Times*, June 13, 2015, https://nyt.com.

7. Francis is, of course, well aware of the long history of environmental thought within the Catholic Church, as is amply demonstrated in the text of *Laudato Si'* and by his decision to take the name of St. Francis of Assisi upon ascending to the papacy.

8. Thomas Burnet, *The Theory of the Earth: Containing an Account of the Original of the Earth, and of All the General Changes Which It Hath Already Undergone, or Is to Undergo, till the Consummation of All Things* (London, 1684), 1:100.

9. On the multiple uses to which the story of Noah's Flood has been put over the centuries, see Alan Dundes, ed., *The Flood Myth* (Berkeley: University of California Press, 1988); and Martin Muslow and Jan Assmann, eds., *Sintflut und Gedächtnis* (Munich: Wilhelm Fink Verlag, 2006). On the place of the Flood in early modern European science, see Don Cameron Allen's classic, *The Legend of Noah: Renaissance Rationalism in Art, Science, and Letters* (Urbana: University of Illinois Press, 1949); and more recently, Paolo Rossi, *The Dark Abyss of Time: The History of the Earth and the History of Nations from Hooke to Vico*, trans. Lydia G. Cochrane (Chicago: University of Chicago Press, 1984); Rhoda Rappaport, *When Geologists Were Historians, 1665–1750* (Ithaca, NY: Cornell University Press, 1997); and William Poole, *The World Makers: Scientists of the Restoration and the Search for the Origins of the Earth* (Oxford: Peter Lang, 2010).

10. I use the term *natural history* both to mean the premodern branch of natural science traditionally devoted to the study of plants, animals, and minerals and also to mean "Nature's history." While historians of science generally use it only to refer to the former, both meanings were routinely ascribed to it in my primary source material. Cf. Thomas Burnet's assertion that the Flood was "a true piece of Natural history" (Burnet, *Theory*, 1:65).

11. Louis Bourguet, "Cours sur la Providence" (1733), BPUN, MS 1243, fol. 65.

12. On confessional division and toleration in early modern Europe, see Benjamin J. Kaplan, *Divided by Faith: Religious Conflict and the Practice of Toleration in Early Modern Europe* (Cambridge, MA: Belknap Press of Harvard University Press, 2007); and Keith P. Luria, *Sacred Boundaries: Religious Coexistence and Conflict in Early Modern France* (Washington, DC: Catholic University of America Press, 2005). On the confessional dimensions of Atlantic world colonialism, see Carla Gardina Pestana, *Protestant Empire: Religion and the Making of the British Atlantic World* (Philadelphia: University of Pennsylvania Press, 2009); and Jorge Cañizares Esguerra, *Puritan Conquistadors: Iberianizing the Atlantic* (Stanford, CA: Stanford University Press, 2006).

13. The impact of the Little Ice Age on European society across the early modern period is discussed in Wolfgang Behringer, *A Cultural History of Climate*, trans. Patrick Camiller (Cambridge: Polity Press, 2010); and Geoffrey Parker, *Global Crisis: War, Climate Change and Catastrophe in the Seventeenth Century* (New Haven, CT: Yale University Press, 2013). The classic work on Atlantic colonialism and disease is Alfred Crosby, *Ecological Imperialism: The Biological Expansion of Europe, 900–1900*, 2nd ed. (Cambridge: Cambridge University Press, 2004).

14. Antonio Barrera-Osorio, *Experiencing Nature: The Spanish American Empire and the Early Scientific Revolution* (Austin: University of Texas Press, 2010); Londa Schiebinger and Claudia Swan, eds., *Colonial Botany: Science, Commerce, and Politics in the Early Modern World* (Philadelphia: University of Pennsylvania Press, 2005).

15. Ann Blair, "Mosaic Physics and the Search for a Pious Natural Philosophy in the Late Renaissance," *Isis* 91, no. 1 (2000): 32–58.

16. The modern coinage is generally attributed to the chemist Paul Crutzen, who first used it in a published article in 2000 and expanded on the term in Crutzen, "Geology of Mankind," *Nature* 415 (2002): 23. A useful overview of the history of the term in recent years is J. R. McNeill, "Introductory Remarks: The Anthropocene and the Eighteenth Century," in "Humans and the Environment," special issue, *Eighteenth-Century Studies* 49, no. 2 (2016): 117–28. James Rodger Fleming begins his multicentury history of climate science with the observation, "The debate over climate change, both from natural causes and human activity, is not new." Fleming, *Historical Perspectives on Climate Change* (New York: Oxford University Press, 1998), 11.

17. Dipesh Chakrabarty, "The Climate of History: Four Theses," *Critical Inquiry* 35 (Winter 2009): 207.

18. Bronislaw Szerszynski, "The End of the End of Nature: The Anthropocene and the Fate of the Human," *Oxford Literary Review* 34, no. 2 (2012): 171. The dating of the so-called golden spike, the world-historical event recorded in the earth's stratigraphic layers that inaugurates the new Anthropocene epoch, has been widely debated across the natural and social sciences and the humanities. Following Crutzen's proposal of the Industrial Revolution, other proposals have included the Neolithic Revolution in agriculture (ca. eight thousand years ago), the rise of anthropogenic soils (ca. two thousand years ago), European colonialism in the Americas (early modern period, specifically 1610), the advent of nuclear testing (1944), and the "Great Acceleration" (1950s–present). A good overview is given in Simon L. Lewis and Mark A. Maslin, "Defining the Anthropocene," *Nature* 519 (March 2015): 171–80. As of the current writing, the Great Acceleration seems poised to win out as the formally designated golden spike. The case for this dating and definition is made in Jan Zalasiewicz et al., "When Did the Anthropocene Begin? A Mid-Twentieth Century Boundary Level Is Stratigraphically Optimal," *Quaternary International* (2014), http://dx.doi.org/10.1016/j.quaint.2014.11.045.

19. Richard H. Grove, *Green Imperialism: Colonial Expansion, Tropical Island Edens and the Origins of Environmentalism, 1600–1860* (Cambridge: Cambridge University Press, 1995), 164.

20. The classic study on theories of degeneration in colonial America is Karen Kupperman, "Fear of Hot Climates in the Anglo-American Colonial Experience," *William and Mary Quarterly* 41, no. 2 (1984): 213–40. More recently, see Jan Golinski, "American Climate and the Civilization of Nature," in *Science and Empire in the Atlantic World*, ed. James Delbourgo and Nicholas Dew (New York: Routledge, 2008), 153–74; and Sam White, "Unpuzzling American Climate: New World Experience and the Foundations of a New Science," *Isis* 106, no. 3 (2015): 544–66. On theories of climate change in the colonies more generally, see Anya Zilberstein, *A Temperate Empire: Making Climate Change in Early America* (Oxford: Oxford University Press, 2016). On theories of local (natural and man-made) climate and environmental change in early modern Europe, see Fredrik Albritton Jonsson, *Enlightenment's Frontier: The Scottish Highlands and the Origins of Environmentalism* (New Haven, CT: Yale University Press, 2013); and Brant Vogel, "The Letter from Dublin: Climate Change, Colonialism, and the Royal

Society in the Seventeenth Century," in "Klima," ed. James Rodger Fleming and Vladimir Jankovic, special issue, *Osiris* 26, no. 1 (2011): 111–28.

21. Tobias Menely and Jesse Oak Taylor describe Crutzen's announcement as "a millennial concept" not only in its timing (2000–2001) but as "a theory of (geo)historical crisis that has followed in the wake of the 'end of history' that Francis Fukuyama proclaimed at the fall of the Berlin Wall in 1989." Menely and Oak Taylor, eds., "Introduction," *Anthropocene Reading: Literary History in Geologic Times* (University Park, PA: Penn State University Press, 2017), 5.

22. Christophe Bonneuil and Jean-Baptiste Fressoz, *The Shock of the Anthropocene: The Earth, History, and Us* (London: Verso, 2017), xiii.

23. Vicky Albritton and Fredrik Albritton Jonsson, *Green Victorians: The Simple Life in John Ruskin's Lake District* (Chicago: University of Chicago Press, 2016); Jesse Oak Taylor, *The Sky of Our Manufacture: The London Fog in British Fiction from Dickens to Woolf* (Charlottesville: University of Virginia Press, 2016); and James Rodger Fleming, *Fixing the Sky: The Checkered History of Weather and Climate Control* (New York: Columbia University Press, 2010).

24. See, e.g., Anne-Marie Mercier-Faivre and Chantal Thomas, eds., *L'invention de la catastrophe au XVIIIe siècle: Du châtiment divin au désastre naturel* (Geneva: Librairie Droz, 2008); and Grégory Quenet, *Les tremblements de terre aux XVIIe et XVIIIe siècles: La naissance d'un risque* (Seyssel: Champ Vallon, 2005). For the argument that risk-based approaches to disaster management were compatible with a Christian providential interpretative framework in premodernity, see Christopher M. Gerrard and David N. Petley, "A Risk Society? Environmental Hazards, Risk and Resilience in the Later Middle Ages in Europe," *Natural Hazards* 69 (2013): 1051–79.

25. For a good overview of these trends in humanist scholarship on eighteenth-century Europe, see Alan Mikhail, "Enlightenment Anthropocene," in "Humans and the Environment," special issue, *Eighteenth Century Studies* 49, no. 2 (2016): 211–31.

26. Craig Martin, *Renaissance Meteorology: Pomponazzi to Descartes* (Baltimore, MD: Johns Hopkins University Press, 2011), 4.

27. Alexandra Walsham, *The Reformation of the Landscape: Religion, Identity, and Memory in Early Modern Britain and Ireland* (Oxford: Oxford University Press, 2011), 379, 393.

28. Denis Cosgrove, *Apollo's Eye: A Cartographic Genealogy of the Earth in the Western Imagination* (Baltimore, MD: Johns Hopkins University Press, 2001); Ayesha Ramachandran, *The Worldmakers: Global Imagining in Early Modern Europe* (Chicago: University of Chicago Press, 2015); Joyce E. Chaplin, *Round about the Earth: Circumnavigation from Magellan to Orbit* (New York: Simon and Schuster, 2012); Alison Bashford, *Global Population: History, Geopolitics, and Life on Earth* (New York: Columbia University Press, 2014); Sebastian Vincent Grevsmühl, *La terre vue d'en haut: L'invention de l'environnement global* (Paris: Editions du Seuil, 2014); and Matthias Dörries, "Krakatau 1883: Die Welt als Labor und Erfahrungsraum," in *Welt-Räume: Geschichte, Geographie und Globalisierung seit 1900*, ed. Iris Schröder and Sabine Höhler (Frankfurt: Campus, 2005), 51–73. See also Lino Camprubí, "The Invention of the Global Environment," *Historical Studies in the Natural Sciences* 46, no. 2 (2016): 243–51; and Alison B. Kavey, ed., *World-Building and the Early Modern Imagination* (New York: Palgrave Macmillan, 2010).

29. Richard Cumberland, *Origines gentium antiquissimae* (London, 1724), 143.

30. Clarence J. Glacken, *Traces on the Rhodian Shore: Nature and Culture in Western Thought*

*from Ancient Times to the End of the Eighteenth Century* (Berkeley: University of California Press, 1967); James Rodger Fleming and Vladimir Jankovic, "Introduction: Revisiting Klima," in "Klima," ed. Fleming and Jankovic, special issue, *Osiris* 26, no. 1 (2011): 1–18.

31. Donald Keene, *The Japanese Discovery of Europe, 1720–1830*, rev. ed. (Stanford, CA: Stanford University Press, 1969), 163. I am grateful to Amy Stanley for this reference.

32. Leydekker's self-identification on the title page as a member "De Republica Hebraeorum" indicates that he viewed his attack on Burnet and defense of Moses as a public intervention he was making on behalf of the Jewish diaspora, who, like the members of the Republic of Letters, formed a virtual community of people distant from one another in space. Melchior Leydekker, *Melchioris Leidekkeri S. S. Theol. D. & P. De Republica Hebraeorum . . . subjicitur Archaeologia Sacra, qua historia creationis et diluvii Mosaica contra Burneti profanam telluris theoriam asseritur* (Amsterdam, 1704).

33. Deborah R. Coen, "Big Is a Thing of the Past: Climate Change and Methodology in the History of Ideas," *Journal of the History of Ideas* 77, no. 2 (2016): 309. See also Coen, *Climate in Motion: Science, Empire, and the Problem of Scale* (Chicago: University of Chicago Press, 2018).

34. See, e.g., Dale Jamieson, *Reason in a Dark Time* (Oxford: Oxford University Press, 2014).

35. Sumathi Ramaswamy, *The Lost Land of Lemuria: Fabulous Geographies, Catastrophic Histories* (Berkeley: University of California Press, 2004), 6.

36. Nodding to Gertrude Stein's famous pronouncement about Oakland, California, Ramaswamy writes: "Lemuria is utterly unavailable outside imagination. There is simply no *there* there." Ramaswamy, *Lemuria*, 6.

37. "If most naturalists discussed the biblical flood, their reasons for doing so should be sought elsewhere than in coercion or timidity in the face of authority." Rappaport, "Geology and Orthodoxy: The Case of Noah's Flood in Eighteenth-Century Thought," *British Journal for the History of Science* 11, no. 1 (1978): 6.

38. Peter Harrison, *The Fall of Man and the Foundations of Science* (Cambridge: Cambridge University Press, 2007); Alexandra Walsham, *The Reformation of the Landscape: Religion, Identity, and Memory in Early Modern Britain and Ireland* (Oxford: Oxford University Press, 2011); and Massimo Mazzotti, *The World of Maria Gaetana Agnesi, Mathematician of God* (Baltimore, MD: Johns Hopkins University Press, 2007).

39. Martin J. S. Rudwick, *Bursting the Limits of Time* (Chicago: University of Chicago Press, 2005); Rappaport, *When Geologists Were Historians*; Rossi, *Dark Abyss of Time*; Anne Goldgar, *Impolite Learning: Conduct and Community in the Republic of Letters, 1680–1750* (New Haven, CT: Yale University Press, 1995); Dániel Margóscy, *Commercial Visions: Science, Trade, and Visual Culture in the Dutch Golden Age* (Chicago: University of Chicago Press, 2014); and Harold J. Cook, *Matters of Exchange: Commerce, Medicine, and Science in the Dutch Golden Age* (New Haven, CT: Yale University Press, 2008). See also Simon Schaffer, Lissa Roberts, Kapil Raj, and James Delbourgo, eds., *The Brokered World: Go-Betweens and Global Intelligence, 1770–1820* (Sagamore Beach, MA: Science History Publications, 2009).

40. See, for example, Poole, *The Worldmakers*; and Roy Porter, *The Making of Geology: Earth Science in Britain, 1660–1815* (Cambridge: Cambridge University Press, 1977).

41. Chakrabarty, "Climate of History"; Chaplin, *Round about the Earth*, xix; Robert O'Neill, "Is It Time to Bury the Ecosystem Concept?," *Ecology* 82, no. 12 (2001): 3275–84.

42. See Andreas Malm and Alf Hornberg, "The Geology of Mankind? A Critique of the

Anthropocene Concept," *Anthropocene Review* 1, no. 1 (2014): 62–69; and Malm, *Fossil Capital: The Rise of Steam Power and the Roots of Global Warming* (London: Verso, 2016).

43. Similar problems arise when we burrow down to the subhuman micro-level. See Julia Adeney Thomas, "History and Biology in the Anthropocene: Problems of Scale, Problems of Value," *American Historical Review* 119, no. 5 (2014): 1587–1607.

44. On the distinctions and possible connections between the global imaginaries of early modern Christian Europe, Ottoman Europe, and the Mughal Empire, see Serge Gruzinski, *What Time Is It There? America and Islam at the Dawn of Modern Times* (Cambridge: Polity Press, 2010); and Ayesha Ramachandran, "A War of Worlds: Becoming 'Early Modern' and the Challenge of Comparison," in *Comparative Early Modernities, 1100–1800*, ed. David Porter (New York: Palgrave Macmillan, 2012), 15–46.

45. As just one potential point of connection, some of the most famous early modern works of earth history, particularly Thomas Burnet's *Sacred Theory of the Earth* (1681–89), were used as textbooks in courses in the natural sciences in Oxbridge and in colleges and universities in the United States well into the nineteenth century, meaning that many of the leading mid-nineteenth-century thinkers on climate and environment in Europe and the United States, such as John Tyndall and John Ruskin, would have been familiar with them. On Tyndall, see Fleming, *Historical Perspectives on Climate Change*, chap. 6. On Ruskin, see Vicky Albritton and Fredrik Albritton Jonsson, *Green Victorians: The Simple Life in John Ruskin's Lake District* (Chicago: University of Chicago Press, 2016).

46. Pope Francis, *Laudato Si'*, sections 113, 222.

47. "There is a huge statement in the film, a strong message about the coming flood from global warming. Noah has been a silly-old-guy-with-a-white-beard story, but really it's the first apocalypse." Darren Aronofsky, quoted in Tad Friend, "Heavy Weather: Darren Aronofksy Gets Biblical," *New Yorker*, March 17, 2014.

### Chapter One · Before the Flood

1. The literature on Erculiani is small but growing and will be immeasurably advanced by the English-language edition and translation of *Letters on Natural Philosophy* and related sources forthcoming from University of Toronto Press as part of the series The Other Voice in Early Modern Europe. I am deeply grateful to Eleonora Carinci, Paula Findlen, and Hannah Marcus for sharing drafts of their essays and translations, and especially to Hannah Marcus for discussing some tricky issues of translation with me. Although I have greatly benefited from reading their translation, all translations from Erculiani's *Lettere* are my own. The secondary literature on Erculiani begins with Eleonora Carinci, "Una 'speziala' padovana: *Lettere di Philosophia Naturale* di Camilla Erculiani (1584)," *Italian Studies* 68, no. 2 (2013): 202–29. See also Meredith K. Ray, *Daughters of Alchemy: Women and Scientific Culture in Early Modern Italy* (Cambridge, MA: Harvard University Press, 2015), chap. 4, "Scientific Circles in Italy and Abroad: Camilla Erculiani and Margherita Sarrocchi"; Sandra Plastina, "'Considerar la mutatione dei tempi e delli stati e degli uomini': Le *Lettere di philosophia naturale* di Camilla Erculiani," *Bruniana & Campanelliana* 20 (2014): 145–58; and Cristina Marcon, "Camilla Gregeta Erculiani, 'scienziata' padovana del Cinquecento," *Padova e il suo territorio* 169 (2014): 37–43. An excerpt from Erculiani's *Letters* recently appeared in Lisa Kaborycha, ed., *A Corresponding Renaissance: Letters Written by Italian Women, 1375–1620* (New York: Oxford University Press, 2016), 248–52.

2. The strong intellectual and political ties between Poland and Italy, especially Padua,

in the fifteenth and sixteenth centuries gives context for Erculiani's decision to publish her book in Kraków. See Harold B. Segel, *Renaissance Culture in Poland: The Rise of Humanism, 1470–1543* (Ithaca, NY: Cornell University Press, 1989); and Joanna Kostylo, "Commonwealth of All Faiths: Republican Myth and the Italian Diaspora in Sixteenth-Century Poland-Lithuania," in *Citizenship and Identity in a Multinational Commonwealth: Poland-Lithuania in Context, 1550–1772*, ed. Karin Friedrich and Barbara M. Pendzich (Leiden: Brill, 2008), 171–205.

3. Carinci, "Una 'speziala' padovana," 221. Specifically, Erculiani was the only Renaissance Italian woman to originate and publish her own philosophical views rather than summarizing the philosophies of others. Paula Findlen, "Aristotle in the Pharmacy: The Ambitions of Camilla Erculiani in Sixteenth-Century Padua," in Erculiani, *Letters on Natural Philosophy*, ed. and trans. Carinci, Findlen, and Marcus (Toronto: University of Toronto Press, forthcoming).

4. George Louis Leclerc, comte de Buffon, *Histoire naturelle, générale et particuliére* (Paris, 1749). *Telliamed* was written in the 1720s and remained unpublished until a decade after the author's death and after having undergone heavy editing by de Maillet's literary executor, who was also a Catholic priest. Benoît de Maillet, *Telliamed, ou Entretiens d'un philosophe indien avec un missionaire francois sur la diminution de la mer, la formation de la terre, l'origine de l'homme, &c.* (Amsterdam, 1748).

5. The chemist Paul J. Crutzen, who coined the term *Anthropocene* in its current usage in 2000, dated the start of the Anthropocene to James Watt's invention of the steam engine in 1784 and the coincidental uptick in emission of greenhouse gases into the atmosphere. Crutzen, "Geology of Mankind," *Nature* 415 (2002): 23. For a good overview of these trends in humanist scholarship on eighteenth-century Europe, see Alan Mikhail, "Enlightenment Anthropocene," in "Humans and the Environment," special issue, *Eighteenth Century Studies* 49, no. 2 (2016): 211–31.

6. Apropos of a sixteenth-century Venetian friar writing about the earth's history, dal Prete observes: "[T]he author cannot provide any other natural evidence to support the Mosaic text, for the simple reason that an 'orthodox' history of the Earth had never existed." Dal Prete thereby calls our attention to the serious challenges faced by natural philosophers, even sincerely pious ones, in determining what an orthodox history of the earth would even look like. Ivano dal Prete, "'Being the World Eternal . . . ': The Age of the Earth in Renaissance Italy," *Isis* 105, no. 2 (2014), 308.

7. Dipesh Chakrabarty, "The Climate of History: Four Theses," *Critical Inquiry* 35 (2009): 206–7.

8. Marcus Frytsche, *Meteorum, hoc est impressionum aerearum et mirabilium naturae operum* (Wittenberg, 1598), A6v. Quoted in Craig Martin, *Renaissance Meteorology: Pomponazzi to Descartes* (Baltimore, MD: Johns Hopkins University Press, 2011), 5. Martin's is the most recent comprehensive survey of early modern meteorology.

9. Martin, *Renaissance Meteorology*, 71–72; Dal Prete, "'Being the World Eternal . . . ,'" 297.

10. Fausto Sebastiano da Longiano, *Meteorologia, cioè discorso de le impressioni humide et secche generate tanto ne l'aria, quanto ne le caverne de la terra* (Venice, 1542), 34r.

11. Jean Buridan, *Quaestiones super tres primos libros metheororum et super majorem partem quarti . . .* ; quoted in Pierre Duhem, *Le système du monde* (Paris: Hermann, 1958), 9:298. This work circulated widely in manuscript in the sixteenth century and was a key point of reference for Renaissance meteorology. Dal Prete, "'Being the World Eternal . . . ,'" 297.

12. Davis A. Young, *The Biblical Flood: A Case Study of the Church's Response to Extrabiblical Evidence* (Grand Rapids, MI: William B. Eerdmans, 1995), chaps. 2–4. Another major problem debated by the commentators, which would rear its head forcefully in the seventeenth century, was the size of the ark and the number and variety of animals on it. See Don Cameron Allen, *The Legend of Noah: Renaissance Rationalism in Art, Science, and Letters* (Urbana: University of Illinois Press, 1949), chap. 4.

13. Anthony Grafton, "Dating History: The Renaissance and the Reformation of Chronology," *Daedalus* 132, no. 2 (2003): 74–85; and Grafton, *New Worlds, Ancient Texts: The Power of Tradition and the Shock of Discovery* (Cambridge, MA: Harvard University Press, 1995).

14. According to Edward Grant, natural philosophers in medieval Europe following the condemnations "would rarely have considered theological issues in their treatises on natural philosophy, largely because they were all too aware that theology was the domain of theologians." Edward Grant, *God and Reason in the Middle Ages* (Cambridge: Cambridge University Press, 2001), 185–86.

15. Ann Blair, "Mosaic Physics and the Search for a Pious Natural Philosophy in the Late Renaissance," *Isis* 91, no. 1 (2000): 32–58; Craig Martin, *Subverting Aristotle: Religion, History, and Philosophy in Early Modern Science* (Baltimore, MD: Johns Hopkins University Press, 2014); Peter Harrison, *The Bible, Protestantism, and the Rise of Natural Science* (Cambridge: Cambridge University Press, 1998), esp. 138–47.

16. Lambert Daneau, *The Wonderfull Workmanship of the World Wherein Is Conteined an Excellent Discourse of Christian Natural Philosophie*, trans. Thomas Twyne (London, 1578), 7r–7v.

17. Sachiko Kusukawa, *The Transformation of Natural Philosophy: The Case of Melanchthon* (Cambridge: Cambridge University Press, 1995).

18. Blair, "Mosaic Physics," 33. In subsequent centuries, the Jesuits would spearhead this effort. See, for example, Marcus Hellyer, *Catholic Physics: Jesuit Natural Philosophy in Early Modern Germany* (Notre Dame, IN: University of Notre Dame Press, 2005).

19. Blair, "Mosaic Physics," 35, 57.

20. Francesco de' Vieri, *Trattato delle Meteore* (Florence, 1573), 4.

21. Virginia Cox, *The Renaissance Dialogue: Literary Dialogue in Its Social and Political Contexts, Castiglione to Galileo* (Cambridge: Cambridge University Press, 1992).

22. On Luther's apocalyptic reading of natural disasters, see Kusukawa, *Transformation*, chap. 4. On the role of the Reformations more generally in encouraging providential readings of natural disasters, see Jennifer Spinks, *Disaster, Death and the Emotions in the Shadow of the Apocalypse, 1400–1700* (London: Palgrave, 2016). Religious interpretations of local natural disasters did not begin in the Reformations, of course. For the medieval period, see Jacques Berlioz, *Catastrophes naturelles et calamités au Moyen Age* ([Florence]: Edizioni del Galluzzo, 1998); and Thomas Labbé, *Les catastrophes naturelles au Moyen Âge* (Paris: CNRS Éditions, 2017).

23. Raingard Esser, "Fear of Water and Floods in the Low Countries," in *Fear in Early Modern Society*, ed. William G. Naphy and Penny Roberts (Manchester: Manchester University Press, 1997), 65. The broader connections between the Little Ice Age, the Dutch revolt, and increased flooding in the Low Countries, are discussed in Dagomar de Groot, *The Frigid Golden Age: Climate Change, the Little Ice Age, and the Dutch Republic, 1560–1720* (Cambridge: Cambridge University Press, 2018), esp. chap. 4.

24. Esser, "Fear of Water and Floods," 67–69.

25. Elaine Fulton, "Acts of God: The Confessionalization of Disaster in Reformation Europe," in *Historical Disasters in Context: Science, Religion, and Politics*, ed. Andrea Janku, Gerrit Jasper Schenk, and Franz Mauelshagen (New York: Routledge, 2012), 54–74.

26. The intellectual and print history of the astrological predictions of a global flood for 1524 is detailed in Paola Zambelli, "Fine del mondo o inizio della propaganda? Astrologia, filosofia della storia e propaganda politico-religiosa nel dibattito sulla congiunzione del 1524," in *Scienze, credenze occulte, livelli di cultura: Convegno internazionale di studi* (Florence: Leo S. Olschki, 1982): 291–368; and Zambelli, "Many Ends for the World: Luca Gaurico Instigator of the Debate in Italy and Germany," in *Astrologi hallucinati: Stars and the End of the World in Luther's Time*, ed. Zambelli (Berlin: Walter de Gruyter, 1986), 239–63. Savonarola's earlier prediction is discussed in Donald Weinstein, *Savonarola: The Rise and Fall of a Renaissance Prophet* (New Haven, CT: Yale University Press, 2011), 103–4. Popular reactions to the predictions are discussed in Ottavia Niccoli, *Prophecy and People in Renaissance Italy*, trans. Lydia Cochrane (Princeton, NJ: Princeton University Press, 1987), chap. 6. Several other apocryphal stories about noblemen building arks in anticipation of the 1524 flood are attested but have so far proved impossible to verify. Starting in the mid-fifteenth century, floods in France and Italy often triggered apocalyptic predictions. See Labbé, *Les catastrophes naturelles*, 175.

27. Steven Vanden Broecke, *The Limits of Influence: Pico, Louvain, and the Crisis of Renaissance Astrology* (Leiden: Brill, 2003), 84.

28. Niccoli, *Prophecy and People*, 134–35, 148–49. At the same time, Craig Martin argues that the backlash against the predictions of a second biblical deluge prompted meteorologists to insist that a natural, global flood was physically impossible. Martin, *Renaissance Meteorology*, 72.

29. On the distinction and collapse between the terms *alluvione* and *Diluvio*, see Niccoli, *Prophecy and People*, 143; and Gerrit Jasper Schenk, "Dis-astri: Modelli interpretativi delle calamità naturali dal Medioevo al Rinascimento," in *Le calamità ambientali nel tardo medioevo europeo: Realtà, percezioni, reazioni*, ed. Michael Matheus, Gabriella Piccinni, Giuliano Pinto, and Gian Maria Varanini (Florence: Firenze University Press, 2010), 45–49. The Greek New Testament likewise uses different terms to distinguish the two: *katakylsomos* for Noah's Flood and *potamos* for all other floods. Young, *Biblical Flood*, 15.

30. Nicolò Vito di Gozze, *Discorsi . . . sopra la Metheore di Aristotele, ridotti in dialogo . . .* (Venice, 1584), 56r.

31. di Gozze, *Discorsi*, 56v–57r.

32. Berzeviczy was clearly comfortable moving between the increasingly divided terrain of Protestant and Catholic Europe, a capacity that reflected the toleration of Poland-Lithuania and served him well on a mission from Báthory to the pope in 1573–74. Felicia Rosu, *Elective Monarchy in Transylvania and Poland-Lithuania, 1569–1587* (Oxford: Oxford University Press, 2018), 84.

33. The professional term used in the title is *speciala*, literally "spicer," or one who sells spices. Given the use of spices and other exotic ingredients in the manufacture of drugs and medicines, the term *speciale* was also used to refer to apothecaries. Ray, *Daughters of Alchemy*, 115.

34. The ways in which Erculiani's family and professional life brought her into regular contact with scholars from the university is discussed in Marcon, "Erculiani, 'scienziata' padovana," 38–40.

35. References in the letters to previous conversations and epistolary communications be-

tween the three authors suggest that Erculiani knew both men personally and continued to correspond with them after they left Padua. Ray raises the tantalizing possibility that Erculiani may have written all of the letters herself, citing the common practice of a single author composing an epistolary exchange specifically for publication (*Daughters of Alchemy*, 123). Carinci doubts this, arguing that Erculiani's surviving correspondence with Sebastiano Erizzo demonstrates Erculiani's real and significant engagement with the Republic of Letters. Carinci, "Introduction," *Letters on Natural Philosophy*. Erculiani nevertheless would have exercised significant control over the selection and editing of the letters for print publication.

36. Marco Sgarbi, "Aristotle and the People: Vernacular Philosophy in Renaissance Italy," *Renaissance and Reformation* 39, no 3 (2016): 59–109.

37. Ray, *Daughters of Alchemy*, 112; Craig Martin, "Meteorology for Courtiers and Ladies: Vernacular Aristotelianism in Renaissance Italy," *Philosophical Readings* 4, no. 2 (2012): 3–14.

38. Erculiani read or at least knew of the meteorological works of her fellow Paduan Alessandro Piccolomini, referencing his *Natural Philosophy* in support of her contention that the flood must have been the result of preexisting water rising up beyond its usual bounds rather than new quantities of water being created, because Piccolomini held that the four elements exist on earth in a fixed quantity. Camilla Erculiani to Márton Berzeviczy, April 9, 1581, in *Lettere di philosophia naturale di Camilla Herculiana, speciala alle tre stelle in Padoua . . . nella quale si tratta la natural causa delli Diluuij, & il natural temperamento dell'huomo, & la natural formatione dell'Arco celeste* (Kraków, 1584), f2v–f3r.

39. Erculiani's decision to include a discussion of the rainbow in her book on the Flood was in line with a Christian philosophical tradition holding that the rainbow that appeared to Noah after the Flood was the first rainbow, ever. Her choice of topics and her insistence on treating them as natural phenomena raises the possibility that she might have had access to Longiano's short, readable 1542 digest of Aristotle's *Meteorology*, whose final paragraph discusses the Flood, the ark, and the rainbow as phenomena that were both natural and supernatural, though he also concludes they were things that "it is not possible for the human intellect to comprehend." Longiano, *Meteorologia*, 44v. Might she have taken this as a challenge?

40. Erculiani to Berzeviczy, April 9, 1581, in *Lettere*, f4r.

41. Plastina, " 'Considerar la mutatione dei tempi e delli stati e degli homini,' " 156.

42. Erculiani to Georges Guarnier, November 9, 1577, in *Lettere*, e3r.

43. On astrology's close ties to religion, especially prophecy and apocalyptic prediction, during the Reformations, see Robin B. Barnes, *Astrology and Reformation* (Oxford: Oxford University Press, 2016); and Barnes, *Prophecy and Gnosis: Apocalypticism in the Wake of the Lutheran Reformation* (Stanford, CA: Stanford University Press, 1988).

44. Galen's theory of mixtures exercised considerable influence in early modern science, medicine, and environmental thought. See Wolfram Schmidgen, *Exquisite Mixture: The Virtues of Impurity in Early Modern England* (Philadelphia: University of Pennsylvania Press, 2013), 25.

45. Erculiani to Guarnier, August 7, 1577, in *Lettere*, b2r.

46. Erculiani to Berzeviczy, April 9, 1581, in *Lettere*, f2r.

47. Erculiani to Guarnier, August 7, 1577, in *Lettere*, b1r–b1v.

48. Erculiani to Guarnier, November 9, 1577, in *Lettere*, e1v–e2r.

49. Erculiani to Berzeviczy, April 9, 1581, in *Lettere*, f2r–f2v.

50. The early modern obsession with antediluvian gigantism is discussed in Nancy G.

Siraisi, *History, Medicine, and the Traditions of Renaissance Learning* (Ann Arbor: University of Michigan Press, 2007), chap. 1.

51. Erculiani to Berzeviczy, April 9, 1581, in *Lettere*, f2v.

52. Findlen, "Aristotle in the Pharmacy."

53. Berns argues that medical practitioners were especially open to the idea of jointly pursuing natural philosophy and religious truth. Crucially, this was an interfaith movement including Jewish as well as Catholic scholars. Andrew Berns, *The Bible and Natural Philosophy in Renaissance Italy: Jewish and Christian Physicians in Search of Truth* (Cambridge: Cambridge University Press, 2015).

54. Erculiani to Guarnier, November 9, 1577, in *Lettere*, e1v.

55. Erculiani to Berzeviczy, April 9, 1581, in *Lettere*, f2r.

56. On Erculiani's awareness as an author of entering a male discursive space, see Carinci, "Una 'speziala' padovana," 221.

57. For example, "All of these things were said by that man and by me. . . ." Erculiani to Guarnier, August 7, 1577, in *Lettere*, b3r.

58. A possible exception is Isotta Nogarola's debate with Ludovico Foscarini, though it was not public in the same way. See the discussion of Nogarola below, note 69. Ray notes that Erculiani "makes use of many topoi of pro-woman *querelle* texts in her letter to Anna Jagiellon," indicating her awareness of this genre, which frequently listed exemplary learned women—modern as well as ancient—as part of broader feminist arguments. Ray, *Daughters of Alchemy*, 119.

59. Erculiani, "A lettori," in *Lettere*, a3v.

60. On the tradition of learned women in the Italian Renaissance, see the rich secondary literature that includes Diana Robin, *Publishing Women: Salons, the Presses, and the Counter-Reformation in Sixteenth-Century Italy* (Chicago: University of Chicago Press, 2007); Virginia Cox, *Women's Writing in Italy, 1400–1650* (Baltimore, MD: Johns Hopkins University Press, 2008); and Sarah Gwyneth Ross, *The Birth of Feminism: Woman as Intellect in Renaissance Italy and England* (Cambridge, MA: Harvard University Press, 2009).

61. Erculiani, "Alla Serenissima Regina Anna," in *Lettere*, a2v.

62. Much of the Paduan Inquisition's archives were destroyed when Napoleon's armies invaded in 1797, so we cannot be certain about the precise date or outcome of her interrogation. However, Carinci has established that Erculiani was not punished harshly or perhaps even at all, nor was *Letters on Natural Philosophy* placed on the *Index of Prohibited Books*. Carinci, "Una 'speziala' padovana," 222–23. On the relationship between the Holy Office in Rome and local inquisitorial tribunals, see Christopher F. Black, *The Italian Inquisition* (New Haven, CT: Yale University Press, 2009). On the workings of the Paduan office in particular, see Antonino Poppi, *Cremonini e Galilei inquisiti a Padova nel 1604: Nuovi documenti d'archivio* (Padua: Editrice Antenore, 1992).

63. Ugo Baldini, "The Roman Inquisition's Condemnation of Astrology: Antecedents, Reasons, and Consequences," in *Church, Censorship and Culture in Early Modern Italy*, ed. Gigliola Fragnito, trans. Adrian Belton (Cambridge: Cambridge University Press, 2001), 91.

64. Erculiani to Guarnier, August 7, 1577, in *Lettere*, b1r.

65. Erculiani to Guarnier, August 7, 1577, in *Lettere*, c2v. See also Erculiani to Guarnier, November 9, 1577, in *Lettere*, f1v.

66. Erculiani to Guarnier, August 7, 1577, in *Lettere*, b2v.

67. Sebastiano Erizzo to Erculiani, February 27, 1584, Biblioteca Bertoliana di Vicenza, MS 277, 190r, 191v. I am grateful to Eleonora Carinci for sharing scans of these letters with me, which are also reproduced and discussed in Maude Vanheulen, "Platonism in Sixteenth-Century Padua: Two Unpublished Letters from Sebastiano Erizzo to Camilla Erculiani," *Bruniana & Campanelliana* 22, no. 1 (2016): 137–47.

68. Erculiani, "A lettori," in *Lettere*, a3v–a4r.

69. Her interlocutor, the Venetian humanist and governor of Verona, Ludovico Foscarini, maintained the misogynist opinion that Eve's relative weakness did not lessen her sin but rather amplified it. See Isotta Nogarola, *Dialogue on the Equal or Unequal Sin of Adam and Eve* (1451), in *Complete Writings: Letterbook, Dialogue on Adam and Eve, Orations*, ed. and trans. Margaret L. King and Diana Robin (Chicago: University of Chicago Press, 2003), chap. 7. Nogarola's views on gender as expressed in this dialogue are discussed in Luka Boršić and Ivana Skuhala Karasman, "Isotta Nogarola: The Beginning of Gender Equality in Europe," *Monist* 98, no. 1 (2015): 43–52.

70. On Marinella in the context of Renaissance Italian debates on women, see Stephen Kolsky, "Moderata Fonte, Lucrezia Marinella, Giuseppe Passi: An Early Seventeenth-Century Feminist Controversy," *Modern Language Review* 96, no. 4 (2001): 973–89; Ross, *The Birth of Feminism*; and Letizia Panizza, "Introduction to the Translation," in Marinella, *The Nobility and Excellence of Women and the Defects and Vices of Men*, ed. and trans. Anne Dunhill (Chicago: University of Chicago Press, 1999), 1–34.

71. Findlen speculates about the possibility that Erculiani knew of the works of Marinella's father, Giovanni Marinelli, and whether Marinella in turn would have encountered Erculiani's book in her father's library as a young woman. Findlen, "Aristotle in the Pharmacy."

72. Lucrezia Marinella, *La nobiltà e l'eccellenza delle donne, co' diffetti, e mancamenti de gli huomini*, 2nd ed. (Venice, 1601).

73. Marinella, *La nobiltà*, 30.

74. Marinella, *La nobiltà*, 38–39.

75. Henry Cornelius Agrippa's *De nobilitate et praecellentia foeminei sexus* [On the nobility and excellence of women, 1529] was an important precursor for Marinella's argument for the superiority of women. It was also translated from Latin into Italian in 1549 by Erculiani's favorite modern meteorologist, Alessandro Piccolomini.

76. Marinella, *La nobiltà*, 9.

77. Marinella, *La nobiltà*, 13. The language of "complexion" appears frequently in early modern scientific texts attempting to explain differences between raced and gendered bodies. See chapter 2, note 60.

78. Marinella, *La nobiltà*, 31. Acknowledging openly that her theories flew in the face of Aristotle and of his modern supporters like Passi, Marinella sought to portray their misogyny as the product of male jealousy and the indirect reflection of female superiority. "This good friend of Aristotle says that women must obey men always and in everything . . . but I desire that we should excuse him, because, being a man, it was fitting that he should desire the greatness and superiority of men over women" (*La nobiltà*, 32). For Marinella's reworking of the Aristotelian theory relating body temperature to gendered capacities, see Marguerite Deslauriers, "Marinella and Her Interlocutors: Hot Blood, Hot Words, Hot Deeds," *Philosophical Studies* 174 (2017): 2525–37.

79. Marinella, *La nobiltà*, 12.

80. Marinella, *La nobiltà*, 79.

81. Marinella, *La nobiltà*, 271.

82. Erculiani to Guarnier, November 9, 1577, in *Lettere*, e4v.

83. On Protestant students at the University of Padua, see Paul F. Grendler, *The Universities of the Italian Renaissance* (Baltimore, MD: Johns Hopkins University Press, 2002), 190–95.

84. Kathleen M. Crowther, *Adam and Eve in the Protestant Reformation* (Cambridge: Cambridge University Press, 2010), 7.

85. The philosophers Pietro Pomponazzi, Jacopo Zabarella, and Cesare Cremonini all spent time at the University of Padua and all considered, through their exploration of the classic Aristotelian question of the immortality of the soul, issues of materialism and embodiment in relation to spiritual states. See Heinrich C. Kuhn, *Venetischer Aristotelismus im Ende der aristotelischen Welt: Aspekte der Welt und des Denkens des Cesare Cremonini (1550–1631)* (Frankfurt: Peter Lang, 1996); Ed Muir, *The Culture Wars of the Late Renaissance: Skeptics, Libertines, and Opera* (Cambridge, MA: Harvard University Press, 2007), chap. 1.

86. On Augustine's popularity across Europe in the Reformations, see Arnoud S. Q. Visser, *Reading Augustine in the Reformation: The Flexibility of Intellectual Authority in Europe, 1500–1620* (Oxford: Oxford University Press, 2011). Augustine's popularity in Renaissance Italy specifically is discussed in Meredith J. Gill, *Augustine in the Italian Renaissance: Art and Philosophy from Petrarch to Michelangelo* (Cambridge: Cambridge University Press, 2005).

87. Young, *Biblical Flood*, 15–18.

88. Menochio's account of the trial references Pelagianism, the heresy of denying original sin or its continued influence on Adam's descendants, indicating that the Inquisitors may have read Erculiani as denying the role of sin in causing the Flood.

89. Giacomo Menochio, *Consiliorum sive responsorum*, vol. 8 (Frankfurt, 1676), 172.

90. Guarnier to Erculiani, September 7, 1577, in *Lettere*, c4v.

91. Carinci, "Una 'speziala' padovana," 228–29.

92. Carinci, "Una 'speziala' padovana," 228.

93. Carinci advances the fascinating argument that Erculiani's case demonstrates that women had more *libertas philosophandi* than men for precisely the reason that their public speech was seen as less serious and less dangerous. Carinci, "Una 'speziala' padovana," 228–29.

94. On the *imbecilitas* defense, see Ian McLean, *The Renaissance Notion of Woman* (Cambridge: Cambridge University Press, 1983), 78–79.

95. Vitale Zuccolo, *Dialogo delle cose meteorologiche* (Venice, 1590), 1612.

### Chapter Two · After the Flood

1. A note on terminology and voice: I have chosen to use the terms *Native American* or *indigenous American* when referring in my scholarly voice to the indigenous peoples and nations of the Americas, but to use the historical terms *Indian* or *peoples of the New World* when inhabiting the voice of my historical actors. This is my uneasy solution to the challenge of trying to decolonize my scholarly account of a colonialist discourse while at the same time calling attention to the denial of indigeneity that (I argue) was a central function and premise of European and Euro-American histories of pre-Colombian migration to the Americas.

2. Henrico Martínez, *Repertorio de los tiempos e historia natural de Nueva España* (Mexico City: Secretaría de Educación Pública, 1948), 120. On Martínez as a Creole natural philosopher, see Jorge Cañizares Esguerra, "New World, New Stars: Patriotic Astrology and the Invention of Indian and Creole Bodies in Colonial Spanish America," *American Historical Review* 104, no. 1 (1999): 60–65. On Martínez and the early modern global imagination,

see Serge Gruzinski, *What Time Is It There? America and Islam at the Dawn of Modern Times* (Cambridge: Polity Press, 2010).

3. Gregorio García, *Origen de los Indios de el Nuevo Mundo e Indias Occidentales* (Valencia, 1607), 13. Italics mine.

4. Further research may of course reveal other lost voices like Erculiani's. Generally speaking, however, advocates for incorporating a global flood into natural philosophy were few and far between prior to the 1660s.

5. Matthew Hale, *The Primitive Origination of Mankind* (London, 1677), 193.

6. Many modern scholars writing about the early modern craze for Noah's Flood have noted that histories of the Flood were often paired with histories of postdiluvian migration, but only a few have noted the key role of land bridges in those debates, or the fact that research on postdiluvian migration was almost always aimed at understanding contemporary human difference in the age of imperial expansion. Exceptions are William Poole, *The World Makers: Scientists of the Restoration and the Search for the Origins of the Earth* (Oxford: Peter Lang, 2010), and Don Cameron Allen, *The Legend of Noah: Renaissance Rationalism in Art, Science, and Letters* (Urbana: University of Illinois Press, 1949), chap. 6.

7. Even the best scholarship on the imbrication of religious and racial thinking in early modern Europe and its empires treats geohistory only in passing. See, for example, David N. Livingstone, *Adam's Ancestors: Race, Religion, and the Politics of Human Origins* (Baltimore, MD: Johns Hopkins University Press, 2008); and Colin Kidd, *The Forging of Races: Race and Scripture in the Protestant Atlantic World, 1600–2000* (Cambridge: Cambridge University Press, 2006). A notable exception is Andrés I. Prieto, "Reading the Book of Genesis in the New World: José de Acosta and Bernabé Cobo on the Origins of the American Population," *Hispanófila* 150 (2010): 1–19, which shows how the need to legitimize American evangelism and find the ancient land bridge combined to produce civilizational, if not racial, hierarchies within the monogenetic human family.

8. See, for example, Londa Schiebinger, *Plants and Empire: Colonial Bioprospecting in the Atlantic World* (Cambridge, MA: Harvard University Press, 2004); and Daniela Bleichmar, *Visible Empire: Botanical Expeditions and Visual Culture in the Hispanic Enlightenment* (Chicago: University of Chicago Press, 2012).

9. Native and indigenous archaeologists have identified lingering racist and colonialist assumptions in contemporary theories of Native American origins, many of which continue to invoke a trans-Pacific land bridge. See Roger Echo-Hawk and Larry J. Zimmerman, "Beyond Racism: Some Opinions about Racialism and American Archaeology," *American Indian Quarterly* 30 (2006): 461–85. I am grateful to Kelly Wisecup for this reference.

10. On the robust borrowing between natural and human history in early modern Europe, see Rhoda Rappaport, *When Geologists Were Historians, 1665–1750* (Ithaca, NY: Cornell University Press, 1997). On their divergence in the eighteenth century, see Robin Valenza, *Literature, Language, and the Rise of the Intellectual Disciplines in Britain, 1680–1820* (Cambridge: Cambridge University Press, 2009). Stephen Gaukroger sees disciplinary separation as well as continued borrowing in the "naturalization of the human and the humanization of nature." Gaukroger, *The Natural and the Human: Science and the Shaping of Modernity, 1739–1841* (Oxford: Oxford University Press, 2016), 9.

11. Fabien Locher and Jean-Baptiste Fressoz, "Modernity's Frail Climate: A Climate History of Environmental Reflexivity," *Critical Inquiry* 38, no. 3 (Spring 2012): 579–98.

12. The classic work on early modern European and Euro-American debates about Native

American origins are discussed in Lee Eldridge Huddleston, *Origins of the American Indians: European Concepts, 1492–1729* (Austin: University of Texas Press, 1967). The most comprehensive and the more analytically sophisticated study is Giuliano Gliozzi, *Adamo e il Nuovo Mondo: La nascita dell'antropologia come ideologia coloniale* (Florence: La nuova Italia, 1977).

13. Martínez, *Repertio*, 120. Key themes and personnel in the debate on the likelihood of postdiluvian migration by land or by sea are surveyed in Gliozzi, *Adamo e il Nuovo Mondo*, 259–73.

14. Walter Raleigh, *History of the World* (London, 1614), 135.

15. Hale, *Primitive Origination*, 202–3. Hale, who had a solid grounding in natural philosophy but was hardly an expert naturalist, was apparently unaware that there were no lions or tigers in the New World.

16. José de Acosta, *Historia natural y moral de las Indias* (Barcelona, 1591), 381. On Acosta's discussion of Indian origins in *Historia natural y moral*, see Gliozzi, *Adamo e il Nuovo Mondo*, 371–81; Saul Jarcho, "Origin of the American Indian as Suggested by Fray Joseph de Acosta (1589)," *Isis* 50, no. 4 (1959): 430–38; and Prieto, "Genesis in the New World."

17. Although they bear different publication dates, *De procuranda Indorum salute* (1588) and *De natura Novi Orbis* (the 1589 Latin edition) were frequently bound together, are continuously paginated, and share the same publisher and place of publication, indicating that Acosta intended them as companion pieces.

18. Paracelsus, *Astronomia Magna, oder die ganze Philosophia sagax der grossen und kleinen Welt (1537/8)*, in *Sämmtliche Werke*, Abteilung 1, Band 12, ed. Karl Sudhoof (Munich: Verlag von K. Oldenbourg, 1929), 35. Paracelsus begins this passage with the declaration, "We are all descended from Adam," by which he could have meant to frame the next sentences as strictly hypothetical, but perhaps "we" was meant to refer only to his European readers, thus leaving open the possibility that the polygenist speculations to come were in fact real possibilities.

19. On European ideas of the presumed otherness or subhumanity of Native Americans, see Patricia Seed, "'Are These Not Also Men?': The Indians' Humanity and Capacity for Spanish Civilisation," *Journal of Latin American Studies* 25, no. 3 (1993): 629–52; and Surekha Davies, *Renaissance Ethnography and the Invention of the Human: New Worlds, Maps and Monsters* (Cambridge: Cambridge University Press, 2017).

20. Felipe Guaman Poma de Ayala, *The First New Chronicle and Good Government Abridged*, ed. and trans. David Frye (Indianapolis, IN: Hackett Publishing, 2006), 294. I am grateful to Paul Ramírez for bringing this source to my attention.

21. Guaman Poma, *The First New Chronicle and Good Government: On the History of the World and the Incas up to 1615*, ed. and trans. Roland Hamilton (Austin: University of Texas Press, 2009), 21.

22. On La Peyrère, see Richard Popkin, *Isaac La Peyrère (1596–1676): His Life, Work, and Influence* (Leiden: Brill, 1987); and Anthony Grafton, "Isaac La Peyrère and the Old Testament," in *Defenders of the Text: The Traditions of Scholarship in the Age of Science, 1450–1800* (Cambridge, MA: Harvard University Press, 1991), 204–13. La Peyrère's complicated religious identity has been memorably characterized as "then a Calvinist, previously perhaps a Jew, and soon to be a Catholic." J. L. Heilbron, "Bianchini, Historian," in *Thinking Impossibilities: The Intellectual Legacy of Amos Funkenstein*, ed. Robert S. Westman and David Biale (Toronto: University of Toronto Press, 2008), 231. La Peyrère's formal conversion from Calvinism to Catholicism is well-documented, but his status as a possible descendant of *conversos*, that is, Iberian Jewish converts to Christianity, is disputed. Popkin reviews the literature on the

question of La Peyrère's possible Jewish heritage and finds the evidence inconclusive, but on balance persuasive (*Isaac La Peyrère*, 21–25).

23. On the pre-Adamic controversy in relation to early modern earth history, see Poole, *The World Makers*, 27–37; and in relation to the longer history of debates about monogenism and polygenism, see Livingstone, *Adam's Ancestors*, 26–51; and Claudine Poulouin, *Le temps des origines* (Paris: Honoré Champion, 1998), 95–144.

24. Isaac La Peyrère, *A Theological Systeme upon the Presupposition, That Men Were before Adam* (London, 1655), 276, 239. *Men before Adam* was published in two parts with differing titles, *Praeadamitae* and *Systema theologicum, ex prae-Adamitarum hypothesi*. They were both published in 1655 and are often found bound together, e.g., Newberry Library, case 4A 1852. English translations of both parts appeared almost immediately as *Men before Adam* (1656) and *A Theological Systeme upon the Presupposition, That Men Were before Adam* (1655). I have chosen to quote from the contemporaneous (and sympathetic) English editions of 1655–56 and have referred to the title uniformly in the main text as *Men before Adam*.

25. Popkin, *Isaac La Peyrère*, 6.

26. Edward Stillingfleet, *Origines Sacrae: Or a Rational Account of the Grounds of Christian Faith*, 3rd ed. (London, 1666), 537.

27. Popkin, *Isaac La Peyrère*, 16.

28. Stillingfleet, *Origines Sacrae*, 534.

29. La Peyrère, *A Theological Systeme upon the Presupposition, That Men Were before Adam*, 8–10.

30. Colin Kidd sums up polygenism's consequences for evangelism nicely: "If the people of America turned out to be an autochthonous race which had sprung up separately from the rest of humanity, then the universality of original sin and of the corresponding gospel promise of redemption was a nonsense." Kidd, *Forging of Races*, 61.

31. On philo-semitism and messianism in La Peyrère's writings, see Ira Robinson, "Isaac de la Peyrère and the Recall of the Jews," *Jewish Social Studies* 40 (1978): 117–30. At the same time, La Peyrère's messianic vision of the newly Christian Jews being led by Jesus and the king of France to reclaim their ancestral homeland indicates that his universalist vision of the future, as opposed to his polycentric vision of the past, partook in some of the same colonialist and evangelical logics as did the universalist histories of his contemporaries.

32. Georg Horn, *Dissertatio de vera aetate mundi* (Leiden, 1659), 61. Voss and Horn's pamphlet war—the original treatise by Voss with the same title, Horn's reply, Voss's counter-reply, and two further counter-counter-replies from Horn, all published over the course of 1659—was bound together by at least one early modern reader in a volume held by the Newberry Library, case BS637.A2 V68 1659.

33. Grandi's treatise was written in the form of a letter and published in Jacopo Grandi and Joannes Quirini, *De testaceis fossilibus Musaei Septalliani et Jacobi Grandii de veritate diluvii universalis, et testaceorum, qua procul a Mari reperiuntur generatione epistolae* (Venice, 1676), 23.

34. See, for example, Joan-Pau Rubiés, "Hugo Grotius's Dissertation on the Origin of the American Peoples and the Use of Comparative Methods," *Journal of the History of Ideas* 52, no. 2 (1991): 221–44; and Benjamin Schmidt, "Space, Time, Travel: Hugo de Groot, Johannes De Laet, and the Advancement of Geographical Learning," *Lias* 25 (1998): 177–99.

35. Pierre Vidal-Naquet, "Atlantis and the Nations," trans. Janet Lloyd, *Critical Inquiry* 18, no. 2 (Winter 1992): 313–14.

36. Antonio de Herrera y Tordesillas, *The General History of the Vast Continent and Islands of America*, trans. John Stevens (London, 1725), 1:22.

37. For an introduction to the theology and politics of Calancha's *Corónica*, see Sabine MacCormack, "Antonio de la Calancha: Un Agostino del siglo XVII en el Nuevo Mundo," *Bulletin Hispanique* 84 (1982): 60–94. Calancha's theory of Native American origins is discussed in Gliozzi, *Adamo e il Nuovo Mondo*, 41–46. For Calancha as a Creole natural philosopher and astrologer, see Cañizares-Esguerra, "New World, New Stars"; and Claudia Brosseder, "Astrology in Seventeenth-Century Peru," *Studies in History and Philosophy of Biological and Biomedical Sciences* 41 (2010): 146–57.

38. On Mendicant orders in the Spanish Americas, see Karen Melvin, *Building Colonial Cities of God: Mendicant Orders and Urban Culture in New Spain* (Stanford, CA: Stanford University Press, 2012). On indigenous Andean religious beliefs and practices and their complex relationship to Catholicism in the colonial era, see Sabine MacCormack, *Religion in the Andes: Vision and Imagination in Early Colonial Peru* (Princeton, NJ: Princeton University Press, 1991); and Claudia Brosseder, *The Power of Huacas: Change and Resistance in the Andean World of Colonial Peru* (Austin: University of Texas Press, 2014).

39. Martínez, *Repertio*, 121.

40. Antonio de la Calancha, *Corónica moralizada del orden de San Augustín en el Perú* (Barcelona, 1638), 43.

41. Calancha, *Corónica*, 42.

42. Calancha, *Corónica*, 42, 44.

43. Edward Brerewood, *Enquiries Touching the Diversity of Languages, and Religions, through the Chief Parts of the World*, 2nd ed. (London, 1622), 96–97.

44. A stimulating recent reappraisal of the significance of the imagined historical relationship between indigenous Americans and Jews by early modern European Christians is Jonathan Boyarin, *The Unconverted Self: Jews, Indians, and the Identity of Christian Europe* (Chicago: University of Chicago Press, 2009).

45. Benjamin Braude, "The Sons of Noah and the Construction of Ethnic and Geographical Identities in the Medieval and Early Modern Periods," *William and Mary Quarterly* 54, no. 1 (1997): 103–42.

46. Calancha, *Corónica*, 35, 42.

47. On anti-Semitism and the Inquisition in Calancha's Peru, see Ana E. Schaposchnik, *The Lima Inquisition: The Plight of Crypto-Jews in Seventeenth-Century Peru* (Madison: University of Wisconsin Press, 2015). On the resistance of conversos, moriscos, and other suspected heretics to the colonial Inquisition, see Stuart B. Schwartz, *All Can Be Saved: Religious Tolerance and Salvation in the Iberian Atlantic World* (New Haven, CT: Yale University Press, 2008).

48. Calancha, *Corónica*, 41–42.

49. Calancha, *Corónica*, 42.

50. Calancha, *Corónica*, 43.

51. Cañizares Esguerra, "New World, New Stars," 58.

52. Cañizares Esguerra, "New World, New Stars," 65. See also James H. Sweet, "The Iberian Roots of American Racist Thought," *William and Mary Quarterly* 54, no. 1 (1997): 143–66.

53. Braude, "Sons of Noah," 127–28.

54. On the Valladolid debate, see Lewis Hanke, *Aristotle and the American Indians: A*

*Study in Race Prejudice in the Modern World* (London: Holis and Carter, 1959). On the theory of natural slavery, see Anthony Pagden, *The Fall of Natural Man: The American Indian and the Origins of Comparative Ethnology* (Cambridge: Cambridge University Press, 1982), chap. 3.

55. On astrology and human diversity, see Cañizares Esguerra, "New World, New Stars"; and Brosseder, "Astrology in Seventeenth-Century Peru." On climatology and human diversity, see Clarence J. Glacken, *Traces on the Rhodian Shore: Nature and Culture in Western Thought from Ancient Times to the End of the Eighteenth Century* (Berkeley: University of California Press, 1967); and David Livingstone, "Race, Space and Moral Climatology: Notes towards a Genealogy," *Journal of Historical Geography* 28, no. 2 (2002): 159–80. On the challenge to classical climatology posed by the European discovery and settlement of the Americas, see Karen Kupperman, "The Puzzle of the American Climate in the Early Colonial Period," *American Historical Review* 87 (1982): 1262–89; and Sam White, "Unpuzzling American Climate: New World Experience and the Foundations of a New Science," *Isis* 106, no. 3 (2015): 544–66. Although the scientific racism may have been new, the idea that human variation, including what we would now call racial characteristics, was produced by diverse climates was not. Steven A. Epstein argues that natural philosophers in the Middle Ages believed that climate "might explain the diversity of living things in this world, down to the ethnicities and even different colors of the people living here now." Epstein, *The Medieval Discovery of Nature* (Cambridge: Cambridge University Press, 2012), 14.

56. "I was moved to do justice to the Indians, who have no defender." Calancha, *Corónica*, 41.

57. Gliozzi, *Adamo e il Nuovo Mondo*, 94–98. Calancha lamented the miserable state of "the Indians who flee from the labor of the mines and the oppression of the colonial governors" (*Corónica*, 44–45). Mining stood at the center of theological, political, and economic debates in the Spanish Empire, especially in Peru; see Orlando Bentancor, *The Matter of Empire: Metaphysics and Mining in Colonial Peru* (Pittsburgh, PA: University of Pittsburgh Press, 2017).

58. Calancha, *Corónica*, 35–39.

59. Calancha, *Corónica*, 68; Martínez, *Repertio*, 176–79. Calancha also cites Galen and Hippocrates along with Aristotle as authorities on the impact of climate on human bodies.

60. The concept of "complexion" was central to racial imaginaries in the early modern Iberian Atlantic. See Rebecca Earle, "Climate, Travel and Colonialism in the Early Modern World," in *Governing the Environment in the Early Modern World*, ed. Sara Miglietti and John Morgan (New York: Routledge, 2017), 24–25. The fact that writers as diverse as Erculiani in late-sixteenth-century Italy, Martínez and Calancha in early-seventeenth-century New Spain and Peru, and Thomas Burnet in late-seventeenth-century England all used the term *complexion* to describe raced and gendered bodies indicates how widespread the concept was across early modern Europe and its Atlantic world colonies.

61. Calancha, *Corónica*, 68.

62. On religion and empire in the Protestant Atlantic World, see Carla Gardina Pestana, *Protestant Empire: Religion and the Making of the British Atlantic World* (Philadelphia: University of Pennsylvania Press, 2009). On race and slavery as it intersected with religion in the Protestant Atlantic, see Kidd, *Forging of Races*; and Katherine Gerbner, *Christian Slavery: Conversion and Race in the Protestant Atlantic World* (Philadelphia: University of Pennsylvania Press, 2018).

63. Louis Bourguet, "Lettre de Mr. L. B.******* à Mr. Bosset de la Rochette, à Neûchâtel;

sur la jonction de l'Amérique à l'Asie," *Mercure Suisse* (July 1735): 67. Contemporary scholarship on Bourguet is scarce and focused on his work in geology. See Kenneth L. Taylor, "Natural Law in Eighteenth-Century Geology: The Case of Louis Bourguet," *Actes du XIIIe Congres International d'Histoire des Sciences* 8 (1971): 72–80; Kennard B. Bork, "The Geological Insights of Louis Bourguet (1678–1742)," *Journal of the Scientific Laboratories, Denison University* 60 (1974): 49–77; and Kerry Magruder, "Understanding a Contested Print Tradition: Bourguet's Mosaic, Platonic, and Aristotelian Theories of the Earth," *Compass* 81 (2008): 9–25.

64. Bourguet, "Lettre . . . à Mr. Bosset," 70.

65. Voltaire, *Traité de métaphysique* (1734) (Manchester, UK: Manchester University Press, 1957), 5. On hair and specifically beards as a marker of racial difference in the eighteenth century, see Londa Schiebinger, *Nature's Body: Gender in the Making of Modern Science* (New Brunswick, NJ: Rutgers University Press, 1993), 120–25.

66. Voltaire's use of bodily differences as racial markers and as evidence of divergent lineages had more in common, in fact, with the racialist monogenism developed in New Spain in the previous century, which offers support to Cañizares Esguerra's argument that Creole philosophers and physicians in the seventeenth century should be considered the true originators of scientific racism. Cañizares Esguerra, "New World, New Stars," 68. On the development of biological accounts of race in the eighteenth century, see Andrew S. Curran, *The Anatomy of Blackness: Science and Slavery in an Age of Enlightenment* (Baltimore, MD: Johns Hopkins University Press, 2012); and Justin E. H. Smith, *Nature, Human Nature, and Human Difference: Race in Early Modern Philosophy* (Princeton, NJ: Princeton University Press, 2015).

67. Bourguet's first *Swiss Mercury* article on the land bridge quotes a letter Bourguet received from the French geographer Joseph Delisle in 1717, indicating that he was pursuing this question at least as far back as 1716. Bourguet, "Lettre . . . à Mr. Bosset," 73–74.

68. Bourguet, "Lettre . . . à Mr. Bosset," 71.

69. Daniela Bleichmar, "Books, Bodies, and Fields: Sixteenth-Century Transatlantic Encounters with New World *Materia Medica*," in *Colonial Botany: Science, Commerce, and Politics in the Early Modern World*, ed. Londa Schiebinger and Claudia Swan (Philadelphia: University of Pennsylvania Press, 2007), 83.

70. Bourguet, "Lettre . . . à Mr. Bosset," 80–81.

71. François Ellenberger, "Bourguet, Louis," *Dictionary of Scientific Biography*, 53.

72. Louis Bourguet to Johann Jakob Scheuchzer, March 23, 1722, ZBZ, MS H 336, no. 76, fols. 277–78. Bourguet and Scheuchzer corresponded extensively over several decades. One hundred thirty letters from Bourguet to Scheuchzer survive, 112 of them in the Zentralbibliothek Zürich, dating from 1704 to 1731. Only two letters from Scheuchzer to Bourguet are known to be extant, so it is unfortunately very difficult to reconstruct Scheuchzer's side of the correspondence. See Simona Boscani Leoni, "Il progetto *Helvetic Networks* e la creazione di un repertorio on line della corrispondenza di Johann Jakob Scheuchzer," in *Le reti in rete: Per l'inventario e l'edizione dell'Archivio Vallisneri*, ed. Ivano dal Prete, Dario Generali, and Maria Teresa Monti (Florence: Leo S. Olschki, 2011), 20.

73. Bourguet, "Lettre . . . à Mr. Bosset," 84–85.

74. Bourguet, "Lettre . . . à Mr. Bosset," 76.

75. Bourguet, "Lettre . . . à Mr. Bosset," 74–75. Delisle would soon accept an invitation from the tsar to join his new observatory in St. Petersburg, making Delisle an even more valuable informant to Bourguet.

76. Philipp Johann von Strahlenberg, *An Histori-geographical Description of the North and Eastern Part of Europe and Asia* (London, 1736), 3.

77. Bourguet, "Lettre à Mr. Engel, Sécrétaire de la Chambre des Orphelins à Berne, servant de réponse à celle qu'il avoit écrite, sur la jonction de l'Amérique avec l'Asie, inserée dans le *Mercure* d'octobre 1735. p. 49," *Mercure Suisse* (February 1736): 54.

78. Bourguet, "Lettre à Mr. Engel," 53–54.

79. Bourguet, "Lettre . . . à Mr. Bosset," 91–92. See also Louis Favre, "Inauguration de l'Académie de Neuchâtel," *Musée Neuchâtelois* 3 (1866): 291.

80. Bourguet, "Lettre à Mr. Engel," 57.

81. Bourguet believed the common idea that California was an island or else a giant peninsula. European theories about the Chinese settlement of the Americas are detailed in Alexander Statman, "Fusang: The Enlightenment Story of the Chinese Discovery of America," *Isis* 107, no. 1 (2016): 1–25.

82. Bourguet, "Lettre . . . à Mr. Bosset," 85.

83. Bourguet to Scheuchzer, March 10, 1722, ZBZ, MS H 336, no. 75, fol. 276.

84. Strahlenberg, *Europe and Asia*, 29; 2.

85. Bourguet to Scheuchzer, August 24 1727, ZBZ MS H 336, no. 92, fol. 353.

86. Bourguet, "Lettre . . . à Mr. Bosset," 83–84.

87. Hale, *Primitive Origination*, 160. Notably, Hale also speculated that Tartar migration to America was relatively recent. Maybe as few as four hundred years had elapsed between the last arrival of Asians to America and the first wave of European settlers, following Columbus (*Primitive Origination*, 197).

88. Bourguet to Scheuchzer, August 24 1727, ZBZ, MS H 336, no. 92, fol. 350. It is interesting to compare Bourguet's fears about the impact of the expansion of the Russian Empire on knowledge of its territories to those of his correspondent Leibniz, whose late-career interest in a position at the court of Peter the Great was motivated in part by a desire to construct "an ethnolinguistic map of all Eurasia" and a sense that gaining access to "a broad portion of the relevant geographical area was through the mediation of the Tsar." Justin E. H. Smith, "Leibniz on Natural and National History," *History of Science* 1 (2012): 390.

89. Bourguet, "Livre d'extrait et remarques," 1695–96, BPUN, MS 1241, p. 385.

90. His financial contributions to these organizations is especially significant considering that he was in financial difficulties for most of his adult life. On Bourguet's ties to British evangelical groups, see his letters to Hans Sloane, 1736–37, BL, MS Sloane 4055, fols. 22–23, 90–91, 108–9, 209–10.

91. Bourguet to Sloane, September [13?] 1730, BL, Sloane MS 4051, fol. 107.

92. Bourguet to Sloane, September [13?] 1730, BL, Sloane 4051, fols. 107–10. Bourguet later established communication with the Mather family directly. Samuel Mather, son of Cotton and grandson of Increase, sent Bourguet an Indian grammar and an Indian Bible. Samuel Mather to Bourguet, June 25, 1740, BPUN, MS 1275, fols. 3–4.

93. "The Roman and Greek Religions displease me almost as much as the rudest, most uncivilized Paganism." Bourguet to Scheuchzer, August 24 1727, ZBZ, MS H 336, no. 92, fol. 350. By "Greek" religion Bourguet appears to have meant all Eastern Orthodox denominations, including the Russian Orthodox Church.

94. Bourguet to Scheuchzer, March 1, 1728, ZBZ, MS H 336, no. 95, fols. 366–67.

95. Naomi Tadmor, *Family and Friends in Eighteenth Century England* (Cambridge: Cambridge University Press, 2001).

96. Samuel Pepys, *The Diary of Samuel Pepys*, transcribed Robert Latham and William Matthews (Berkeley: University of California Press, 1970), 1:291 (November 12, 1660), and 2:4 (January 2, 1661).

97. Samuel Purchas, *Purchas His Pilgrimage: Or Relations of the World and the Religions Observed in all Ages and Places Discovered, from the Creation unto This Present* (London, 1613), 546.

98. Poole, *World Makers*, 32.

99. Jean Bodin, *Oeuvres philosophiques de Jean Bodin*, ed. and trans. Pierre Mesnard (Paris: Presses Universitaires de France, 1951), 241, 448.

100. "The onely probable reason, which induced these Nations to make themselves Aborigines, was, because they supposed themselves to be the first Inhabitants of the Countryes they lived in" (Stillingfleet, *Origines Sacrae*, 558–59).

101. Colin Kidd, for example, contends: "During the early modern era theological concerns helped to inhibit—and at very least to circumscribe—the articulation of racial prejudices and the formulation of identities based upon race" (*Forging of Races*, 57). Even scholars who note correctly that religious belief in common ancestry did not preclude extraordinarily cruel treatment of non-European peoples nevertheless often espouse this notion that monogenism led to softer attitudes and better treatment than polygenism would have. Braude, for example, notes that "belief in common Noachian descent gave no guarantee of human compassion" but argues, counterfactually, that treatment of Jews, Black Africans, and Indians would have been even worse in the absence of that belief ("Sons of Noah," 104).

102. Influential arguments for race's essential modernity are Ivan Hannaford, *Race: The History of an Idea in the West* (Baltimore, MD: Johns Hopkins University Press, 1996); and Brian Niro, *Race* (New York: Palgrave Macmillan, 2003).

103. George M. Fredrickson, *Racism: A Short History* (Princeton, NJ: Princeton University Press, 2002); Joyce E. Chaplin, "Natural Philosophy and an Early Racial Idiom in North America: Comparing English and Indian Bodies," *William and Mary Quarterly* 54 (1997): 229–52; Cassander L. Smith, *Black Africans in the English Imagination: English Narratives of the Early Atlantic World* (Baton Rouge: Louisiana State University Press, 2016).

### Chapter Three · Protestant Climate Change

1. Bill McKibben, *The End of Nature* (New York: Doubleday, 1989); Thomas Burnet, *The Theory of the Earth: Containing an Account of the Original of the Earth, and of All the General Changes Which It Hath Already Undergone, or Is to Undergo, till the Consummation of All Things* (London, 1684), 1:249. Although the first English edition omitted the word *Sacred* from the title, it remained in the frontispiece, and subsequent editions restored it to the title page as well. The original Latin title, *Telluris theoria sacra*, translates as "sacred theory of the earth," so although all quotations are taken from the first English edition of 1684, I refer to the book throughout by that title.

2. Dipesh Chakrabarty, "The Climate of History: Four Theses," *Critical Inquiry* 35 (Winter 2009): 206–7.

3. Fabien Locher and Jean-Baptiste Fressoz, "Modernity's Frail Climate: A Climate History of Environmental Reflexivity," *Critical Inquiry* 38 (Spring 2012): 581.

4. Alexandra Walsham, *The Reformation of the Landscape: Religion, Identity, and Memory in Early Modern Britain and Ireland* (Oxford: Oxford University Press, 2011), 379, 393.

5. William Poole, *The World Makers: Scientists of the Restoration and the Search for the Origins of the Earth* (Oxford: Peter Lang, 2010).

6. Burnet has been the subject of only one full-length monograph, by Mirella Pasini, who also reads *Sacred Theory* as an attempt to forge "una nuova unità tra storia naturale ed umana." Pasini, *Thomas Burnet: Una storia del mondo tra ragione, mito e rivelazione* (Florence: La nuova Italia, 1981), 150.

7. For a prolific writer and cosmopolitan man of letters of the late seventeenth century, Burnet left behind surprisingly little of what must have been a sizable correspondence and set of personal papers; at least, precious few such sources are known to survive. His will makes no specific mention of his letters or papers, though there is a tantalizing suggestion in the preface to a 1729 edition of one of his other printed works that Burnet privately disposed of them before his death or had his executors do so shortly thereafter. ["Philalethes"], preface in Burnet, *Archaelogiae philosophicae: Or, the Ancient Doctrine concerning the Originals of Things . . . Faithfully Translated into English, with Remarks Thereon, by Mr. Foxton* (London, 1729).

8. One of the first works of humanistic scholarship to try to resuscitate Burnet's reputation as a singularly influential thinker was Marjorie Hope Nicolson, *Mountain Gloom and Mountain Glory: The Development of the Aesthetics of the Infinite* (New York: W. W. Norton and Cornell University Press, 1959), a trend continued in Stephen Jay Gould, *Time's Arrow, Time's Cycle: Myth and Metaphor in the Discovery of Geological Time* (Cambridge, MA: Harvard University Press, 1987). Once this rehabilitation of Burnet's reputation had begun, it was possible to revisit the imaginative and religious aspects of his history of the earth in a more positive light. See, for example, Al Coppola, "Imagination and Pleasure in the Cosmography of Thomas Burnet's *Sacred Theory of the Earth*," in *World-Building and the Early Modern Imagination*, ed. Allison B. Kavey (New York: Palgrave Macmillan, 2010), 119–39; and Kerry V. Magruder, "Thomas Burnet, Biblical Idiom, and Seventeenth-Century Theories of the Earth," in *Nature and Scripture in the Abrahamic Religions: Up to 1700*, ed. Scott Mandelbrote (Leiden: Brill, 2008), 451–90.

9. Walsham, *Reformation of the Landscape*, 393.

10. Burnet studied with the Cambridge Platonists Henry More and Ralph Cudworth, who attempted to "Christianize" Descartes in the mid-seventeenth century. Burnet drew inspiration from Descartes's history of the world in *Principles of Philosophy* (1644) while also trying to make it more "Mosaic." Kerry Magruder, "Global Visions and the Establishment of Theories of the Earth," *Centaurus* 48 (2006): 247–51.

11. Burnet, *Theory*, 1:22.

12. Burnet, *Theory*, 1:25.

13. Burnet, *Theory*, 1:9.

14. John Woodward, *An Essay toward a Natural History of the Earth . . . with an Account of the Universal Deluge* (London, 1695), 82.

15. On the concept of the golden spike in planetary history, see note 18 to the introduction in this book.

16. "[T]he face of the Earth before the Deluge was smooth, regular, and uniform, without Mountains, and without a Sea." Burnet, *Theory*, 1:52.

17. Burnet, *Theory*, 1:176.

18. Burnet, *Theory*, 1:22, 175.

19. Nicholson's classic study *Mountain Gloom, Mountain Glory* documents the popular appeal of Burnet's vivid and poetic descriptions of "a World lying in its rubbish." Burnet, *Theory*, 1:110.

20. Burnet, *Theory*, 1:140.

21. Burnet, *Theory*, 1:110.

22. Burnet, *Theory*, 1:128.

23. Burnet, *Theory*, 1:132.

24. Burnet, *Theory*, 1:132.

25. Burnet, *Theory*, 1:185.

26. Burnet, *Theory*, 1:132. Cf. "the dissolution of the first Earth, and its fall in to the Abysse" (Burnet, *Theory*, 1:111).

27. Burnet, *Theory*, 1:107. Burnet's theory of "natural providence" is discussed in Rienk Vermij, "The Flood and the Scientific Revolution: Thomas Burnet's System of Natural Providence," in *Interpretations of the Flood*, ed. Florentino García Martínez and Gerardus Petrus Luttikhuizen (Leiden: Brill, 1998), 150–66.

28. Burnet, *Theory*, 1:151; Andrew Marvell, "Upon Appleton House, to My Lord Fairfax," in *Miscellaneous Poems* (London, 1681), 103. I am grateful to David Simon for this reference.

29. Gilbert Burnet, *Travels* (Amsterdam, 1687), 11.

30. Peter Dear, "Totius in verba: Rhetoric and Authority in the Early Royal Society," *Isis* 76 (1985): 145–61.

31. Edmond Halley, "Some Considerations about the Cause of the Universal Deluge, Laid before the Royal Society, on the 12th of December 1694," *Philosophical Transactions* 33 (1724–25): 122. In a subsequent *Philosophical Transactions* article, Halley prevaricated about whether the "ruins" he described were caused by the Flood or referred instead to some other catastrophe that befell the earth "before the Creation."

32. Woodward, *Essay*, 82.

33. Matthew Mackaile, *Terrae prodromus theoricus: . . . Animadversions upon Mr. Thomas Burnet's Theory of His Imaginary Earth, &c.* (Edinburgh, 1691), 7.

34. William Whiston, *New Theory of the Earth* (London, 1696), 100.

35. The classic work on the Renaissance idea of the world's decay is Victor Harris, *All Coherence Gone: A Study of the Seventeenth Century Controversy over Disorder and Decay in the Universe* (Chicago: University of Chicago Press, 1949). On the Calvinist sympathies of declensionism, see Gordon L. Davies, *The Earth in Decay: A History of British Geomorphology, 1578–1878* (New York: American Elsevier Publishing, 1969), 6.

36. George Hakewill, *An Apologie of the Power and Providence of God in the Government of the World* (Oxford, 1627), 1. Hakewill himself disputed the idea of the world's decay.

37. As Nancy Siraisi has argued, "The belief that the human body had changed since early times belonged to the larger pattern of thought that viewed all of nature—indeed, the world itself—as subject to aging and deterioration." Nancy G. Siraisi, *History, Medicine, and the Traditions of Renaissance Learning* (Ann Arbor: University of Michigan Press, 2007), 26. Gigantism was attributed to causes both moral and climactic; it could signify evil, but height was also understood as an index of general good health such that tall races of people were thought to live in Edenic climates. Amy Morris, "Geomythology on the Colonial Frontier: Edward Taylor, Cotton Mather and the Claverack Giant," *William and Mary Quarterly* 70, no. 4 (October 2013): 701–24; Surekha Davies, *Renaissance Ethnography and the Invention of the Human: New Worlds, Maps and Monsters* (Cambridge: Cambridge University Press, 2017), chap. 5.

38. Walter Raleigh, *The History of the World* (London, 1614), 83.

39. See F. Egerton, "The Longevity of the Patriarchs: A Topic in the History of Demography," *Journal of the History of Ideas* 27 (1966): 575–84.

40. Richard Cumberland, *Origines gentium antiquissimae* (London, 1724), 143.

41. Burnet, *Theory*, 1:67–68.

42. Burnet mentions Aristotle's *Meteorology* several times, usually critically, placing him in the same tradition of critical engagement with Aristotelian natural philosophy that inspired Erculiani's reworking of the *Meteorology* in her account of the natural causes of the Flood. See, for example, Burnet, *Theory*, 1:158, 226.

43. Burnet, *Theory*, 1:224.

44. "[T]he posture of the Earth to the Sun, was such, that there was no diversity or alternation of seasons in the Year, as there is now . . ." (Burnet, *Theory*, 1:71).

45. Burnet, *Theory*, 1:226.

46. Burnet, *Theory*, 1:243.

47. Burnet, *Theory*, 1:242.

48. Burnet, *Theory*, 1:196. Burnet projected the temperate climate onto the whole planet several times. Cf. "[F]rom this immediately follow'd a perpetual Æquinox all the earth over, or, if you will, a perpetual Spring" (*Theory*, 1:194). Elsewhere, though, he carefully distinguished the planet's orientation from the springlike climate of the midlatitudes; for example, "If there was a perpetual Spring and perpetual Æquinox in Paradise, there was at the same time a perpetual Æquinox all the Earth over" (*Theory*, 1:186).

49. Burnet, *Theory*, 1:186. Earlier in the century, the Peruvian friar Antonio de la Calancha modified the Hippocratic dictum of universal causes producing universal effects in formulating his astro-climatological theory of racial difference in the Americas, speculating about the differential effects of the stars and climate on different kinds of human bodies (as discussed in chapter 2). Burnet likewise appealed to this Hippocratic principle in order to explain a universal human condition as the product of universal natural causes (air and soil) that he elsewhere specified were particular to the temperate zones.

50. "Climate literally produced seasons and endemic diseases, vegetation and diet, soil and vernacular architecture, customs and political organization." James Rodger Fleming and Vladimir Jankovic, "Introduction: Revisiting Klima," in "Klima," ed. Fleming and Jankovic, special issue, *Osiris* 26, no. 1 (2011), 2. The classic intellectual history of early modern concepts of climate is Clarence J. Glacken, *Traces on the Rhodian Shore: Nature and Culture in Western Thought from Ancient Times to the End of the Eighteenth Century* (Berkeley: University of California Press, 1967). Recent scholars seek to tie these concepts more firmly to their social and political contexts and uses. See Jan Golinski, *British Weather and the Climate of Enlightenment* (Chicago: University of Chicago Press, 2010); and Anya Zilberstein, *A Temperate Empire: Making Climate Change in Early America* (Oxford: Oxford University Press, 2016).

51. On the Hippocratic tradition in climate and medicine, see Andrew Wear, "Place, Health, and Disease: The Airs, Waters, Places Tradition in Early Modern England and North America," *Journal of Medieval and Early Modern Studies* 38 (2008): 443–65. On environmental medicine in the Enlightenment, see Vladimir Jankovic, *Confronting the Climate: British Airs and the Making of Environmental Medicine* (New York: Palgrave Macmillan, 2010); and L. J. Jordanova, "Earth Science and Environmental Medicine: The Synthesis of the Late Enlightenment," in *Images of the Earth*, ed. Jordanova and Roy Porter (Chalfont St. Giles: British Society for the History of Science, 1979), 119–39.

52. Tobias Venner, *Via recta ad vitam longam . . .* (London, 1620), 7.

53. Burnet, *Theory*, 1:251.

54. Burnet, *Theory*, 1:194.

55. Burnet, *Theory*, 1:206.

56. Burnet, *Theory*, 1:180.

57. Burnet, *Theory*, 1:186.

58. Burnet, *Theory*, 1:201.

59. On hurricanes as evidence of the wildness of New World nature, see Matthew Mulcahy, *Hurricanes and Society in the Greater British Caribbean, 1624–1783* (Baltimore, MD: Johns Hopkins University Press, 2006). The religious politics of the idea of American nature as Edenic are explored in Zachary Mcleod Hutchins, *Inventing Eden: Primitivism, Millennialism, and the Making of New England* (Oxford: Oxford University Press, 2014). Dueling English views of American climate as temperate or extreme are detailed in Zilberstein, *Temperate Empire*.

60. Further evidence that Burnet saw the New World in particular as less important to sacred and natural history are his quasi-polygenist speculations that the Flood never reached the New World and the modern inhabitants of the New World were descended from Adam but not from Noah (*Theory*, 1:271–73). He developed this theory further in *Archaelogiae philosophicae* (1692), arguing that God must have provided some other means for the salvation of the American peoples, thereby further excluding them from his supposedly universal history (Pasini, *Thomas Burnet*, 92–93).

61. Burnet, *Theory*, 1:191. The theory of acclimatization and the practice of bioprospecting jointly developed within the expansion of European empires and commercial networks. On acclimatization of people and plants, see Lisbet Koerner, *Linnaeus: Nature and Nation* (Cambridge, MA: Harvard University Press, 1999); and Mark Harrison, "'The Tender Frame of Man': Disease, Climate, and Racial Difference in India and the West Indies, 1760–1860," *Bulletin of the History of Medicine* 70, no. 1 (1996): 68–93. On bioprospecting, see Londa Schiebinger, *Plants and Empire: Colonial Bioprospecting in the Atlantic World* (Cambridge, MA: Harvard University Press, 2004).

62. Roxann Wheeler, *The Complexion of Race: Categories of Difference in Eighteenth-Century British Culture* (Philadelphia: University of Pennsylvania Press, 2000), 2.

63. Burnet, *Theory*, 1:191.

64. The importance of Europe's American colonies as a site for theorizing the intersections of climate, health, and race is increasingly being recognized. The classic study is Karen Kupperman, "Fear of Hot Climates in the Anglo-American Colonial Experience," *William and Mary Quarterly* 41, no. 2 (1984): 213–40. More recently, see Jan Golinski, "American Climate and the Civilization of Nature," in *Science and Empire in the Atlantic World*, ed. James Delbourgo and Nicholas Dew (New York: Routledge, 2008), 153–74; Sam White, "Unpuzzling American Climate: New World Experience and the Foundations of a New Science," *Isis* 106, no. 3 (2015): 544–66; and Golinski, "Debating the Atmospheric Constitution: Yellow Fever and the American Climate," *Eighteenth Century Studies* 49, no. 2 (2016): 149–66.

65. Woodward, *Brief Instructions for Making Observations in All Parts of the World* (London, 1696), 9.

66. One of Cavert's key contributions is to demonstrate not only that London's air pollution worsened significantly over the early modern period but that the city's inhabitants and visitors were very much aware of this change. William M. Cavert, *The Smoke of London: Energy and Environment in the Early Modern City* (Cambridge: Cambridge University Press, 2016). The nine hundred- and ninety-year figures are in Burnet, *Theory*, 1:192.

67. John Ray, *Three Physico-Theological Discourses . . .* (London, 1693), 282.

68. Robert St. Clair, *The Abyssinian Philosophy Confuted* . . . (London, 1697), "To the Reader," [lxi].

69. Whiston, *New Theory of the Earth*, 181.

70. Brian Fagan, *The Little Ice Age: How Climate Made History, 1300–1850* (New York: Basic Books, 2000), 113; Burnet, *Theory*, 1:200.

71. Raleigh, *History of the World*, 131.

72. Burnet, *Theory*, 1:200. The controversy surrounding Geoffrey Parker's *Global Crisis: War, Climate Change and Catastrophe in the Seventeenth Century* (New Haven, CT: Yale University Press, 2013) has prompted both excitement in revisiting key themes and events of early modern history in light of the Little Ice Age as well as concerns about environmental determinism. Cavert offers a useful caveat against climate determinism with specific reference to the English winter of 1683–84 in Cavert, "Winter and Discontent in Early Modern England," in *Governing the Environment in the Early Modern World: Theory and Practice*, ed. Sara Miglietti and John Morgan (New York: Routledge, 2017), 114–33.

73. William Whiston, for example, reported having "an exceeding liking of the main part of Dr. Burnet's Theory of the Earth" during his college years at Cambridge "and thought my self never more pleas'd than in a repeated perusal of so ingenious and remarkable a Book." William Whiston, *A Vindication of the New Theory of the Earth from the Exceptions of Mr. Keill and Others* (London, 1698), preface, A2r.

74. Burnet claimed to have done the English translation at the king's behest (*Theory*, "Epistle Dedicatory").

75. After gaining his post at the Charterhouse, Burnet made his debut on the political stage in 1686–87 by spearheading the resistance to the decision by King James II to appoint a Catholic pensioner to the Charterhouse in violation of the school's charter, which required oaths of office. T. Carte, *The Life of James, Duke of Ormond* (Oxford: Oxford University Press, 1851), 4:682–83.

76. This figure comes from James E. Force, "Some Eminent Newtonians and Providential Geophysics at the Turn of the Seventeenth Century," *History of Geology* 2, no. 1 (1983): 5.

77. Halley, "Some Considerations about the Cause of the Universal Deluge," 120–21.

78. Herbert Croft, *Some Animadversions upon a Book Intituled, The Theory of the Earth* (London, 1685), preface.

79. Thomas Robinson, *New Observations on the Natural History of This World of Matter, and This World of Life* (London, 1696), 16.

80. John Beaumont, *Considerations on a Book, Entituled The theory of the Earth, Publisht Some Years since by the Dr. Burnet* (London, 1693), 186.

81. John Keill, *Examination of Dr. Burnet's Theory of the Earth Together with Some Remarks on Mr. Whiston's New Theory of the Earth* (Oxford, 1698), 178.

82. Archibald Lovell, *A Summary . . . Which May Be Enlarged and Improved into a Compleat Answer to Dr. Burnet's Theory of the Earth* (London, 1696), 23.

83. Burnet, *Theory*, 1:106.

84. Halley, "Some Considerations about the Cause of the Universal Deluge," 121.

85. Burnet, *A Review of the Theory of the Earth . . . Especially in Reference to Scripture* (London, 1697), 183.

86. Burnet to Newton, January 13, 1680/1, in Isaac Newton, *Correspondence*, ed. H. W. Turnbull (Cambridge: Cambridge University Press, 1960), 2:323. The philosophical and exegetical issues raised in this correspondence are discussed at length in Scott Mandelbrote, "Isaac

Newton and Thomas Burnet: Biblical Criticism and the Crisis of Late Seventeenth-Century England," in *The Books of Nature and Scripture*, ed. Richard H. Popkin and James E. Force (Dordrecht: Kluwer, 1994), 149–78.

87. Newton to Burnet, January 1680/1, in Newton, *Correspondence*, 2:331.

88. "The 'atheist' denied something so basic and fundamental to the notion of God or of faith that its denial amounted to no God or belief at all from the accuser's point of view." Sachiko Kusukawa, *The Transformation of Natural Philosophy: The Case of Philip Melanchthon* (Cambridge: Cambridge University Press, 1995), 128.

89. Croft, *Some Animadversions*, 18–19.

90. *Two Essays Sent in a Letter from Oxford, to a Nobleman in London. The First concerning Some Errors about the Creation, General Flood, and the Peopling of the World. . . . By L. P.* (London, 1695), 13–14. Rappaport declines to positively identify the author of *Two Essays*, rumored then and now to be John Toland, whose *Christianity Not Mysterious* (1696) was widely denounced as deistic. Rhoda Rappaport, "Questions of Evidence: An Anonymous Tract Attributed to John Toland," *Journal of the History of Ideas* 58, no. 2 (1997): 339–48.

91. John Keill, *An Examination of the Reflections on the Theory of the Earth Together with a Defence of the Remarks on Mr. Whiston's New theory* (Oxford, 1699), 4.

92. This is one of the most significant and underappreciated aspects of the theology of *Sacred Theory*. Martin Rudwick does flag this remarkable fact in *The Meaning of Fossils: Episodes in the History of Paleontology*, 2nd ed. (Chicago: University of Chicago Press, 1985), 78; as does Poole, *World Makers*, 59.

93. Burnet, *Theory*, 1:99–100.

94. Augustinianism was influential across religious divides in early modern Europe. Arnoud S. Q. Visser, *Reading Augustine in the Reformation: The Flexibility of Intellectual Authority in Europe, 1500–1620* (Oxford: Oxford University Press, 2011). It exerted considerable influence in Britain at the height of the Scientific Revolution, among Burnet's generation. Peter Harrison, *The Fall of Man and the Foundations of Science* (Cambridge: Cambridge University Press, 2007), especially chap. 2.

95. The literature on the relationship between Calvinism and Anglicanism is vast and itself deeply contentious. Recent scholarship relevant to the present discussion posits neo-Augustinianism as a link between the two and as an important factor in the formation of a post-Restoration confessional state. See Jean-Louis Quantin, *The Church of England and Christian Antiquity: The Construction of a Confessional Identity in the Seventeenth Century* (Oxford: Oxford University Press, 2009), 170–90.

96. These accusations were enough to dislodge another one of the world historians, William Whiston, from his chair at Cambridge. Stephen Snobelen, "Caution, Conscience, and the Newtonian Reformation: The Public and Private Heresies of Newton, Clarke, and Whiston," *Enlightenment and Dissent* 16 (1997): 151–84.

97. Constructing orthodoxy in Anglican Britain was very different but no less challenging than in post-Tridentine Italy. The challenge of determining a "predestinarian orthodoxy" within Anglicanism and with respect to Calvinism is discussed in David Como, "Puritans, Predestination and the Construction of Orthodoxy in Early Seventeenth-Century England," in *Conformity and Orthodoxy in the English Church, c. 1560–1660*, ed. Peter Lake and Michael Questier (Suffolk, UK: Boydell Press, 2000), 64–87.

98. Lovell, *Summary*, 20.

99. Burnet, *Theory*, 1:246. As startlingly Marxist as the term may sound to us in the twenty-

first century, the *OED* records several instances of the word *superstructure* in seventeenth-century English.

100. Burnet, *Theory*, vol. 2, "Preface to the Reader," [R3c].

101. Harrison, *The Fall of Man*.

102. Burnet, *Theory*, vol. 2, "Preface to the Reader," [R3c].

103. Golinski, *British Weather*.

104. Burnet, *Theory*, vol. 2, "Preface to the Reader," [R3c].

105. The best account of Woodward's life and work is Joseph M. Levine, *Dr Woodward's Shield: History, Science, and Satire in Augustan England* (Berkeley: University of California Press, 1977), which showcases his career as both a naturalist and antiquarian, someone interested equally in natural history and human history.

106. Michael Heyd, "Original Sin, the Struggle for Stability, and the Rise of Moral Individualism in Late Seventeenth-Century England," in *Early Modern Europe: From Crisis to Stability*, ed. Philip Benedict and Myron P. Gutmann (Newark: University of Delaware Press, 2005), 197–233.

107. Woodward, *Essay*, preface [A7r], 160–66.

108. John Fisher to John Woodward, August 19, 1718, CUL, MS Add 7647/126.

109. Edward Lhwyd to Martin Lister, March 28, 1695, BOD, MS Lister 36, fol. 118.

110. Tancred Robinson to Lhwyd, April 11, 1696, BOD, MS Ashmole 1817a [fol. 328].

111. Woodward, *Essay*, 8.

112. Woodward, *Essay*, 98.

113. Woodward, *Essay*, 85–87.

114. Woodward, *Essay*, 87–90.

115. Woodward, *Essay*, 91.

116. Woodward, *Essay*, 91–93.

117. The political and religious stakes of "improvement" in Britain are explored in Simon Schaffer, "The Earth's Fertility as a Social Fact in Early Modern Britain," in *Nature and Society in Historical Context*, ed. Mikulas Teich, Roy Porter, and Bo Gustafsson (Cambridge: Cambridge University Press, 1997), 124–47. On the imperial dimensions of the politics of improvement, see Richard Drayton, *Nature's Government: Science, Imperial Britain, and the "Improvement" of the World* (New Haven, CT: Yale University Press, 2000).

118. "Enquiries concerning Agriculture," *Philosophical Transactions* 1, no. 5 (1665): 92.

119. John Evelyn, *A Philosophical Discourse of Earth, relating to the Culture and Improvement of It for Vegetation, and the Propagation of Plants* (London, 1676), 100.

120. Economic historians have since confirmed the perception of popular resisters that enclosure and related "improvements" benefited landowners almost exclusively. Roderick Floud and D. McCloskey, *The Economic History of Britain since 1700*, vol. 1, *1700–1860* (Cambridge: Cambridge University Press, 1981), 118.

121. Henry Rowlands, *Idea agriculturae, or the Principles of Vegetation Asserted and Defended, Being an Essay on the Theory and Practice of Husbandry* (Dublin, 1764), 71–72.

122. Efforts to re-create Eden were popular among aristocrats, manual laborers, and political dissidents alike. Joanna Picciotto, "Reforming the Garden: The Experimentalist Eden and *Paradise Lost*," *English Literary History* 72, no. 1 (2005): 23–78; Sandra Sherman, "Replanting Eden: John Evelyn and His Gardens," *Endeavour* 26, no. 3 (2002): 113–17; Ariel Hessayon, "Restoring the Garden of Eden in England's Green and Pleasant Land: The Diggers and the Fruits of the Earth," *Journal for the Study of Radicalism* 2, no. 2 (2009): 1–25.

123. Woodward, "Two Discourses: I. An Account of the Ores of Metalls . . . II. The Art of Assaying" (ca. 1724), BL Add. MS 25096, fol. 18r.

124. On field tests, see Schaffer, "Earth's Fertility," 135–36.

125. Woodward, "Two Discourses," BL Add. MS 25096, fol. 18r. Woodward may have intended a dig at the Spanish, who were dealing with massive inflation as a result of the influx of specie into Spain from the gold and silver mines in their American colonies.

126. Woodward, "Two Discourses," BL Add. MS 25096, fol. 18v.

127. Woodward, *Essay*, 154–55.

128. Woodward, "Two Discourses," BL Add. MS 25096, fol. 22r.

129. Burnet, *Theory*, 1:185.

130. Burnet, *Theory*, 1:247, 248.

131. On Woodward's relationship with Italian fossil collectors and fossil literature and with his often-hostile English colleagues, see Paula Findlen, "The Specimen and the Image: John Woodward, Agostino Scilla, and the Depiction of Fossils," *Huntington Library Quarterly* 78, no. 2 (2015): 217–61.

132. Tancred Robinson to Lhwyd, April 11, 1696, BOD, MS Ashmole 1817a [fol. 328].

133. Walsham, *Reformation of the Landscape*, 389.

134. Burnet, *Theory*, 1:100.

135. Burnet, *Theory*, 1:132.

136. Mackaile, *Terrae prodromus theoricus*, 4; Croft, *Some Animadversions*, 138.

137. Woodward, *Essay*, 147–50.

138. St. Clair, *Abyssinian Philosophy*, "To the Reader," [xlviii].

139. Burnet, *Theory*, 1:144. Burnet may have had in mind the "maps of Philipp Clüver (1580–1622), which represented regions of the Earth as they existed in antiquity before human settlement changed the face of the Earth" (Magruder, "Global Visions," 251).

140. Lorraine Daston and Katherine Park, *Wonders and the Order of Nature, 1150–1750* (New York: Zone Books, 2001), chap. 7, "Wonders of Art, Wonders of Nature."

141. Ursula K. Heise, "Terraforming for Urbanists," *Novel* 49, no. 1 (2016): 10–25.

### Chapter Four · The Flood and the Apocalypse

1. John Woodward, *Brief Instructions for Making Observations in All Parts of the World . . . Being an Attempt to Settle an Universal Correspondence for the Advancement of Knowledg Both Natural and Civil* (London, 1696), 1.

2. Travel across and beyond Europe became an increasingly vital part of natural history collecting and of geopolitical place-making by the eighteenth century. Sverker Sörlin, "National and International Aspects of Cross-Boundary Science: Scientific Travel in the Eighteenth Century," in *Denationalizing Science: The Contexts of International Scientific Practice*, ed. Elisabeth Crawford, Terry Shinn, and Sverker Sörlin (Dordrecht: Springer, 1993), 43–72; Alix Cooper, "From the Alps to Egypt (and Back Again): Dolomieu, Scientific Voyaging, and the Construction of the Field in Eighteenth-Century Europe," in *Making Space: Territorial Themes in the History of Science*, ed. Crosbie Smith and Jon Agar (London: Macmillan, 1998), 39–63; and David Philip Miller and Peter Hanns Reill, eds., *Visions of Empire: Voyages, Botany, and Representations of Nature* (Cambridge: Cambridge University Press, 2011).

3. The query list genre as a means of information gathering and social network building is discussed in Elizabeth Yale, "Making Lists: Social and Material Technologies in the Making of Seventeenth-Century British Natural History," in *Ways of Making and Knowing: The Mate-

*rial Culture of Empirical Knowledge*, ed. Pamela H. Smith, Amy R. W. Meyers, and Harold J. Cook (Ann Arbor: University of Michigan Press, 2014), 280–301. The social and political contexts of the related genre of "wish lists" are discussed in Vera Keller, *Knowledge and the Public Interest, 1525–1725* (Cambridge: Cambridge University Press, 2015).

4. Woodward, *Brief Instructions*, 7, 1, 10, 8. Like José de Acosta's *Natural and Moral History of the Indies* (1590), Woodward's *Brief Instructions* was concerned to document foreign cultures as well as foreign nature, as indicated in the subtitle.

5. Woodward, *Brief Instructions*, 6.

6. The intellectual history of the early modern fossil debate is richly described in Martin J. S. Rudwick, *The Meaning of Fossils: Episodes in the History of Paleontology*, 2nd ed. (Chicago: University of Chicago Press, 1985), chaps. 1–2; and Nicoletta Morello, "The Question on the Nature of Fossils in the 16th and 17th Centuries / La questione della natura dei fossili nel cinquecento e seicento," in *Four Centuries of the Word Geology: Ulisse Aldrovandi 1603 in Bologna / Quadricentenario della parola Geologia: Ulisse Aldrovandi 1603 Bologna*, ed. Gian Battista Vai and William Cavazza (Bologna: Minerva Edizioni, 2003), 127–52.

7. John Woodward, *An Essay toward a Natural History of the Earth . . . with an Account of the Universal Deluge* (London, 1695), A5r–A6r. The somewhat odd phrasing of Woodward's insistence that fossils were found on continents other than Asia was intended to refute the idea that the Flood itself was limited to Asia, as discussed in chapter 2.

8. Mordechai Feingold, ed., *Jesuit Science and the Republic of Letters* (Cambridge, MA: MIT Press, 2003); Steven J. Harris, "Confession-Building, Long-Distance Networks, and the Organization of Jesuit Science," *Early Science and Medicine* 1, no. 3 (October 1996): 287–318.

9. On the European drive to envision the globe before it was possible to really see or know it empirically, see Denis Cosgrove, *Apollo's Eye: A Cartographic Genealogy of the Earth in the Western Imagination* (Baltimore, MD: Johns Hopkins University Press, 2001); and Ayesha Ramachandran, *The Worldmakers: Global Imagining in Early Modern Europe* (Chicago: University of Chicago Press, 2015). As Joyce Chaplin reminds us, humans have been circumnavigating the globe, or coming close to it, for centuries, which forms part of "a longer history of planetary consciousness." Chaplin, *Round about the Earth: Circumnavigating from Magellan to Orbit* (New York: Simon and Schuster, 2012), xix.

10. Paul N. Edwards, *A Vast Machine: Computer Models, Climate Data, and the Politics of Global Warming* (Cambridge, MA: MIT Press, 2010), 8. Edwards's definition of a knowledge infrastructure ("Knowledge infrastructures comprise robust networks of people, artifacts, and institutions that generate, share, and maintain specific knowledge about the human and natural worlds" [17]) is also a useful way of understanding the exchange of knowledge about the human and natural worlds in the seventeenth and eighteenth centuries. The institutional basis was less robust, of course, but it was there by the end of the seventeenth century; see, for example, Marie Boas Hall, "The Royal Society's Role in the Diffusion of Information in the Seventeenth Century," *Notes and Records of the Royal Society* 29 (1975): 173–92.

11. The literature on the early modern Republic of Letters has grown rapidly in recent decades. An excellent introduction is Anthony Grafton, "A Sketch Map of a Lost Continent: The Republic of Letters," *Republics of Letters* 1, no. 1 (2008): https://arcade.stanford.edu/rofl /sketch-map-lost-continent-republic-letters. Of special relevance to the present discussion are Anne Goldgar, *Impolite Learning: Conduct and Community in the Republic of Letters, 1680–1750* (New Haven, CT: Yale University Press, 1995); and Lorraine Daston, "The Ideal and Reality of the

Republic of Letters in the Enlightenment," *Science in Context* 4, no. 2 (1991): 367–86, which offer important insights into the social composition, values, and dynamics of this scholarly community. Claudine Poulouin argues that the concept of "universalism" was a key link between the Republic of Letters and the study of the Universal Deluge in *Le temps des origines* (Paris: Honoré Champion, 1998), pt. 1, chap. 2.

12. Scholarly gift exchange was a crucial force of social cohesion in the Republic of Letters; see Paula Findlen, "The Economy of Scientific Exchange in Early Modern Italy," in *Patronage and Institutions: Science, Technology, and Medicine at the European Court, 1500–1750*, ed. Bruce T. Moran (Woodbridge, UK: Boydell Press, 1991), 5–24; and Franz Mauleshagen, "Networks of Trust: Scholarly Correspondence and Scientific Exchange in Early Modern Europe," *Medieval History Journal* 6, no. 1 (2003): 1–32. The long-distance exchange of gift-objects played an important role in elaborating global imaginaries. Anya Zilberstein, "Objects of Distant Exchange: The Northwest Coast, Early America, and the Global Imagination," *William and Mary Quarterly*, 3rd ser., 64, no. 3 (2007): 591–620.

13. Woodward to Scheuchzer, July 21, 1702, ZBZ, MS H 294, no. 20, p. 87.

14. Woodward and Scheuchzer's relationship unfolded within Anglo-Swiss networks of exchange in the early eighteenth century; see Michael Kempe, "Die 'Anglo-Swiss Connection': Zur Kommunikationskultur der Gelehrtenrepublik in der Frühaufklärung," in *Wissen und Wissensvermittlung im 18. Jahrhundert*, ed. Robert Seidel (Heidelberg: Palatina Verlag, 2001), 71–91; and Simona Boscani Leoni, "La ricerca sulla montagna nel Settecento sotto nuove prospettive: Il 'network' anglo-elvetico-alpino," in *Histoire des Alpes/Storia delle Alpi/ Geschichte der Alpen* 12 (2007): 201–13. The best study of Scheuchzer's scientific career, which was dominated by his study of fossils and the Flood, is Kempe, *Wissenschaft, Theologie, Aufklärung: Johann Jakob Scheuchzer (1672–1733) und die Sintfluttheorie* (Epfendorf, DE: Bibliotheca Academica Verlag, 2003).

15. Woodward to Scheuchzer, July 20, 1703, ZBZ, MS H 294, no. 23, p. 97.

16. Woodward to Scheuchzer, January 20, 1712, ZBZ, MS H 294, no. 54, pp. 229–32.

17. David Price, "John Woodward and a Surviving British Geological Collection from the Early Eighteenth Century," *Journal of the History of Collections* 1 (1989): 79–95.

18. Woodward, "A Catalogue of the Foreign Fossils in the Collection of J. Woodward M.D.," in *An Attempt towards a Natural History of the Fossils of England* (London, 1729), 2:7. Woodward's extensive network of admirers, correspondents, and fossil suppliers is well documented in Joseph M. Levine, *Dr. Woodward's Shield: History, Science, and Satire in Augustan England* (Berkeley: University of California Press, 1977), 96–100.

19. Scheuchzer's earliest publications espoused a chemical theory of the genesis and growth of figured stones. Woodward warned him away from this view and toward his own in their early letters. See, for example, Woodward to Scheuchzer, March 8, 1705/6, ZBZ, MS H 294, no. 32, pp. 143–46.

20. John Ray to Edward Lhwyd, July 26, 1704, BOD, MS Ashmole 1817a, fol. 228a.

21. Gisbert Cuper to Woodward (copy), April 14, 1708, CUL, Add MS 7647/96.

22. Woodward to Scheuchzer, May 20, 1709, ZBZ, MS H 294, no. 46, p. 199.

23. See Woodward to Bethell, CUL, MS Add 7647, nos. 47–49.

24. Woodward to Scheuchzer, July 21, 1702, ZBZ, MS H 294, no. 20, p. 87.

25. Walter Tega, ed. *Anatomie accademiche* (Bologna: Il Mulino, 1986), 1:70.

26. Marta Cavazza, "Bologna and the Royal Society in the Seventeenth Century," *Notes*

*and Records of the Royal Society of London* 35, no. 2 (1980): 105–23. The definitive study of the Bologna Institute and its ties to other parts of Europe is Cavazza, *Settecento inquieto: Alle origini dell'Istituto delle Scienze di Bologna* (Bologna: Il Mulino, 1990).

27. Scheuchzer's younger brother, for example, served as Marsigli's secretary and translator. John Stoye, *Marsigli's Europe, 1680–1730: The Life and Times of Luigi Ferdinando Marsigli, Soldier and Virtuoso* (New Haven, CT: Yale University Press, 1994), 242.

28. Gian Battista Vai, "Liberal Diluvialism / Un diluvianismo liberale," in *Four Centuries of the Word Geology: Ulisse Aldrovandi 1603 in Bologna / Quadricentenario della parola Geologia: Ulisse Aldrovandi 1603 Bologna*, ed. Gian Battista Vai and William Cavazza (Bologna: Minerva Edizioni, 2003), 237.

29. Monti's "Museum Diluvianum" is partially preserved, along with the original labels, in the University of Bologna's museum of natural history, and I am grateful to Carlo Sarti for his guided tour. It is cataloged and contextualized in Carlo Sarti, *I fossili e il Diluvio Universale* (Bologna: Pitagora Editrice, 1988). Scheuchzer published a fossil catalog by the same name in 1716.

30. Giuseppe Monti, *De monumento diluviano nuper in agro Bononiensi detecto dissertatio* (Bologna, 1719), 6.

31. This letter also mentions an anticipated shipment of "books from England." Giuseppe Monti to Gian Giacinto Vogli, August 21, 1719, BUB, MS 2086, vol. 2, no. 285, fol. 7.

32. Monti to Bourguet, March 12, 1715, BPUN, MS 1276, fol. 108.

33. Monti to Vogli, undated, BUB, MS 2086, vol. 2, no. 286, fol. 8.

34. Woodward, *Essay*, 182–88.

35. The controversy in Britain over Woodward's alleged plagiarism of Italian authors reveals much about the close yet fraught scholarly ties between Britain and Italy around the turn of the eighteenth century. See Paula Findlen, "The Specimen and the Image: John Woodward, Agostino Scilla, and the Depiction of Fossils," *Huntington Library Quarterly* 78, no. 2 (2015): 217–61.

36. Monti to Bourguet, June 5, 1714, BPUN, MS 1276, fol. 85v. While Grandi's name has been mostly lost to history, the powerful combination of these two ideas that he helped to engineer—the notion that fossils were organic remains and that their global distribution proved the universality of Noah's Flood—would reverberate throughout European and Euro-American science for the next several decades and, indeed, centuries. He is mentioned briefly in Morello, "Steno, the Fossils, the Rocks and the Calendar of the Earth," in *The Origins of Geology in Italy*, ed. Gian Battista Vai and W.G.E. Caldwell (Boulder, CO: Geological Society of America, 2006), 90–91; and in André Robinet, *G. W. Leibniz iter Italicum* (Florence: Leo S. Olschki, 1988), 414–18.

37. The pathbreaking work in paleontology by Scilla and Steno, jointly and singly, is documented in Bruno Accordi, "Agostino Scilla, Painter from Messina (1629–1700), and His Experimental Studies on the True Nature of Fossils," *Geologica Romana* 17 (1978): 129–44; Morello, "Steno"; and Findlen, "Specimen and the Image."

38. Morello, "Question on the Nature of Fossils," 129; Davis A. Young, *The Biblical Flood: A Case Study of the Church's Response to Extrabiblical Evidence* (Grand Rapids, MI: William B. Eerdmans, 1995), 26–27, 34–36.

39. The popular Renaissance notion that fossils were "jokes of Nature" invited people to appreciate fossils aesthetically without inviting scrutiny as to the exact natural processes involved in their production. Paula Findlen, "Jokes of Nature and Jokes of Knowledge: The

Playfulness of Scientific Discourse in Early Modern Europe," *Renaissance Quarterly* 43 (1990): 292–331.

40. Nicolaus Steno, *The Prodromus to a Dissertation on a Solid Naturally Contained within a Solid*, ed. and trans. Troels Kardel and Paul Maquet (Heidelberg; Berlin: Springer, 2013), 625.

41. Woodward to Scheuchzer, June 22, 1717, ZBZ, MS H 294, no. 67, p. 287.

42. John Harris, *Remarks on Some Late Papers, Relating to the Universal Deluge: And to the Natural History of the Earth* (London, 1697), 102. One of Woodward's staunchest defenders in Britain, Harris was attempting to counter charges like the one from Tancred Robinson, who alleged that Woodward "hath transcribed Columna, Steno, Scilla, Boccone & others." Robinson to Edward Lhwyd, April 11, 1696, BOD, MS Ashmole 1817a [fol. 328].

43. Neither Burnet nor Whiston appealed to paleontological evidence in their demonstrations of the Flood's universality. Ray, the only one of the four to travel to Italy and to meet Steno, proposed that marine fossils could provide evidence for the Universal Deluge in *Three Physico-Theological Discourses* (1693), but his ambivalence about the nature of fossils likely explains his choice not to make them central to his discussion of the Flood, as Woodward did. See, for example, Ray to Lhwyd, November 7, 1690, BOD, MS Eng hist c 11, fol. 43.

44. On the history of scaling, see Deborah R. Coen, "Big Is a Thing of the Past: Climate Change and Methodology in the History of Ideas," *Journal of the History of Ideas* 77 (2016): 305–21; and Coen, *Climate in Motion: Science, Empire, and the Problem of Scale* (Chicago: University of Chicago Press, 2018).

45. Some of these religious, colonial, and commercial networks that provided the infrastructure for scholarly exchange are described in Feingold, ed., *Jesuit Science and the Republic of Letters*; Goldgar, *Impolite Learning*; and Kathleen S. Murphy, "Collecting Slave Traders: James Petiver, Natural History, and the British Slave Trade," *William and Mary Quarterly* 70, no. 4 (2013): 637–70.

46. The earliest attempts to record, for example, weather and climate data in a standardized and systematic fashion date to the eighteenth century for national scales and the nineteenth for global ones. James Rodger Fleming, *Historical Perspectives on Climate Change* (New York: Oxford University Press, 1998).

47. Antoni Maczak, *Travel in Early Modern Europe* (Cambridge: Polity Press, 1995).

48. Kircher perhaps came closest to achieving an empirically grounded portrait of the earth based on long-distance data collection. Drawing on the far-flung network of Jesuit missionaries to which he belonged, Kircher collected information on magnetic variation from Asia, Africa, Europe, and the Americas in order to create a composite picture of the magnetic planet. Martha Baldwin, "Kircher's Magnetic Investigations," in *The Great Art of Knowing: The Baroque Encyclopedia of Athanasius Kircher*, ed. Daniel Stolzenberg (Stanford, CA: Stanford University Libraries, 2001), 32.

49. Brian W. Ogilvie, *The Science of Describing: Natural History in Renaissance Europe* (Chicago: University of Chicago Press, 2006).

50. The choice of scale—local or national—in early modern natural history was exceedingly complex, driven by social, political, and intellectual forces. Alix Cooper, *Inventing the Indigenous: Local Knowledge and Natural History in Early Modern Europe* (Cambridge: Cambridge University Press, 2007); Elizabeth Yale, *Sociable Knowledge: Natural History and the Nation in Early Modern Britain* (Philadelphia: University of Pennsylvania Press, 2016).

51. Ray to Lhwyd, April 6, 1691, BOD, MS Eng hist c 11, fol. 47.

52. Ray to Lhwyd, May 27, 1691, BOD, MS Eng hist c 11, fol. 48.

53. Ray to Lhwyd, June 1, 1694, BOD, MS Ashmole 1817a, fol. 224.

54. Woodward to Lhwyd, January 19, 1691, BOD, MS Eng hist c 11, fol. 104.

55. Steno, *Prodromus*, 654. Italics mine.

56. On the publication history of the early papers, see Rhoda Rappaport, "Leibniz on Geology: A Newly Discovered Text," *Studia Leibnitiana* 29 (1997): 6–11.

57. Gottfried Wilhelm Leibniz, *Protogaea*, ed. and trans. Claudine Cohen and André Wakefield (Chicago: University of Chicago Press, 2008), 3.

58. Claudine Cohen, "Leibniz's *Protogaea*: Patronage, Mining, and the Evidence for a History of the Earth," in *Proof and Persuasion: Essays on Authority, Objectivity, and Evidence*, ed. Suzanne Marchand and Elizabeth Lunbeck (Turnhout, BE: Brepols, 1996), 124–43; E. P. Hamm, "Knowledge from Underground: Leibniz Mines the Enlightenment," *Earth Sciences History* 16 (1997): 77–99.

59. The traveler was assumed to be an Englishman, insofar as he was instructed to observe "what kinds of Trees, shrubs, and herbs it produceth that we have, and what kinds that we have not in England" (Woodward, *Brief Instructions*, 8). Levine also notes the Anglocentrism of Woodward's network: "Gifts of fossils came from all over the earth, wherever England traded or colonized" (*Dr. Woodward's Shield*, 42).

60. Woodward to Lhwyd, January 19, 1691, BOD, MS Eng hist c 11, fols. 104–5. This letter suggests that the *Brief Instructions* was a formalized version of an informal manuscript query list he had already been sending out for several years before publishing it in 1696.

61. [Noël Bonaventure d'Argonne], *Mélanges d'histoire et de littérature* (Rotterdam, 1700), 2:62.

62. The complicated ways in which marginalized groups were partially integrated into the Republic of Letters are described in Carol Pal, *Republic of Women: Rethinking the Republic of Letters in the Seventeenth Century* (Cambridge: Cambridge University Press, 2012); and Alexander Bevilacqua, *The Republic of Arabic Letters: Islam and the European Enlightenment* (Cambridge, MA: Harvard University Press, 2018).

63. Petiver to Lhwyd, n.d. (between 1701 and 1703), BOD, MS Eng hist c 11, fol. 33.

64. Scheuchzer to Kaspar Wetstein, May 20, 1724, BL, Add. Ms 32414, fol. 331.

65. On colonial informants in the transatlantic British Empire, see Susan Scott Parrish, *American Curiosity: Cultures of Natural History in the Colonial British Atlantic World* (Chapel Hill: University of North Carolina Press, 2006). Sloane and Petiver's collecting depended on colonial infrastructures generally and slaving infrastructures in particular. James Delbourgo, *Collecting the World: Hans Sloane and the Origins of the British Museum* (Cambridge, MA: Harvard University Press, 2017); Murphy, "Collecting Slave Traders."

66. Mather to Woodward, November 17, 1712, RS, MS EL/M2/21. On Mather's correspondence with Woodward and the Royal Society on the subject of fossils, see David Levin, "Giants in the Earth: Science and the Occult in Cotton Mather's Letters to the Royal Society," *William and Mary Quarterly* 45, no. 4 (1988): 751–70; and Lydia Barnett, "Giant Bones and the Taunton Stone: American Antiquities, World History, and the Protestant International," in *Empires of Knowledge: Scientific Networks in the Early Modern World*, ed. Paula Findlen (London: Routledge, 2018). Jones sent the fossil to Lhwyd, who was closely aligned with Petiver's circle and who had by this time fallen out with Woodward. Levine, *Dr. Woodward's Shield*, 99.

67. Magnus von Bromell to James Petiver, April 18, 1709, BL, Sloane MS 4064, fols. 198–201.

68. Bromell wrote in French but asked Petiver (not without reason) to excuse the poorness of it, which might be seen as symptomatic of his desire to join the Republic of Letters—which was in the early eighteenth century, turning increasingly from Latin to French—while at the same time recognizing his own marginality in relation to it. Bromell to Petiver, April 18, 1709, BL, Sloane MS 4064, fols. 198–201. Sweden's peripheral position in pre-Linnaean European science is discussed in Göran Rydén, ed., *Sweden in the Eighteenth-Century World: Provincial Cosmopolitanisms* (Surrey, UK: Ashgate, 2013) and J.F.C. Danneskiold-Samsøe, *Muses and Patrons: Cultures of Natural Philosophy in Seventeenth-Century Scandinavia* (Lund: Lund University Press, 2004).

69. Woodward to Scheuchzer, July 3, 1710, ZBZ, MS H 294, no. 49, pp. 213–14.

70. Scheuchzer, *Piscium querelae et vindiciae* (Zurich, 1708), 12.

71. In this respect, fossils functioned in a similar fashion to the medical facts that circulated across cultures under the guise of being theory-free. Harold J. Cook, *Matters of Exchange: Commerce, Medicine, and Science in the Dutch Golden Age* (New Haven, CT: Yale University Press, 2008). The utility of theory-free facts as a mechanism of social cohesion in late-seventeenth-century science is influentially described by Lorraine Daston, "Baconian Facts, Academic Civility, and the Prehistory of Objectivity," *Annals of Scholarship* 8 (1991): 337–65.

72. The symmetry of Flood and Apocalypse in early modernity has not received serious or sustained attention by modern scholars, though it has been noted by Mirella Pasini, *Thomas Burnet: Una storia del mondo tra ragione, mito e rivelazione* (Florence: La nuova Italia, 1981), 44; Young, *Biblical Flood*, 13; Michael Kempe, "Noah's Flood: The Genesis Story and Natural Disasters in Early Modern Times," *Environment and History* 9 (2003): 154; and William Poole, *The World Makers: Scientists of the Restoration and the Search for the Origins of the Earth* (Oxford: Peter Lang, 2010), chap. 12.

73. Henry More, *An Explanation of the Grand Mystery of Godliness* (London, 1660), 233.

74. Johann Jakob Scheuchzer, *Kupfer-Bibel, in welcher die Physica Sacra, oder Geheiligte Natur-wissenschafft derer in Heil: Schrifft vorkommenden natürlichen Sachen* (Augsburg, 1735), 4:1405.

75. Matthew Mackaile, *Terrae prodromus theoricus . . . by Way of Animadversions upon Mr. T. Burnet's Theory of His Imaginary Earth* (Aberdeen, 1691), 7.

76. Luke 3:16–17, *New Oxford Annotated Bible, New Revised Standard Version*, ed. Michael D. Coogan, 4th ed. (Oxford: Oxford University Press, 2010).

77. Matthew 24:37, *New Oxford Annotated Bible*.

78. *The New Testament, with Moral Reflections upon Every Verse* (London, 1719–25), 475–76.

79. Nicolò Vito di Gozze, *Discorsi . . . sopra la Metheore di Aristotele, ridotti in dialogo* (Venice, 1584), 57r.

80. See, for example, Matthew Hale, who references "the Covenant that God made never to bring a Flood again," in *The Primitive Origination of Mankind* (London, 1677), 223.

81. Jacopo Grandi and Joannes Quirini, *De testaceis fossilibus Musaei Septalliani et Jacobi Grandii de veritate diluvii universalis, et testaceorum, qua procul a Mari reperiuntur generatione epistolae* (Venice, 1676), 28.

82. Hale, *Primitive Origination of Mankind*, 217.

83. Hale, *Primitive Origination of Mankind*, 229.

84. Thomas Burnet, *The Theory of the Earth: Containing an Account of the Original of the Earth, and of All the General Changes Which It Hath Already Undergone, or Is to Undergo, till the Consummation of All Things* (London, 1690), 2:60–64.

85. Burnet, *Theory*, 2:63.

86. Two hundred years later, the American geologist William Denton would favorably compare coal-heated buildings to the projected human-engineered warming of the entire planet, arguing with great optimism that humans' invention of the former in recent centuries augured well for humanity's ability to figure out how to engineer the latter in centuries to come. Denton, *Our Planet, Its Past and Future: Lectures on Geology*, 2nd ed. (Boston: William Denton, 1869), 308–11.

87. Tobias Menely, "The Rise of Coal and the Narrativization of Geo-history," talk given at Society of Literature, Science, and the Arts annual meeting, Dallas, TX, October 2015.

88. As Poole argues, "belief in the Conflagration itself was universal and unproblematic" (*World Makers*, 155).

89. Herbert Croft, *Some Animadversions upon a Book Intituled, The Theory of the Earth* (London, 1685), 38–39.

90. Ray to Lhwyd, November 25, 1691, BOD, MS Ashmole 1817a, fol. 214. In a letter to Lhwyd the following year, Ray wrote: "I am solicited upon that account to put my Physico-Theological Discourses into Latine. but they are not particularly directed ag[ain]st Mr Burnets theory, though I look upon it as no more or better th[a]n a meer Chimaera or Romance." Ray to Lhwyd, March 22, 1692, BOD, MS Ashmole 1817a, fol. 221.

91. Ray to Lhwyd, November 25, 1691, BOD, MS Ashmole 1817a, fol. 214.

92. Ray to Lhwyd, November 7, 1692, BOD, MS Ashmole 1817a, fol. 219. By contrast, when he first told Lhwyd about his intention to turn the sermon into a short treatise, he called it "a Short Discourse concerning ye Dissolution of ye World." Ray to Lhwyd, copy, August 17, 1691, BOD, MS Eng hist c 11, fol. 49. The published titles also reflect this shift of emphasis. The main title of the first edition is *Miscellaneous Discourses concerning the Dissolution and Changes of the World*, with "the primitive chaos and creation" and "the general deluge" relegated to the subtitle. The full title of the second and all subsequent editions signal that the three topics are now on a level playing field: *Three Physico-Theological Discourses, concerning I. The Primitive Chaos, and Creation of the World. II. The General Deluge, Its Causes and Effects. III. The Dissolution of the World, and Future Conflagration.*

93. The 1713 edition was prepared by Ray's literary executors based on the expanded manuscript Ray himself prepared prior to his death in 1705, suggesting that the changing balance of coverage from the second to the third edition was an authorial decision rather than an editorial one, in spite of its posthumous publication. Geoffrey Keynes, *John Ray: A Bibliography* (London: Faber and Faber, [1951]), 109.

94. Ray, *Three Physico-Theological Discourses* . . . (London, 1693), 331.

95. Ray, *Three Physico-Theological Discourses*, 332, 331.

96. Ray, *Three Physico-Theological Discourses*, 342.

97. Ray, *Three Physico-Theological Discourses*, 343.

98. M. C. Jacob and W. A. Lockwood, "Political Millenarianism and Burnet's Sacred Theory," *Science Studies* 2, no. 3 (1972): 269. Beverley actually published a defense of Burnet and his treatment of the Apocalypse, making him one of the few people to do so publicly. Thomas Beverley, *Reflections upon The Theory of the Earth* (London, 1699). This tract is sometimes attributed to Burnet himself.

99. Mackaile, *Terrae prodromus theoricus*, 3.

100. Daniel Rosenberg and Anthony Grafton, *Cartographies of Time* (New York: Princeton Architectural Press, 2010).

101. On early modern perceptions of the obscurity of the ancient past, see Paolo Rossi, *The Dark Abyss of Time: The History of the Earth and the History of Nations from Hooke to Vico*, trans. Lydia G. Cochrane (Chicago: University of Chicago Press, 1984); and Arthur B. Ferguson, *Utter Antiquity: Perceptions of Prehistory in Renaissance England* (Durham, NC: Duke University Press, 1993).

102. Isaac Newton, "Fragments from a Treatise on Revelation," Jewish National and University Library, Yahuda MS 1.1, 16r. Reproduced in Frank E. Manuel, *The Religion of Isaac Newton* (Oxford: Oxford University Press, 1974), 122.

103. Newton's understanding of prophecy is discussed in Sarah Hutton, "More, Newton, and the Language of Biblical Prophecy," in *The Books of Nature and Scripture: Recent Essays on Natural Philosophy, Theology, and Biblical Criticism in the Netherlands of Spinoza's Time and the British Isles of Newton's Time*, ed. Richard H. Popkin and James E. Force (Dordrecht: Kluwer, 1994), 39–54. Newton's lifelong, frustrated project of constructing a universal chronology provides a counterpoint to his youthful distinction about the relative certainty of past history. Jed Z. Buchwald and Mordechai Feingold, *Newton and the Origin of Civilization* (Princeton, NJ: Princeton University Press, 2012).

104. Samuel Catherall, *An Essay on the Conflagration in Blank Verse* (Oxford, 1719), 58, 57.

105. Catherall, *Essay on the Conflagration*, 26–27.

106. H.G.I.D.P.E.C.D.R. [Henri Gautier], *Nouvelles conjectures sur le globe de la terre* (Paris, 1721), 34.

107. Gautier, *Nouvelles conjectures*, 40.

108. Johann Beringer, *The Lying Stones of Dr. Johann Bartholomew Beringer, Being His Lithographiae Wirceburgensis*, trans. and annotated Melvin E. Jahn and Daniel J. Woolf (Berkeley: University of California Press, 1963), 18. Given that all the fossil specimens described in *Lithographiae Wirceburgensis* turned out to be fakes manufactured by his enemies at the university in an effort to discredit him, Beringer's sober words of warning now ring somewhat hollow.

109. Nicolas Malebranche, *Méditations chrétiennes*, in *Oeuvres complètes*, ed. Henri Gouthier and André Robinet, 2nd ed. (Paris: J. Vrin, 1967), 10:80; l'Abbé de la Pluche, *Le spectacle de la nature* (Paris, 1764), 1:516–17. Burnet's observation that the Italian peninsula was already prone to earthquakes and was well stocked with volcanoes, several of which had been active in the past century, only served as further proof that the Apocalypse would begin in the seat of Roman Catholicism. Burnet, *Theory*, 2:56–59, 84–85.

110. Michael Heyd, *"Be Sober and Reasonable": The Critique of Enthusiasm in the Seventeenth and Early Eighteenth Centuries* (Leiden: Brill, 1995).

111. On the deliberate sidelining of phenomena deemed too "supernatural," see William E. Burns, *An Age of Wonders: Prodigies, Politics, and Providence in England, 1658–1727* (Manchester, UK: Manchester University Press, 2002).

### Chapter Five • Catholic Climate Change

1. In spite of being one of the most celebrated authors of earth history in the first half of the eighteenth century, Vallisneri's name was not routinely included, as Burnet's and Woodward's were, in the histories of the field that prefaced works of earth history in the second

half of the century. As a consequence, his influence on the field went largely unrecognized until the 1990s, when he was discovered by historians of science like Dario Generali and Rhoda Rappaport. See Generali, *Antonio Vallisneri: Gli anni della formazione e le prime ricerche* (Florence: Leo S. Olschki, 2007); and Rappaport, "Italy and Europe: The Case of Antonio Vallisneri (1661–1730)," *History of Science* 29 (1991): 73–98. Recent studies of Vallisneri's work in earth history are Michael Cunningham, "Seashells on the Mountains: Antonio Vallisneri, Fossils, and the Republic of Letters" (PhD diss., University of Connecticut, 2005); and Francesco Luzzini, *Il miracolo inutile: Antonio Vallisneri e le scienze della terra in Europea tra XVII e XVIII secolo* (Florence: Leo S. Olschki, 2013).

2. Vallisneri to Bourguet, January 29, 1721, BPUN, MS 1282, fol. 243.

3. The word *isporcato*, meaning "dirty," "foul," or "impure," conveys a dual sense of spiritual and physical pollution. Florio's 1611 Italian-English dictionary gives several English synonyms for *sporcare*, including "to pollute," "to defile," and "to make impure." John Florio, *Queen Anna's New World of Words* (London, 1611), 526. As when describing the Flood, the words Vallisneri used to describe the causes and effects of the plague could connote both physical and moral states or processes, and were almost certainly intended to connote both.

4. Daniel Gordon, "Confrontations with the Plague in Eighteenth-Century France," in *Dreadful Visitations: Confronting Natural Catastrophe in the Age of Enlightenment*, ed. Alessa Johns (New York: Routledge, 1999), 3–30.

5. See, for example, Bourguet to Vallisneri, April 12, 1712, BACR, MS Conc. 328/67, no. 5.

6. Vallisneri to Bourguet, August 14, 1719, BPUN, MS 1282, fols. 231–32.

7. Vallisneri to Bourguet, January 29, 1721, BPUN, MS 1282, fol. 244.

8. Vallisneri to Bourguet, August 30, 1721, BPUN, MS 1282, fol. 250. In spite of their Woodwardian leanings, Vallisneri respected Monti and especially Zannichelli. He corresponded with both of them, and he named them both in his list of Italy's greatest natural philosophers in Vallisneri to Bourguet, February 12, 1729, BPUN, MS 1282, fols. 307–8.

9. The eulogy for Zannichelli that appeared in the *Bibliothèque Italique*, likely written by Bourguet, claimed he never made up his mind about the origin of figured stones. *Bibliothèque Italique* 6 (September–December 1729): 152–69.

10. Brendan Dooley, *Science, Politics, and Society in Eighteenth-Century Italy: The "Giornale de' letterati d'Italia" and Its World* (New York: Garland, 1991); Dario Generali, "Il *Giornale de' letterati d'Italia* e la cultura veneta del primo Settecento," *Rivista di storia della filosofia* 2 (1984): 243–81.

11. Vallisneri, *Che ogni Italiano debba scrivere in lingua purgata italiana, o toscana . . .* , in Antonio Vallisneri, *Opere fisico-mediche* (Venice: Sebastiano Coleti, 1733), 3:257.

12. Vallisneri, *Che ogni Italiano*, 257.

13. Francesco Bianchini to Lodovico Antonio Muratori, February 7, 1705, in Gian-Francesco Soli Muratori, *Vita del proposto Lodovico Antonio Muratori* (Naples: Giovanni Gravier, 1773), 203–4.

14. Giovanni Giacinto Vogli to Vallisneri, June 17, 1721, BE, MS It. 588 α.H.4.3, no. 124.

15. I describe the social forces that prompted Swiss and northern Italian scholars to establish epistolary relationships in greater detail in Lydia Barnett, "Strategies of Toleration: Talking across Confessions in the Alpine Republic of Letters," *Eighteenth-Century Studies* 48, no. 2 (2015): 141–57. The dynamics of the Italo-Swiss network are further described in Walter Kurmann, *Presenze italiane nei giornali elvetici del primo Settecento* (Bern: Herbert Lange, 1976), which focuses on the triangular relationship between Bourguet, Scheuchzer, and Vallisneri,

and, more generally, Clorinda Donato, "Illustrious Connections: The Premises and Practices of Knowledge between Switzerland and the Italian Peninsula," in *Scholars in Action: The Practice of Knowledge and the Figure of the Savant in the Eighteenth Century*, ed. André Holenstein, Hubert Steinke, Martin Stuber, and Philippe Rogger (Leiden: Brill, 2013), 535–67.

16. Lorraine Daston, "The Ideal and Reality of the Republic of Letters in the Enlightenment," *Science in Context* 4, no. 2 (1991): 379.

17. The archives of the Royal Society contain numerous manuscripts translated and sent by the Scheuchzers and authored by Italian savants, including Eustachio Manfredi, Michele Pinelli, Dominico Bottoni, Antonio Benevoli, and Vallisneri. The Scheuchzers' connections to Britain were many and deep, forging a scholarly network connecting England and Switzerland. Urs B. Leu, "Swiss Mountains and English Scholars: Johann Jakob Scheuchzer's Relationship to the Royal Society," *Huntington Library Quarterly* 78, no. 2 (2015): 329–48. See also chapter 4, note 14.

18. Francesca Bianca Crucciti Ullrich, *La Bibliothèque Italique: Cultura "italianisante" e giornalismo letterario* (Milan: Riccardo Riccardi, 1974).

19. Jean Le Clerc to Vallisneri, September 1721, BE, MS Italiani 588.α.H.3, no. 11.

20. The classic study of Enlightenment Italy is Vincenzo Ferrone, *The Intellectual Roots of the Italian Enlightenment*, trans. Sue Brotherton (Atlantic Highlands, NJ: Humanities Press, 1995), focused on the spread of Newtonianism as well as more radical philosophies of deism and materialism. More recent scholarship on the Catholic Enlightenment in Italy shows that many Italians viewed the pursuit of knowledge, political and ecclesiastical reform, and a vibrant spiritual life as compatible pursuits. See Massimo Mazzotti, *The World of Maria Gaetana Agnesi, Mathematician of God* (Baltimore, MD: Johns Hopkins University Press, 2007); and Rebecca Messbarger, Christopher M. S. Johns, and Philip Gavitt, eds., *Benedict XIV and the Enlightenment: Art, Science, and Spirituality* (Toronto: University of Toronto Press, 2016). For a provocative discussion of the methodological value of taking religion seriously when writing intellectual history, see Alister Chapman, John Coffey, and Brad S. Gregory, eds., *Seeing Things Their Way: Intellectual History and the Return of Religion* (Notre Dame, IN: University of Notre Dame Press, 2009).

21. Monti to Vallisneri, June 13, 1719, BACR, MS Conc. 344/75, no. 2. The letters preserved in BACR, MS Conc. 344/75 include nine letters from Monti to Vallisneri over the span of ten years; all subsequent ones stick to neutral, shared topics of interest such as the description and exchange of fossil, animal, and plant specimens.

22. The critical reception of Newton in Italy could take on nationalist overtones. Giovanni Rizzetti, the Venetian author of a famous essay on light and color theory who took issue with Newton's *Optics*, wrote disparagingly of "the foreigners [*Oltramontani*], and especially the English" in a letter to Vallisneri. Rizzetti to Vallisneri, November 17, 1727, ASRE, Fondo Brunelli II C, fol. 18.

23. Vallisneri to Guido Grandi, March 10, 1729, in Antonio Vallisneri, *Epistolario 1714–1729*, ed. Dario Generali (Florence: Leo S. Olschki, 2006), 1685 [CD-ROM]. The word Vallisneri used to describe Burnet was *visionario*, somebody who has visions, sees ghosts, or claims to see into the future. It was not, in other words, the compliment that the modern English "visionary" would convey; instead, it suggested that Burnet was a fabulist and a crank.

24. Vallisneri to Bourguet, August 30, 1721, BPUN, MS 1282, fol. 249.

25. Vallisneri was "one of the most overtly Galilean scientists of his generation." Paula Findlen, "Founding a Scientific Academy: Gender, Patronage and Knowledge in Early Eighteenth-

Century Milan," *Republics of Letters* 1, no. 1 (2008): https://arcade.stanford.edu/rofl/founding -scientific-academy-gender-patronage-and-knowledge-early-eighteenth-century-milan.

26. Sebastiano Rotari to Vallisneri, July 22, 1721, BE, MS It. 588 α.H.4.3, no. 54.

27. Mazzotti, *The World of Maria Gaetana Agnesi*, 103. Scholars have tended to regard Vallisneri's religious orientation as unorthodox, perhaps verging on irreligious. See, for example, Ferrone, *Intellectual Roots of the Italian Enlightenment*, 105–11, which places Vallisneri in the milieu of Venetian "libertinism." More recently, historians have stressed that Vallisneri saw himself as a good Catholic, not a secularizer. Cunningham, "Seashells," 37, 190; Francesco Luzzini, "Flood Conceptions in Vallisneri's Thought," in *Geology and Religion: A History of Harmony and Hostility*, ed. Martina Kölbl-Ebert (London: Geological Society, 2009), 80. They have also emphasized that he could be quite inconsistent on questions of science and religion. Paolo Rossi, *The Dark Abyss of Time: The History of the Earth and the History of Nations from Hooke to Vico*, trans. Lydia G. Cochrane (Chicago: University of Chicago Press, 1984), 78–79; Cunningham, "Seashells," 185–90. Most recently, Francesco Luzzini's excellent monograph on Vallisneri, *Il miracolo inutile*, offers a nuanced and detailed account of his views on science and religion, which takes seriously Vallisneri's faith as well as his anticlerical streak and efforts to dodge the Inquisition's censors.

28. Vallisneri to Lioni, "Seconda lettera all'Illustriss. Sig. Abate Girolamo Conte Lioni . . . intorno le produzioni marine, che si trovano su' monti, agli effetti del Diluvio; e all'annosa vita degli uomini innanzi 'l medesimo," July 12, 1719, in *De' corpi marini, che su' monti si trovano; dello loro origine, e dello stato del mondo avanti il Diluvio, nel Diluvio, e dopo il Diluvio: Lettere critiche* (hereafter *DCM*), 2nd ed. (Venice, 1728), 107–8.

29. Paola Potestà, "La questione del 'contagium vivum' e la genesi del concetto di 'ambiente' nella medicina tra XVII e XVIII secolo," *Nuncius* 6, no. 1 (1991): 49–67.

30. Giovanni Maria Lancisi, *De noxiis paludum effluviis, eorumque remediis* . . . (Rome, 1717), 7–8.

31. Vallisneri to Lioni, July 12, 1719, in *DCM*, 107–8.

32. Vallisneri to Lioni, July 12, 1719, in *DCM*, 111.

33. Vallisneri to Lioni, July 12, 1719, in *DCM*, 97, 108.

34. Paul-Gabriel Boucé, "Imagination, Pregnant Women, and Monsters in Eighteenth-Century England and France," in *Sexual Underworlds of the Enlightenment*, ed. G. S. Rousseau and Roy Porter (Chapel Hill: University of North Carolina Press, 1988), 86–100.

35. Lorraine Daston and Katherine Park, *Wonders and the Order of Nature, 1150–1750* (New York: Zone Books, 2001), 197.

36. Fernando Vidal, "Onanism, Enlightenment Medicine, and the Immanent Justice of Nature," in *The Moral Authority of Nature*, ed. Vidal and Lorraine Daston (Chicago: University of Chicago Press, 2004), 272, 266.

37. Gary P. Cestaro, *Dante and the Grammar of the Nursing Body* (Notre Dame. IN: University of Notre Dame Press, 2003). I am grateful to Caterina Mongiat-Farina for this reference.

38. Vallisneri to Lioni, July 12, 1719, in *DCM*, 108.

39. John Woodward, *An Essay toward a Natural History of the Earth . . . with an Account of the Universal Deluge* (London, 1695), 91; Vallisneri to Sebastiano Rotari, "Risposta del Sig. Vallisneri al Sig. Rotari," n.d., in *DCM*, 49.

40. Thomas Burnet, *Theory of the Earth*, vol. 2 (London, 1690), "Preface to the Reader," [R3c].

41. Burnet, *Theory of the Earth* (London, 1684), 1:65.

42. The debate on the nature of generation, a contest between "ovists" and "spermists," was one of the biggest philosophical debates of the late seventeenth and early eighteenth centuries, in part because of the thorny medical and theological issues involved. See Clara Pinto-Correia, *The Ovary of Eve: Egg and Sperm and Preformation* (Chicago: University of Chicago Press, 1997). Vallisneri's participation in the debates about parasites in Eden is discussed in John Farley, "The Spontaneous Generation Controversy (1700–1860): The Origin of Parasitic Worms," *Journal of the History of Biology* 5, no. 1 (1972): 101–2.

43. D. Antonio Maria Borromeo to Vallisneri, September 21, 1711, in Vallisneri, *Nuove osservazioni ed esperienze intorno all'ovaia scoperta ne' vermi tondi dell'uomo* (Padua, 1713), 106–7.

44. Vallisneri to Borromeo, "Risposta alla suddetta lettera," n.d., in *Nuove osservazioni*, 115.

45. See, for example, Vallisneri to Lodovico Antonio Muratori, October 8, 1721, in Muratori, *Edizione nazionale del carteggio di L. A. Muratori*, ed. Michela L. Nichetti Spanio (Florence: Leo S. Olschki, 1978), 44:243: "My book about the Deluge is, as you have seen, skeptical, because no opinion satisfies me."

46. Vallisneri to Lioni, July 12, 1719, in *DCM*, 106, 101.

47. Vallisneri to Lioni, July 12, 1719, in *DCM*, 105.

48. Cunningham, "Seashells," 77–80.

49. Vallisneri to Bourguet, May 27, 1727, BPUN, MS 1282, fols. 289–90. The practice of circulating private correspondence locally recalls Daston's argument that "the scholarly letter of this period was a peculiar hybrid of the personal and the public" ("Ideal and Reality," 371).

50. Vallisneri to Scheuchzer, January 10, 1705, ZBZ, MS H 312, fol. 61.

51. "Bourguet," in *La France protestante*, ed. Eugène Haag (Paris: Sandoz & Fischbacher, 1881), 3:485.

52. "Préface," *Bibliothèque Italique* 1 (January–April 1728): xix–xx.

53. Bourguet to Polier, May 30, 1728, BPUN, MS 1261, fol. 67.

54. Woodward, *Essay*, 82.

55. Vallisneri to Scheuchzer, August 26, 1721, ZBZ, MS H 312, fol. 242.

56. Vallisneri to Bourguet, August 30, 1721, BPUN, MS 1282, fol. 250.

57. Vallisneri to Riccati, September 7, 1721, in Jacopo Riccati and Antonio Vallisneri, *Carteggio (1719–1729)*, ed. Maria Laura Soppelsa (Florence: Leo S. Olschki, 1985), 123.

58. Vallisneri to Bourguet, August 30, 1721, BPUN, MS 1282, fol. 249.

59. Vallisneri to Riccati, September 7, 1721, *Carteggio (1719–1729)*, 123. Emphasis mine.

60. On Jesuit networks, see Mordechai Feingold, *Jesuit Science and the Republic of Letters* (Cambridge, MA: MIT Press, 2003). Gendered divisions (and also collaborations) are highlighted in Carol Pal, *Republic of Women: Rethinking the Republic of Letters in the Seventeenth Century* (Cambridge: Cambridge University Press, 2012).

61. Louis Bourguet to Georges Polier, January 28, 1728, BPUN, MS 1261, fol. 43.

62. Bayle immediately followed this statement of a general principle with a reference to religious differences in particular, observing that it would hardly be appropriate to use "the same style with Protestants, as with good Roman Catholics." [Pierre Bayle], "Lettres choisies de seu Monsieur Guy Patin . . . ," *Nouvelles de la République des Lettres* 1 (April 1684): 107–8.

63. C. Scott Dixon, "Introduction," *Living with Religious Diversity in Early-Modern Europe*, ed. Dixon, Dagmar Freist, and Mark Greengrass (Farnham, UK: Ashgate, 2009), 19. See also Benjamin J. Kaplan, *Divided by Faith: Religious Conflict and the Practice of Toleration in Early Modern Europe* (Cambridge, MA: Belknap Press, 2010), 336, which describes premodern toleration as "a pragmatic arrangement for the limited accommodation of regrettable realities."

64. Vallisneri to Bourguet, December 20, 1710, BPUN, MS 1282, fol. 32.

65. Vallisneri to Bourguet, January 1, 1722, BPUN, MS 1282, fol. 251.

66. Johann Jakob Scheuchzer, *Deum ex terrae structura aliisque affectionibus demonstratum* (Zurich, 1715), 8.

67. Vallisneri to Bourguet, March 19, 1715, BPUN, MS 1282, fols. 190–91.

68. Bourguet to Vallisneri, April 2, 1715, ASRE, Archivio Vallisneri 4/I, 1, fol. 109.

69. Vallisneri to Bourguet, April 5, 1715, BPUN, MS 1282, fol. 192.

70. Many more of Vallisneri's letters to Bourguet survive than Bourguet's to Vallisneri, so it is not always possible to reconstruct a complete exchange.

71. Vallisneri to Bourguet, November 23, 1718, BPUN, 1282, fol. 226.

72. Vallisneri to Bourguet, March 3, 1720, BPUN, MS 1282, fol. 235.

73. Bourguet to Vallisneri, February 4, 1722, BE, MS Italiani 588 α.H.4.3, no. 6.

74. Vallisneri to Ubertino Landi, May 11, 1726, in Vallisneri, *Epistolario 1714–1729*, 1335. I am grateful to Brad Bouley for this reference.

75. Bourguet to Polier, January 28, 1728, BPUN, MS 1261, fol. 42.

76. Bourguet to Polier, May 30, 1728, BPUN, MS 1261, fol. 67.

77. Bourguet to Jallabert, "Lettre sur l'origine des petrifications qui ressemblent aux corps marins," January 5, 1741, in *Traité des petrifications* (Paris, 1742), 53–94.

78. Vallisneri to Muratori, May 6, 1726, in *Carteggio di L. A. Muratori*, 289–90. Vallisneri's difficulties with his role as the medical examiner of an alleged saintly corpse were faced by other Catholic physicians as well. Bradford A. Bouley, *Pious Postmortems: Anatomy, Sanctity, and the Catholic Church in Early Modern Europe* (Philadelphia: University of Pennsylvania Press, 2017), 85–88.

79. Nor was he alone in worrying about this. Catholic officials also "feared that false miracles could discredit Catholicism as easily as true ones could support it." Kaplan, *Divided by Faith*, 33.

80. Anne Goldgar, *Impolite Learning: Conduct and Community in the Republic of Letters, 1680–1750* (New Haven, CT: Yale University Press, 1995), 213.

81. He closed the above statement with the common Ockhamian refrain of philosophers defending themselves against charges of impiety since the fourteenth century: "I'm not among those who like to multiply miracles without necessity." Riccati hoped perhaps that invoking a philosophical principle popular in the Renaissance from an intellectual tradition that both men knew well might serve as an alternative source of legitimacy for his controversial opinion. Riccati to Vallisneri, n.d. [sometime between June 14 and July 13, 1719], in *Carteggio (1719–1729)*, 66. A very similar conversation about the Flood's universality transpired between Bourguet and the Veronese scholar Ottavio Alecchi, who compared the needless miracle of a Universal Deluge to flooding the entire Mediterranean basin just to destroy one little village on the Lago di Garda. Ivano dal Prete, *Scienza e società nel Settecento Veneto* (Milan: FrancoAngeli, 2008), 223.

82. Vallisneri to Riccati, July 13, 1719, *Carteggio (1719–1729)*, 70.

83. Riccati to Vallisneri, n.d. [sometime between July 13 and 26, 1719], in *Carteggio (1719–1729)*, 75–77.

84. Vallisneri to Riccati, June 14, 1719, *Carteggio (1719–1729)*, 63. He used nearly the same words—Galileo's words—to make the same point to Bourguet the following year: "Holy Scripture teaches natural philosophers nothing, and only fills the head with prejudices. Scripture teaches the ways of Heaven, not the phenomena of the Earth." Vallisneri to Bourguet,

September 23, 1720, BPUN, MS 1282, fol. 238. Galileo's formulation was "The intention of the Holy Spirit is to teach us how one goes to heaven and not how heaven goes." Galileo Galilei, *Letter to the Grand Duchess Christina*, in *The Galileo Affair: A Documentary History*, ed. Maurice Finocchiaro (Berkeley: University of California Press, 1989), 96.

85. Riccati referred to the "supposedly mobile earth . . . demonstrated by Huygens and Newton," when discussing the shape and size of the earth and the possibility of its being completely flooded. Riccati to Vallisneri, n.d. [sometime between June 14 and July 13, 1719], in *Carteggio (1719–1729)*, 69.

86. Vallisneri to Riccati, July 26, 1719, in *Carteggio (1719–1729)*, 78.

87. Vallisneri to Riccati, September 7, 1721, *Carteggio (1719–1729)*, 123.

88. Vallisneri to Muratori, October 8, 1721, in *Carteggio di L. A. Muratori*, 243.

89. Vallisneri to Rotari, n.d., in *DCM*, 24. His statement of faith comes in the next letter, Vallisneri to Lioni, July 12, 1719, in *DCM*, 82.

90. Conti, who exchanged letters with Vallisneri on the problems of a too-literal interpretation of Genesis, was one of several in Vallisneri's circle of correspondents who believed that the Flood's miraculous nature placed it beyond the pale of philosophy. Rhoda Rappaport, *When Geologists Were Historians, 1665–1750* (Ithaca, NY: Cornell University Press, 1997), 169.

91. Ferrone, *Intellectual Roots of the Italian Enlightenment*, 103.

92. Giovanni Giacomo Spada, *Dissertazione, ove si prova, che li petrificati corpi marini, che nei (1719–1729) monti adiacenti a Verona si trovano, non sono scherzi di natura, nè diluviani; ma antediluviani* (Verona, 1737), 15.

93. Spada, *Dissertazione . . . petrificati corpi marini*, 17.

94. Antonio Lazzaro Moro, *De' crostacei e degli altri marini corpi che si truovano su' monti* (Venice, 1740), 211.

95. Moro, *De' crostacei*, 15–16.

### Epilogue · The Flood Subsides

1. Galileo was not a Paduan by birth; he took up a professorship at the University of Padua in 1592, around the same time as Erculiani's death.

2. Theriac, which Erculiani mentions having prepared in her 1577 letter to Guarnier, was a highly prized remedy for plague. Camilla Erculiani to Georges Guarnier, August 7, 1577, in *Lettere di philosophia naturale* (Kraków, 1584), c3r. Paula Findlen, "Aristotle in the Pharmacy: The Ambitions of Camilla Erculiani in Sixteenth-Century Padua," in Erculiani, *Letters on Natural Philosophy*, ed. and trans. Eleonora Carinci, Paula Findlen, and Hannah Marcus (Toronto: University of Toronto Press, forthcoming). Vallisneri's research on the microbial causes of contagious disease is described in Dario Generali, *Antonio Vallisneri: Gli anni della formazione e le prime ricerche* (Florence: Leo S. Olschki, 2007), 300–307.

3. Richard J. Blackwell, *Galileo, Bellarmine, and the Bible* (Notre Dame, IN: University of Notre Dame Press, 1991).

4. Antonio Vallisneri to Sebastiano Rotari, n.d., in *DCM*, 24.

5. Vallisneri to Bourguet, January 1, 1722, BPUN, MS 1282, fols. 251–52. See Francesco Luzzini, *Il miracolo inutile: Antonio Vallisneri e le scienze della terra in Europea tra XVII e XVIII secolo* (Florence: Leo S. Olschki, 2013).

6. Vallisneri to Rotari, n.d., in *DCM*, 53.

7. Rotari to Vallisneri, July 22, 1721, BE, MS It. 588 α.H.4.3, no. 54.

8. Antonio Lazzaro Moro, *De' crostacei e degli altri marini corpi che si truovano su' monti* (Venice, 1740), 15 (italics mine).

9. Vallisneri to Rotari, n.d., in *DCM*, 34.

10. Vallisneri to Bourguet, November 23, 1718, BPUN, MS 1282, fols. 225–26 (italics mine).

11. Vallisneri to Bourguet, November 9, 1710, BPUN, MS 1282, fol. 17.

12. Vallisneri to Bourguet, January 2, 1714, BPUN, MS 1282, fol. 146.

13. Bernadino Ramazzini, *De fontium Mutinensium admiranda* (Modena, 1691), 48–49, 55.

14. Giovanni Giacomo Spada, *Dissertazione, ove si prova, che li petrificati corpi marini, che nei monti adiacenti a Verona si trovano, non sono scherzi di natura, nè diluviani; ma antediluviani* (Verona, 1737), 1:15.

15. M. [Antoine] de Jussieu, "Examen des causes des impressions de plantes marquées sur certaines pierres des environs de Saint-Chaumont dans le Lionnois," *Mémoires de l'Académie Royale des Sciences* (1718): 291–93.

16. Jussieu, "Examen des causes des impressions," 287.

17. René Antoine Ferchault de Réaumur, "Remarques sur les coquilles fossiles de quelques cantons de la Touraine, & sur les utilités qu'on en tire," *Histoire de l'Académie Royale des Sciences* (1720): 400–416. Rappaport notes that Jussieu, Réaumur, and Fontenelle were unique among their contemporaries in divorcing natural and human history. Rhoda Rappaport, *When Geologists Were Historians, 1665–1750* (Ithaca, NY: Cornell University Press, 1997), 92.

18. Scheuchzer, *Charta invitatoria* (Zurich, 1699), 2. Scheuchzer's religious, political, and intellectual commitments dovetailed in his pursuit of patriotic Swiss natural history and positive rehabilitation of both the Flood and the Alps. Michael Kempe, "Noah's Flood: The Genesis Story and Natural Disasters in Early Modern Times," *Environment and History* 9 (2003): 162–63; Alix Cooper, *Inventing the Indigenous: Local Knowledge and Natural History in Early Modern Europe* (Cambridge: Cambridge University Press, 2007), 131–39.

19. Kirwan and Whitehurst receive notices and Buckland considerable attention in Martin J. S. Rudwick, *Bursting the Limits of Time* (Chicago: University of Chicago Press, 2005); and Rudwick, *Worlds before Adam* (Chicago: University of Chicago Press, 2008).

20. "The same subterraneous fires (which originally raised the continents and islands that now appear, and have ever since been making great changes in the bowels of the earth, and producing those tremendous earthquakes, which have happened from time to time) may in the end break forth with redoubled violence, and destroy it, in the manner foretold in Scripture." Edward King, "An Attempt to Account for the Universal Deluge," *Philosophical Transactions of the Royal Society* 57 (1767): 56.

21. Rudwick, "Biblical Flood and Geological Deluge: The Amicable Dissociation of Geology and Genesis," in *Geology and Religion: A History of Harmony and Hostility*, ed. Martina Kölbl-Ebert (Bath, UK: Geological Society of London, 2009), 103–10.

22. On the emergence of the Romantic conception of nonhuman nature, see the classic essay by William Cronon, "The Trouble with Wilderness; or, Getting Back to the Wrong Nature," in *Uncommon Ground: Toward Reinventing Nature*, ed. William Cronon (New York: W. W. Norton, 1995), 69–90.

23. Martin J. S. Rudwick, *Earth's Deep History: How It Was Discovered and Why It Matters* (Chicago: University of Chicago Press, 2014).

Page numbers in *italics* refer to illustrations.

acclimatization, 103–5, 221n61

Acosta, José de, 57–58, 64, 74

Adam, 39, 41–42, 45, 58–64, 94, 173–75, 209n88, 211n18; and Eve, 4, 50, 58, 60–62, 94, 173–74; and Noah, 12, 51–55, 58, 60, 62–64, 175, 221n60. *See also* pre-Adamism

Agnesi, Maria Gaetana, 166

air, 98, 101; antediluvian, 94, 98, 102, 105–6; degradation of, 3, 17, 93, 100–101, 105, 120, 167; in London, 92, 105, 107, 118, 124, 221n66; as vector of disease, 101, 105–6, 167–68

Albritton, Vicky, 9, 202n45

Alecchi, Ottavio, 238n81

All Saints' Flood (1570), 28–30, *29*

Anthropocene, 8–10, 18, 21, 199nn16–18, 200n25, 203n5; and deep time, 21–22, 194; early modern origins of, 8–9, 17, 21–22; and environmental reflexivity, 9, 90; and geological agency, 8, 22, 90; and theology, 9

anti-Semitism, 13, 72, 111, 201n32, 213n47

Apocalypse, 15, 148–50; as global natural disaster, 133, 149, 153–54

argument from design, 126–27

Aristotelian natural philosophy, 10, 22, 43–44, 101; Burnet and, 220n42; meteorological traditions, 98, 191, 206n37; power of water and fire, 150–51, 189

Aristotle, 23, 25–27, 37, 39, 48, 208n78, 209n85; on climate's effects on human bodies and intellect, 74, 214n59; Erculiani

and, 23, 36–37, 39, 203n3, 206nn37–39; and gender, 43–44, 208n78; *Meteorology*, 24, 27, 31, 33, 48–49, 98, 160, 206nn38–39, 220n42; misogyny of, 208n78; and natural slavery, 73, 213n54; silence on New World people, 54

ark, Noah's, 3, 6, 19, 28, 30, 34, 40, 74, 114, 170, 205n26; and monogenism, 50, 52, 54; size of, 61, 64, 204n12

astrology, 39, 74, 137; and Calancha, 200n25, 213n37; and disaster prediction, 29–30, 34–35, 205n26, 214n55; and Erculiani, 33–36, 40, 49; Noah as astrologer, 34–35

Augustine, Saint, 45–46, 114, 166; Augustinianism, 110, 223n4

Bacon, Francis, 121, 231n71

Bashford, Alison, 11

Báthory, Stefan, 32, 38, 205n32

Bayle, Pierre, 180, 237n62

Beaumont, John, 109

Becher, Johann Joachim, 138

Benevoli, Antonio, 235n17

Bering, Captain Vitus, 79

Beringer, Johann, 156, 233n108

Berns, Andrew, 37, 207n53

Bertrand, Élie, 193

Berzeviczy, Márton, 32–33, 34–38, 42, 45–46, 205n32

Bethell, Hugh, 134–35

Beverley, Thomas, 154, 232n98

Bianchini, Francesco, 164

Bible, 13, 31, 36–37, 52; biblical history, 21–23, 31, 34, 37, 46, 49–64, 66, 68, 72–73, 76, 81, 96–97, 174–75, 189; early modern editions of, 25, 37, 78, 84, 148, 216; and natural philosophy, 26–27, 31, 61–64; Scheuchzer and, 148. *See also* scripture

*Bibliothèque Italique*, 165, 177, 183, 234n9

Bignon, Abbé Jean-Paul, 135

Blair, Ann, 5, 26–27

Boccone, Paolo, 164, 229n42

Bodin, Jean, 87

Bonaventure d'Argonne, Noël, 142

Bonneuil, Christophe, 9

Borromeo, Antonio Maria, 173

Borromeo, Clelia Grillo, 177

Bottoni, Domenico, 235n17

Bourguet, Louis, 3–4, 7, 66, 75–83, 182–83, 214n63, 215n67, 216n81; and *Bibliothèque Italique*, 165, 171, 177, 183, 234n9; on the Flood, 10, 178; *Mercure Suisse*, 76–79, 81, 84, 177, 215n67; and Native Americans, 66, 83–84, 178, 216n90; and postdiluvian migration, 66, 77, 81–82; and Scheuchzer, 83, 84, 180–83, 215n72, 216n88; and Tartary theory, 52, 66, 76–84; and Vallisneri, 160, 163, 176–83, 189–90

Boyle, Robert, 121–22, 130, 139

Brerewood, Edward, 69–70

Buckland, William, 193, 240n19

Buffon, George Louis Leclerc, comte de, 21, 194, 203n4

Buridan, Jean, 24–25, 203n11

Burnet, Gilbert, 95

Burnet, Thomas, 3, 7, 12, 21, 89–90, 94, 104, 137, 222n75; and Aristotle's meteorology, 98, 220n42; critical responses to, 91–92, 95, 107–12, 126–27, 152–53, 167, 232n98; environmental determinism, 91–92, 115, 124–25; Eurocentrism of, 102, 221n60; Mosaic natural philosophy, 92, 110–11; natural vs. artificial earth, 89, 126–27, 194; and Newton, 110; on New World weather, 103, 106; providence and ruined earth, 94–95, 110, 126, 219n27; salvation, 112–15; and sin, 89–90, 93, 126; synchrony of human and natural history, 93–94, 103, 108, 112, 161, 167; and Universal Conflagration, 151, 157, 232n98; universality of the Flood, 93–94

Calancha, Antonio de la, 12, 52, 66–68, 73, 213n37; climate's impact on human body, 74, 214n59; and evangelism, 66, 68, 70–71, 73; and imperial expansion, 66, 68, 73; and racial divisions, 12, 52, 73–75, 214n60, 220n49; and Tartary theory, 66, 68–74

Calvin, John, 44; and Church of England, 114, 126, 223nn95–97; and declensionism, 219n35; depravity, 45, 114, 126; salvation, 114, 126

Calvinists, 13, 26, 60, 114, 126, 211n22

Carinci, Eleanora: on Erculiani and female liberty of philosophizing, 209n93; and Inquisition, 47–48, 207n62; and male discursive space, 207n56; and Republic of Letters, 205n35

Catherall, Samuel, 155–56

Catholic Reformation, 2, 31, 173; and learned women, 207n60; and pious natural philosophy, 3, 25–56

Cavert, William, 105, 221n66, 222n72

Chakrabarty, Dipesh, 8, 22, 90

Chaplin, Joyce, 11, 226n9

Charles II, King, 108, 114

China, 25; fossils from, 184; and New World migration, 55, 70, 77, 81

Church of England, 91–92, 115, 126, 153, 157; and Calvinism, 114, 126, 158, 223n95, 223n97

climate, 11, 214n59, 220nn50–51; and degeneration, 8, 103–4, 199n20; Edenic climactic zones, 98, *99*, 100–101; effect on humans, 102, 116

climate change: anthropogenic, 1–2, 9, 11, 18–19, 90–91, 107, 172, 199n16; and degeneration, 8; denialism, 2, 197n5; diluvial, 11–12, 19, 101–2, 120, 167, 170, 175

coal, 135, 152, 232n86

Coen, Deborah R., 14, 229n44

colonialism, 5, 103, 198n1, 221n64; Columbian expeditions, 50, 58, 76, 77; and monogenism, 63

complexion, 104, 208n77, 214n60; and gender, 44, 208n77; and race, 74–75, 104, 214n60

Conti, Antonio, 177, 186, 239n90

Cook, Harold J., 17, 225n3, 231n71

Cosgrove, Denis, 8, 11, 226n9

Cremonini, Cesare, 46, 209n85

Croft, Herbert: denunciation of Burnet, 109,
    152; on Flood as a miracle, 112, 152–53; on
    ruined world, 110, 112, 127
Crutzen, Paul, 9, 199n16, 199n18, 200n21, 203n5
cultural nationalism, 16, 161, 164, 181
Cumberland, Richard, 11, 97
Cuper, Gisbert, 134

dal Prete, Ivano, 22, 203n6, 203n11, 238n81
Daneau, Lambert, 26
Dante Alighieri, 171, 174
de Laet, Johannes, 66
decay: of world, 96, 106–7, 120–21, 219n35,
    219n37
declensionism, 90, 96, 101; and Calvinism,
    219n35; human and natural decline, 97, 101
deep time, 193–94; and Anthropocene, 10,
    21–23; as Enlightenment discovery, 10, 21;
    geological, 14, 194–95
deism: charges of, 16, 91–92, 108–12, 116–17,
    152, 223n90; and earth's origin, 21, 46, 91,
    235n20
Delisle, Joseph-Nicolas, 79, 80, 215n67, 215n75
de Luc, Jean-André, 193
Descartes, René, 10, 138, 218n10; Cartesian
    rationality, 92
de' Vieri, Francesco, 27
di Gozze, Nicolò Vito, 27, 31–33, 150; and
    Gondola preface on learned women, 34; on
    Noah as astrologer, 34–35; and Zuzori, 34
disasters, natural, 9–10, 13, 15, 24, 31, 119, 133;
    and Apocalypse, 15, 133, 148, 152, 157–58,
    167, 204n22, 231n72; as divine punishment,
    3–4, 10, 28–29; humans as passive victims
    of, 8; providentializing, 27–28
Dixon, C. Scott, 180
Dörries, Matthias, 11
double truth, doctrine of, 25–26; Erculiani
    and, 47–48; and liberty of philosophizing,
    25–26, 47
Dryden, John, 122

Eden, 11, 15, 96, 173, 219n37; climate of, 11, 98,
    *99*, 100–101, 175; original state of perfection,
    3, 11; paradox of human parasites in, 173;
    recreating, 122, 224n122
Edenocene: characteristics of, 94, 98–101; ends
    with Flood, 94, 98, 100, 105; vs. Fallocene,

95, 98, 100, 102–4, 119–20, 124–25, 169;
    natural abundance in, 102, 116, 121
empiricism, 14–15, 37; Baconian emphasis
    on practical natural knowledge, 121; and
    empirical fieldwork, 190; evidence of fossils,
    131–32; scaling and scalar imagination, 14,
    229n44, 229n50
environmental consciousness, 8–9, 18, 51,
    226n9
environmental degradation: anthropogenic, 3,
    18, 90, 125, 169; as sin, 1–4, 19
environmental determinism, 92, 222n72;
    Burnet and, 92, 115, 124
environmental reflexivity, 9, 90–91
Erculiani, Camilla, 2, 7, 12, 20–21, 23, 35–36,
    38, 202n1, 205n35, 210n4; and Aristotle,
    33–34, 36–37, 39, 206n39; on astrology
    33–35, 40; the Bible and earth's history,
    27, 37, 188; de-emphasis on sin as cause of
    Flood, 39; and empirical authority, 37; and
    entry into male discursive space, 38, 48,
    207n56; free will vs. determinism, 39–40;
    and Galen, 34, 35–36, 37, 39, 206n44; and
    gender, 37–39, 42, 47; on human embodi-
    ment and harm to nature, 20, 23, 34; and
    Inquisition, 39, 46–48, 51; on learned
    women, 38, 207n60; on male vs. female
    sin, 41; and Mosaic natural philosophy, 36;
    on Noah's Flood and natural causes, 20–21,
    32, 34–36, 40–41, 188; on overpopulation,
    35–36, 45; publication in Kraków, 32, 46,
    188, 203n2; and rainbow, 34, 206n39; on
    sin, suffering, and salvation, 33, 40–41; and
    University of Padua, 20, 33, 45–64, 205n34;
    on women subject to men's laws, 41–42
Erizzo, Sebastiano, 41–42, 206n35
evangelism, 7, 58, 77, 90, 115; and Bourguet,
    83, 180; and imperialism, 49, 60, 66–67,
    72, 87; and monogenism, 12, 52, 59, 64, 66,
    68, 210n7; and Native Americans, 58, 60,
    64, 68–69, 75, 210n7; and polygenism, 58,
    60–61, 63–64, 212n30; race and, 52, 53, 58,
    70, 75, 210n7
Evelyn, John, 121

Fallocene, 89, 125; climate of, 104,
    108; vs. Edenocene, 89, 101–2, 126; and
    environmental determinism, 115, 124;

Fallocene (*cont.*)
and impoverished soil, 102, 120–21; labor as
means to salvation, 116, 120–21; materialism
and greed in, 125; persistence of, 114, 118;
racial transformations in, 104–5; variability
of weather and seasons during, 98, 100
Fausto da Longiano, Sebastiano, 24, 206n39
Fisher, John, 117
Fleming, James Rodger, 9, 199n16
Flood, the (Noah's/Universal): and Apocalypse, 15, 231n72; and astrology, 34; as
catastrophe punishing human sin, 3–5,
30, 49; and end of Edenocene, 100; and
geological change, 94; as historical event,
27; less than planetary, 13, 49, 60–61, 111;
and monogenism, 51, 85–88; natural vs.
supernatural causes of, 20, 31–32, 48, 63–64,
121; planetary changes, 3, 22, 48–49, 94,
100; and problem of New World animals,
57; secularization of, 194–95
floods, global, 31–32, 51; *Diluvio* vs. *alluvione*,
31, 205n29; physical impossibility of, 24;
predictions of, 29–32, 34, 45
floods, local: All Saints' Flood, 28–30; *alluvione* vs. *Diluvio*, 31; astrological causes,
34; comparisons to Noah's Flood, 28–30;
miraculous, 31; non-miraculous, 190
Foscarini, Ludovico, 207n58, 208n69
fossils, 147, 157, 190; Beringer and, 156; Bourguet and, 181–83; early modern debate on,
130, 134, 157, 226n6; as empirical evidence,
22, 131–32, 141, 146, 191; exchange of, 14, 116,
131–47; Italian scholarship on, 164, 186; Jussieu and, 190–93; Lhwyd and, 146; Mather
and, 144; Monti and, 135–36; sacralization
of, 131; Scheuchzer and, 192; Vallisneri
on, 166, 181, 183; Woodward and, 129–42,
225n131, 226n7
Francis, Pope, *Laudauto Si'*: call to environmental action, 1–2; and consumerism,
19; critics of, 1–2; and history of Catholic
environmental thought, 198n7
free will, 4, 45; Erculiani and, 39, 41, 46; and
sin, 168–69
Fressoz, Jean-Baptiste, 9, 90
Frytsche, Marcus, 24
Fulton, Elaine, 28

Galen, 23, 42, 295n44; Calancha and, 214n59;
Erculiani and, 35–37, 39, 42; Galenic medicine, 34–35, 206n44
Galileo, 1–2, 166, 185, 188, 239n1; and scripture
in support of science, 46, 185, 189; Vallisneri
and, 238n84
García, Gregorio, 50–51, 62
Gaurico, Luca, 30
Gautier, Henri, 156
gender: belief in Eve's inferiority to Adam,
42; and body temperature, 208n78; and
complexion, 208n77; effects of sin on male
health and bodies, 11; Erculiani in dialogue
with men, 38–39; female embodiment and
spirituality, 43–45; and Galen, 42; and
male discursive space, 38, 207n56; male vs.
female sin, 41–42; Marinella on, 43–45; and
Republic of Letters, 179
Genesis, 23, 96–97; Edenic human-nature
harmony, 1, 96; and Golden Age, 96; and
history of nature, 31, 34, 36–37, 50, 54, 108,
110, 239n90
Genghis Khan, 82
geology, 16, 53; and Burnet, 97–98, 193;
diluvial formation of France, 191–92;
earthquakes, 24, 27–28, 105, 114, 151, 191,
233n109, 240n20; geological agency, 8, 22,
90–91, 115; geological determinism, 54, 57,
77; golden spike, 8, 94, 199n18, 218n15;
secularization of, 194–95; volcanic activity,
151, 187, 233n109
gigantism, antediluvian, 94, 96–97, 206n50,
219n37; Burnet on, 101, 107; Erculiani on,
35–36, 38, 97
*Giornale de' Letterati d'Italia*, 164
Golden Age, the, 96, 122; and Eden, 11, 96–97,
102; and lost longevity, 174; and world's
decay, 107
Goldgar, Anne, 17, 226n11, 229n45
Gondola, Maria, 34
Grandi, Jacopo, 136–37, 228n36; on the Flood
and Apocalypse as global, 63–64, 150; Vallisneri and, 166
Grevsmühl, Sebastian, 11
Grove, Richard, 8–9
Grotius, Hugo, 66, 77, 212n34
Guaman Poma de Ayala, Felipe, 59–60

Guarnier, Georges: and Erculiani, 32–33, 35–36, 38, 39, 42, 45, 47; objection to contradicting Aristotle and Galen, 36–37

Hakewill, George, 96, 219n26
Hale, Matthew, 51, 57, 151, 211n15; no second Flood, 231n80; and Tartary theory, 82, 216n87
Halley, Edmond, 96, 108–9, 219n31
Harris, John, 137
Harrison, Peter, 16, 116, 223n94
Hartnett White, Kathleen, 1–2, 197nn4–5
Heise, Ursula, 128
Herrera y Tordesillas, Antonio de, 68
Hippocrates: and astrology, 34, 74–75; on climate's effect on human bodies, 101, 168, 214n59, 220n51; universal causes and effects, 34, 74, 220n49
Hirata Atsutane, 13
historiographic narratives, 81: chronology of past and future, 155; declensionist view of environmental harm, 90, 96; end of history, 200n21; interdependence of human and natural history, 96; polycentric, 63, 212n31; prophecy vs. unknowability of future, 155–56; provincializing Mosaic history, 13, 49, 61; religion and the secularization of science, 17, 21; re-universalizing biblical history, 64; synchrony of human and natural history, 93–94, 103, 108, 112, 161
Hobbes, Thomas, 97
Horn, Georg: *Arca Noae*, 6; and pre-Adamism, 63; and Tartary theory, 77; and Voss, 63, 212n32
Hornberg, Alf, 17
Huguenots: 83, 114; and interconfessional scholarly discourse, 177, 179, 182
human agency: and free will, 9–10, 39; monolithic fallacy, 18
human embodiment: damaged by Fall, 172–73; Erculiani on, 20, 35, 39, 42, 45–46; and morality, 43; of sin, 39, 42, 46, 169
humanity: antediluvian longevity and gigantism, 35–36, 94; as planetary force, 9, 10; postdiluvian changes to, 3, 11–12, 104
human mortality, 17, 33, 41–42, 118, 167; and original sin, 39, 45, 173

human origins and diversity: monogenism, 12, 49, 50–52, 59, 75–76, 86–87, 214n55; polygenism, 54, 57–59, 62; and pre-Adamism, 52, 58–61; quasi-polygenism, 60, 221n60
human reproduction: ovist-spermist debates, 170, 237n42; postdiluvian debility, 170–71
hurricanes, 103, 221n59
Huygens, Christiaan, 239n85

imperialism, 5, 73, 210n6, 214n62; and evangelism, 13, 49, 52–53, 60, 64, 66; and monogenism, 60, 73; and racial categories, 52–53, 76; and Republic of Letters, 138
Incas, 59–60
indigeneity, 87, 209n1
Iroquois, 81–82, 84
Istituto delle Scienze (Bologna), 135
*Italian Library. See Bibliothèque Italique*

Jews: conversion of, 212n31; Jewish-Christian relations in Spain, 71–72; Jewish "contumacy" and New World evangelism, 72; Native Americans as Semitic, 72
Jones, Hugh, 144, 230n66
Jonsson, Fredrik Albritton, 9
Jussieu, Antoine de, 190–91, 240n17

Keill, John, 109, 112
Kidd, Colin, 212n30, 214n62, 217n101
King, Edward, 193, 240n20
Kircher, Athanasius: and empiricism, 109, 138, 229n48; on the Flood and earth's topography, 64, 65
Kirwan, Richard, 193
knowledge networks, 6–7, 23, 131–33, 138, 147, 160, 226n19; global networks, 140–42; Jesuit knowledge networks, 130, 237n60; rhetoric of global participation, 142–45; toleration, 176
Kusukawa, Sachiko, 110, 204n22, 223n88

labor, necessity of, 3, 42, 102; and theology of improvement, 122–25, 224n117, 224n120
La Hontan, Louis Armand, Baron de, 82
Lancisi, Giovanni Maria, 168

land bridge, 51, 55, 57–58, 65, 66–68, 81, 210n6;
  Bourguet on, 79–81; and debate between
  Grotius and de Laet, 66; and Kamchatka,
  79, 83; and racism, 210n9
La Peyrère, Isaac, 13, 52, 111, 211n22; and assertion that Flood affected only Jewish lands,
  13, 60–61; and autochthonous indigenous
  accounts, 87; Catholic and Protestant
  condemnation of, 61; and evangelism, 60,
  211n22; on persecution of Jews, 63; and
  polygenism, 60–62; and provincialization
  of Mosaic history, 13, 49, 61; and writers
  against pre-Adamism, 63–65
Las Casas, Bartolomé de, 73
learned women, 34; Erculiani's awareness of,
  38; Marinella's examples, 43
Le Clerc, Jean, 165
Leibniz, Gottfried Wilhelm, 10, 191, 216n88;
  and Steno's methodology, 140–41
Lemuria (lost continent), 15, 201n36
Leydekker, Melchior, 13, 201n32
Lhwyd, Edward, 117, 125, 191; and British
  fossils, 139, 143–44; opposition to idea
  of fossils as Flood relics, 146; and Ray,
  139, 153, 232n92; and Woodward, 140,
  142
liberty of philosophizing, 209n93; double
  truth and, 25–26, 47–48; and gender, 48
Lioni, Girolamo, 167, 174–75
Lister, Martin, 117
Little Ice Age, 5, 28, 92, 103, 107, 198n13,
  204n23, 222n72; and acclimatization, 103;
  and environmental determinism, 222n72;
  and flooding, 30, 53; and idea of frozen land
  bridge, 69, 107
Locher, Fabien, 90
longevity, antediluvian, 11, 97, 169, 173, 175;
  Burnet on, 100–104; decline of, 3, 11, 97,
  104–5, 169; and Edenic climate, 100, 102,
  175; Erculiani on, 36–38; Methuselah, 97,
  107; and overpopulation, 36; Ray on, 105;
  and virility, 11, 97
Lovell, Archibald, 15
Luther, Martin, 30, 44; on Adam's embodiment and sin, 45; on frequency of natural
  disasters, 28
Lutheran Reformation, 5, 26, 30, 45, 46
Lyell, Charles, 92

Machiavelli, Niccolò, 30
Mackaile, Matthew, 96, 126; on Burnet, 96,
  126; and coming Apocalypse, 149–50, 154–55
Maffei, Scipio, 164
Maillet, Benoît de, 21, 303n4
Malebranche, Nicolas, 157
Malm, Andreas, 18
Manfredi, Eustachio, 235n17
Margóscy, Dániel, 17
Marinella, Lucrezia, 208n70; and Ercu-
  liani, 43; opposition to misogyny, 43–44,
  208nn77–78; on women's embodiment and
  spirituality, 42–43, 208nn70–71, 208n75,
  208n78
Marinelli, Giovanni, 208n70
Marsigli, Luigi Ferdinando, 135, 228n27
Martin, Craig, 10
Martínez, Henrico (Heinrich Martin), 50, 51,
  55, 58, 209n2; and racial divisions, 74
Marvell, Andrew, 95
Mather, Cotton, 78–79, 84; and Bourguet,
  216n92; and fossils, 144; and Royal Society,
  78, 144, 230n66; and Taunton boulder,
  78–79, 230n66; and Woodward, 144–45
Mather, Increase, 83–84, 216n92
Mather, Samuel, 216n92
Mazzotti, Massimo, 16, 166, 236n27
McKibben, Bill, 89, 127
Medici, Ferdinando II de, 137
Melanchthon, Philip, 26
Menochio, Giacomo, 46–48, 209n88
*Mercure Suisse*, 76–79, 81, 84, 177, 215n67
meteorology, 24; Aristotle's *Meteorologica*, 24,
  27, 31, 33, 48–49, 98, 160, 189, 206nn38–39;
  Catholic, 31, 192; return to Renaissance
  theory, 189
monogenism, 12, 49, 50–52; and denial of
  indigeneity, 87, 209n1; and imperialism and
  evangelism, 59, 75; and Noah, 51, 76, 87;
  and racism, 85–88
Monti, Giuseppe, 135–36; and "Woodwardian-
  ism," 137, 163, 165, 187, 191, 228n29, 234n8,
  235n21
Moro, Antonio Lazzaro: the Flood as matter
  of faith not philosophy, 189; refutation of
  Burnet and Woodward, 187
Mosaic natural philosophy, 31, 36, 49, 157, 162,
  189, 203n6; and ancient history, 109–10;

Aristotle supplanted by Moses, 26, 36; Bourguet and, 76; and Catholic learning, 26, 31, 36; and double truth, 26–27; Erculiani and, 23, 36, 42, 49; as global history, 94; as irenic, 27, 131, 157–59; and monogenetic human origins, 76, 87; Monti and, 135–36; and Newtonianism, 165; transnational character of, 131, 136, 148, 187; unifying science and religion, 5, 91–92, 108, 144; uniting Catholics and Protestants, 131, 136; and Vallisneri, 165, 175, 187, 189; Zuccolo and, 49

Moses, 23, 26, 110; and La Peyrère, 49, 61

Muratori, Lodovico Antonio, 164; on separation of scripture and natural philosophy, 186; and Vallisneri, 183, 186, 237n45, 238n78

Native American origins, European theories of: autochthony and indigeneity, 63, 87, 209n1, 217n100; Hamitic descent, 68, 70, 73–75; Japhetic descent, 68, 70–74; migration by sea, 55–56, 63, 81; and monogenism, 55, 59, 62; Moses and Aristotle, 54; as "natural slaves," 73; and polygenism, 54, 59, 60, 64; and post-Flood migration, 63; racial politics of origin debate, 66, 75, 210n12; Semitic descent, 68, 70, 72, 73, 213n44; Tartar descent, 52, 68–70, 75; and universal Flood, 50, 64, 66

Native American origins, indigenous theories of, 59–60, 211nn20–21

natural history, 109, 198n10; and Baconian empiricism, 121; and Calvinism, 219n35; and declensionism, 90, 101; and Enlightenment vision of progress, 96; and improvement of soil as a means of grace, 121–22; and local and global natural histories, 138–39; and manifestation of sin in nature, 3, 17; and query lists, 7, 129–30, 139–40, 192–93, 225n3, 230n59, 230n60; and scripture as source of knowledge, 26, 46; and separation of faith and philosophy, 176, 187, 204n14

nature: Book of Nature as revelation, 166; Edenic perfection of, 3; environmental degradation and sin, 1–3; Romantic conception as non-human, 240n22

neo-Augustinianism, 46, 209n86, 223n95

Newton, Isaac: and Burnet, 110, 222n86; on history and prophecy, 155, 233n103; lack of evidence for future, 157; Newtonianism, 165, 168; reception in Italy, 165–66, 235n22, 239n85; synthesis of science and religion, 165–66

Nicholson, Marjorie Hope, 218n7, 218n19

Nile, flooding of, 24

Noah: as astrologer, 34–35; and ark, 6; post-diluvian trauma and heritable problems, 170, 175. *See also* Adam; ark, Noah's; Flood, the

Nogarola, Isotta, 42, 207n58, 208n69

Oldenburg, Henry, 121, 137

original sin: and compromise of cognition and bodily integrity, 115; and human mortality, 39, 173; and Native Americans, 54; and polygenic provincialization, 62

Orsato, Giovanni Antonio, 175

overpopulation, 2, 35, 45

paleontology, 125, 191, 223n92, 226n6; absence in Burnet and Whiston, 229n43; in works of Scilla and Steno, 136–37, 164, 229n37; and Woodward, 137, 225n131, 228n35, 229n42

Paracelsus, 59, 211n18

Passi, Giuseppe, 42–43

Pelagianism, 209n88

Pepys, Samuel, 86

Petiver, James, 134, 143; and competition for fossils, 144–46; and von Bromell, 146–47

Pflaum, Jacob, 29–30

Phillip II (king of Spain), 59

philo-semitism, 13, 212n31

*Philosophical Transactions*, 78–79

Piccolomini, Alessandro, 206n38

Pinelli, Michele, 235n17

plague, 2, 105, 119, 160, 161, 169, 171, 188, 239n2; as divine punishment, 28, 160; parallel to the Flood, 160, 167, 171

Plato, 31, 151

Plot, Robert, 139

Pluche, Abbé Noël-Antoine de la, 157

Polier, Georges, 178, 182–83

polygenism, 54, 57–59; denies universal original sin, 62; threat to Christian doctrine, 131, 212n30

Pomponazzi, Pietro, 209n85

Poole, William, 15, 91, 232n88
pre-Adamism, 49, 52, 54, 58–64, 212nn23–24
providence: Burnet on, 95, 97, 107–8, 110, 118, 125, 219n27; natural disaster as sign of, 27–28, 64, 91, 178, 204n22; and Noah's ark, 50; Woodward on, 117, 123–24, 126
Purchas, Samuel, 86

race: acclimatization and, 104; bodily markers of, 76, 86; and complexion, 208n77; Creole philosophy of, 71–75, 87, 169, 215n66; European assertion of superiority, 51, 57, 68, 86; and migration theory of Native American origins, 68, 73; modernity of, 217n102; monogenism and scientific racism, 85–88, 217n101; and religion, 210n7
Raleigh, Walter, 56, 58; on contemporary vs. biblical weather, 106–7; on gigantism, 97
Ramachandran, Ayesha, 11, 202n44, 226n9
Ramaswamy, Sumathi, 15, 201n36, 210n10
Ramazzini, Bernardino, 190
Rappaport, Rhoda, 17, 201n37, 223n90, 230n56, 233n1, 239n90, 240n17
Ray, John, 105, 117, 134, 136–37, 232n93; on Apocalypse, 153–54; and Burnet, 153; and Lhwyd, 139–40, 153, 232n92
Réaumur, René Antoine Ferchault de, 191–92, 240n17
Republic of Letters, 13–14, 48, 225n11, 231n68; Alpine, 162, 165, 176–80, 187, 234n15; collecting correspondents, 147; colonial entry into, 145, 230n65; confessional dialogue and divisions in, 162, 172, 178–79, 187; cosmopolitanism, 178–79; exchange of specimens, 131–33, 143, 145–46, 227n12; excluded non-Christians, 13; and gender, 179, 237n60; and imperial expansion, 132, 138; language of respect, 178; and local/global empirical projects, 141; and other networks, 179, 229n45; pluralistic community of inquirers, 131, 138; rhetoric of inclusivity, 142–44; and scholarly sociability, 16, 227n12; social utility of Universal Deluge studies, 132, 148
Revolution of 1688–89, 91, 114, 126
Riccati, Jacopo, 184–85, 238n81; and Vallisneri, 179, 184–85, 239n85
Richelieu, Cardinal Armand Jean du Plessis, Duc de, 61

Rizzetti, Giovanni, 235n22
Robinson, Tancred, 117, 125
Robinson, Thomas, 109
Roman Catholic Church, 7, 46; Catholic Enlightenment, 166, 187, 235n20, 236n27; Council of Trent, 25, 45; divine causes of natural disasters, 28, 30–32; Fifth Lateran Council, 26; the Flood as supernatural, 163, 186; *Index of Prohibited Books*, 47, 188, 207n62; Inquisition, 1–2, 20, 25, 37, 39, 49, 175–76, 207n62, 213n47; Jesuit missionary/knowledge network, 130; and Mosaic natural philosophy, 27, 36, 136, 187; natural philosophy and faith, 25–26, 176, 187; New World evangelism, 68, 70–72, 77, 115; and Republic of Letters, 17, 134–35, 137, 175–76; and Universal Conflagration, 150, 156
Rossi, Paolo, 17, 198n9, 233n101, 236n27
Rotari, Sebastiano, 166
Rowlands, Henry, 121
Royal Society, the, 95–96, 135, 144, 165, 235n17
Rudwick, Martin J. S., 194, 226n6
ruined earth, the, 2–3, 5, 15, 100–101, 118, 171, 219n31; and argument from design, 125–28; arguments against, 111, 120; Burnet on, 93–98; caused by human sin, 3, 11, 90, 95; postdiluvian ruin of humanity and nature, 7, 90, 95, 98, 118, 120, 160, 171; and potential renovation of the planet, 116, 172; and providence, 94–95; and ruin of humanity, 161, 170, 172–74; sin and salvation in, 4, 95, 116, 120, 172
Russiliano, Tiberio, 34

Sarmiento de Gamboa, Pedro: on continental drift, 67; Spanish origin of Native Americans, 72
Savonarola, Fra Girolamo, 30, 205n26
Scheuchzer, Johann Jakob, 7, 81–82, 132, 165, 191, 240n19; and Bourget, 79, 179–81, 215n72; and Burnet, 181; exchange of specimens, 134–35, 144, 146, 227n14; on the Flood, 148–50, 178, 182; fossil catalogues, 146, 147, 192; and Swiss local geohistory, 192–93; Universal Conflagration, 149; and Vallisneri, 146, 160, 163, 165, 177–81; and Woodward, 133–35, 191, 192, 224n19, 227n14

Scilla, Agostino, 136, 164; and Woodward, 136–37, 225n131, 228nn35

scripture: accusations of misinterpretation, 61, 108, 110, 112, 117, 166; and confessional polemics, 166, 179–80, 182–83, 238n84; interpretation of, 3, 27, 37, 46, 118, 166, 189; as source of knowledge about nature and history, 22, 26–27, 31, 58, 64, 91, 148–50, 154, 178, 240n20; as source of religious authority, 47, 50, 58, 185. *See also* Bible

Sepúlveda, Juan Ginés de, 73

settler colonialism, 53, 63, 70, 82, 87

Sforza, Bona, 38

sin: and anthropogenic damage to nature, 1–2; connection between the Flood and universal sin, 49, 64; doctrine of depravity, 45–46, 114; legal vs. natural sin, 62; and unintended consequences, 10, 161

Siraisi, Nancy G., 218n37

slavery: Aristotle and "natural slaves," 73, 213n54; Christian injunction against, 82; race and, 104; Spanish justifications for, 73

Sloane, Hans, 84, 134, 144, 230n65

Society for the Propagation of the Gospel in Foreign Parts, 83

Spada, Giovanni Giacomo, 186, 190

Spinoza, Baruch, 10, 16

St. Clair, Robert, 105, 127

Steno, Nicolaus (Niels Stensen), 136–38, 164, 228n37; and Leibniz, 140–41; and scaling, 140–42

Stillingfleet, Edward, 61–62, 87, 111, 217n100

Stöffler, Johannes, 29–30

Strahlenberg, Philipp Johann von, 79, 81–82

*Swiss Mercury*. See *Mercure Suisse*

Szerszynski, Bronislaw, 8, 199n18

Tartars, 72; European ideas about, 52, 66, 68–75, 78–79, 81–82, 216n87; Strahlenberg on, 79–82

Taunton stone, the, 78–79, 81–82, 84, 230n66

Taylor, Jesse Oak, 9

Toland, John, 223n90

toleration, 46, 61, 63, 180; and confessional divisions, 198n12; in correspondence networks, 176, 179–80; racial, 85; and Republic of Letters, 132; rhetoric of, 179; and secularization, 180

*Two Essays Sent in a Letter from Oxford* (by L. P.), 111

Universal Conflagration, 132, 146, 152, 154, 232n88, 240n20; Burnet on, 151–52, 157; compared with Universal Deluge, 148–51, 193, 231n72; religious divisions over, 157–58; unsuited to inquiry, 157

universalism, Christian, 4, 5, 11; La Peyrère's rejection of, 63

Ussher, James, 23

Vallisneri, Antonio, 8, 233n1; on antediluvian gigantism and longevity, 174; on antediluvian vices, 170; and Antonio Maria Borromeo, 173; and Clelia Grillo Borromeo, 177; and Bourguet, 160, 163, 176–80, 183, 189; and Burnet, 136, 160–61, 163, 171; on Burnet, 165, 235n23; Catholicism of, 176, 179, 236n27; climate degraded by Flood, 161, 167, 172; damage to human bodies, 167–69, 171–73; and empirical fieldwork, 190; and Erculiani, 162, 188–89; fears of Protestant mockery, 183; fossil exchange, 180–81; and Galileo, 166, 184–85, 235n25, 238n84; on geology and the Flood, 161, 169, 172, 184; and Hippocrates, 169; imagery from Dante, 171; and Giovanni Maria Lancisi, 168; letter-writing, 166–67, 176, 180, 189; and Girolamo Lioni, 167, 174–75; on local climate and floods, 167–68, 190, 192; and miracles, 182, 189; and Monti, 165, 187; and Moro, 187, 189; and Mosaic natural philosophy, 165, 175, 187, 189; and Muratori, 183, 186, 237n45, 238n78; nationalism of, 164, 180; *Of Marine Bodies*, 7, 8, 136, 146, 147, 161–63, 180–81; and Polier, 178, 182–83; and Riccati, 179, 184; and Scheuchzer, 160, 163, 165, 177–80, 192; separating faith and philosophy, 184–89; separating human and natural history, 162, 171–72, 189; on sin as cause of Flood and plagues, 60–87, 160–61, 167, 171, 188; skeptical mode, 174, 176, 185–86; on theology of improvement, 172; three versions of biblical history, 174–76; on undue attention to scripture in natural history, 166, 179; and Whiston, 136, 163, 166; and Woodward, 136, 160–61, 163, 171–72, 178–79, 185; on "Woodwardians," 163–65, 178, 187

Venner, Tobias, 101

Vidal, Fernando, 171

Virgil, 102, 107; *Georgics*, 97, 122; and Golden Age, 96, 102, 107, 121–22

Vogli, Giovanni Giacinto, 164

Voltaire, 76, 215nn65–66

von Bromell, Magnus, 145–47, 231n68

Voss, Isaak (Vossius), 63, 111, 212n32

Walsh, Francis, 193

Walsham, Alexandra, 10, 16, 91–92

Wars of Religion, 25, 27, 28

water: and astrology, 29, 40; fogs and rain, 106; postdiluvian degradation of, 3, 17; vapor and disease, 98; as vital element, 35–36, 74, 101–2

Whiston, William, 90, 137, 223n96; on Burnet, 222n73; Newtonianism of, 166; postdiluvial air and disease, 106–7; Vallisneri on, 166

White, Lynn, Jr., 18

Whitehurst, John, 193

women: education of, 33; superior spiritual state and embodiment, 42–43; supposedly less sinful than men, 42

Woodward, John, 8–9, 90, 92, 96, 116, 133–35, 191, 224n105; and Burnet, 116–18; charges of deism against, 116–18; charges of plagiarism against, 25, 136–37; climate and acquired racial characteristics, 104; and fossils, 129–30, 133–36, 143–44, 225n131; health, longevity, and postdiluvian soil, 118, 120; and Italian work on fossils, 136, 164; and Lhwyd, 140, 142; and Mather, 144–45; and Mosaic history, 117; overcoming original sin, 116–17; query list, 129–30; reception of, 117, 120, 125, 136; and ruined earth, 96, 120; and Scheuchzer, 7, 130–33, 227n14, 227n19; and Scilla, 136–37, 225n131, 228nn35–37, 229n42; and theology of improvement, 120–24, 172; and theory of soil sterility, 120–21; Universal Deluge supernatural, 117; version of Edenocene, 118; on Woodward and "Woodwardians," 136, 160–61, 163–64, 171–72

Zabarella, Jacopo, 209n85

Zanichelli, Gian Girolamo, 163

Zuccolo, Vitale, 48–49

Zuzori, Fiore, 34